Titles in This Series

Volume

1 **Markov random fields and their applications,** Ross Kindermann and J. Laurie Snell

2 **Proceedings of the conference on integration, topology, and geometry in linear spaces,** William H. Graves, Editor

3 **The closed graph and P-closed graph properties in general topology,** T. R. Hamlett and L. L. Herrington

4 **Problems of elastic stability and vibrations,** Vadim Komkov, Editor

5 **Rational constructions of modules for simple Lie algebras,** George B. Seligman

6 **Umbral calculus and Hopf algebras,** Robert Morris, Editor

7 **Complex contour integral representation of cardinal spline functions,** Walter Schempp

8 **Ordered fields and real algebraic geometry,** D. W. Dubois and T. Recio, Editors

9 **Papers in algebra, analysis and statistics,** R. Lidl, Editor

10 **Operator algebras and K-theory,** Ronald G. Douglas and Claude Schochet, Editors

11 **Plane ellipticity and related problems,** Robert P. Gilbert, Editor

12 **Symposium on algebraic topology in honor of José Adem,** Samuel Gitler, Editor

13 **Algebraists' homage: Papers in ring theory and related topics,** S. A. Amitsur, D. J. Saltman, and G. B. Seligman, Editors

14 **Lectures on Nielsen fixed point theory,** Boju Jiang

15 **Advanced analytic number theory. Part I: Ramification theoretic methods,** Carlos J. Moreno

16 **Complex representations of GL(2, K) for finite fields K,** Ilya Piatetski-Shapiro

17 **Nonlinear partial differential equations,** Joel A. Smoller, Editor

18 **Fixed points and nonexpansive mappings,** Robert C. Sine, Editor

19 **Proceedings of the Northwestern homotopy theory conference,** Haynes R. Miller and Stewart B. Priddy, Editors

20 **Low dimensional topology,** Samuel J. Lomonaco, Jr., Editor

21 **Topological methods in nonlinear functional analysis,** S. P. Singh, S. Thomeier, and B. Watson, Editors

22 **Factorizations of $b^n \pm 1$, b = 2, 3, 5, 6, 7, 10, 11, 12 up to high powers,** John Brillhart, D. H. Lehmer, J. L. Selfridge, Bryant Tuckerman, and S. S. Wagstaff, Jr.

23 **Chapter 9 of Ramanujan's second notebook—Infinite series identities, transformations, and evaluations,** Bruce C. Berndt and Padmini T. Joshi

24 **Central extensions, Galois groups, and ideal class groups of number fields,** A. Fröhlich

25 **Value distribution theory and its applications,** Chung-Chun Yang, Editor

26 **Conference in modern analysis and probability,** Richard Beals, Anatole Beck, Alexandra Bellow, and Arshag Hajian, Editors

27 **Microlocal analysis,** M. Salah Baouendi, Richard Beals, and Linda Preiss Rothschild, Editors

28 **Fluids and plasmas: geometry and dynamics,** Jerrold E. Marsden, Editor

29 **Automated theorem proving,** W. W. Bledsoe and Donald Loveland, Editors

30 **Mathematical applications of category theory,** J. W. Gray, Editor

31 **Axiomatic set theory,** James E. Baumgartner, Donald A. Martin, and Saharon Shelah, Editors

32 **Proceedings of the conference on Banach algebras and several complex variables,** F. Greenleaf and D. Gulick, Editors

33 **Contributions to group theory,** Kenneth I. Appel, John G. Ratcliffe, and Paul E. Schupp, Editors

34 **Combinatorics and algebra,** Curtis Greene, Editor

Titles in This Series

Volume

35 **Four-manifold theory,** Cameron Gordon and Robion Kirby, Editors

36 **Group actions on manifolds,** Reinhard Schultz, Editor

37 **Conference on algebraic topology in honor of Peter Hilton,** Renzo Piccinini and Denis Sjerve, Editors

38 **Topics in complex analysis,** Dorothy Browne Shaffer, Editor

39 **Errett Bishop: Reflections on him and his research,** Murray Rosenblatt, Editor

40 **Integral bases for affine Lie algebras and their universal enveloping algebras,** David Mitzman

41 **Particle systems, random media and large deviations,** Richard Durrett, Editor

42 **Classical real analysis,** Daniel Waterman, Editor

43 **Group actions on rings,** Susan Montgomery, Editor

44 **Combinatorial methods in topology and algebraic geometry,** John R. Harper and Richard Mandelbaum, Editors

45 **Finite groups–coming of age,** John McKay, Editor

46 **Structure of the standard modules for the affine Lie algebra $A_1^{(1)}$,** James Lepowsky and Mirko Primc

47 **Linear algebra and its role in systems theory,** Richard A. Brualdi, David H. Carlson, Biswa Nath Datta, Charles R. Johnson, and Robert J. Plemmons, Editors

48 **Analytic functions of one complex variable,** Chung-chun Yang and Chi-tai Chuang, Editors

49 **Complex differential geometry and nonlinear differential equations,** Yum-Tong Siu, Editor

50 **Random matrices and their applications,** Joel E. Cohen, Harry Kesten, and Charles M. Newman, Editors

51 **Nonlinear problems in geometry,** Dennis M. DeTurck, Editor

52 **Geometry of normed linear spaces,** R. G. Bartle, N. T. Peck, A. L. Peressini, and J. J. Uhl, Editors

53 **The Selberg trace formula and related topics,** Dennis A. Hejhal, Peter Sarnak, and Audrey Anne Terras, Editors

54 **Differential analysis and infinite dimensional spaces,** Kondagunta Sundaresan and Srinivasa Swaminathan, Editors

55 **Applications of algebraic K-theory to algebraic geometry and number theory,** Spencer J. Bloch, R. Keith Dennis, Eric M. Friedlander, and Michael R. Stein, Editors

56 **Multiparameter bifurcation theory,** Martin Golubitsky and John Guckenheimer, Editors

57 **Combinatorics and ordered sets,** Ivan Rival, Editor

58.I **The Lefschetz centennial conference. Proceedings on algebraic geometry,** D. Sundararaman, Editor

58.II **The Lefschetz centennial conference. Proceedings on algebraic topology,** S. Gitler, Editor

58.III **The Lefschetz centennial conference. Proceedings on differential equations,** A. Verjovsky, Editor

59 **Function estimates,** J. S. Marron, Editor

60 **Nonstrictly hyperbolic conservation laws,** Barbara Lee Keyfitz and Herbert C. Kranzer, Editors

61 **Residues and traces of differential forms via Hochschild homology,** Joseph Lipman

62 **Operator algebras and mathematical physics,** Palle E. T. Jorgensen and Paul S. Muhly, Editors

63 **Integral geometry,** Robert L. Bryant, Victor Guillemin, Sigurdur Helgason, and R. O. Wells, Jr., Editors

64 **The legacy of Sonya Kovalevskaya,** Linda Keen, Editor

65 **Logic and combinatorics,** Stephen G. Simpson, Editor

66 **Free group rings,** Narian Gupta

67 **Current trends in arithmetical algebraic geometry,** Kenneth A. Ribet, Editor

Titles in This Series

Volume

68 Differential geometry: The interface between pure and applied mathematics, Mladen Luksic, Clyde Martin, and William Shadwick, Editors

69 Methods and applications of mathematical logic, Walter A. Carnielli and Luiz Paulo de Alcantara, Editors

70 Index theory of elliptic operators, foliations, and operator algebras, Jerome Kaminker, Kenneth C. Millett, and Claude Schochet, Editors

71 Mathematics and general relativity, James A. Isenberg, Editor

72 Fixed point theory and its applications, R. F. Brown, Editor

73 Geometry of random motion, Rick Durrett and Mark A. Pinsky, Editors

74 Geometry of group representations, William M. Goldman and Andy R. Magid, Editors

75 The finite calculus associated with Bessel functions, Frank M. Cholewinski

76 The structure of finite algebras, David C. Hobby and Ralph Mckenzie

77 Number theory and its applications in China, Wang Yuan, Yang Chung-chun, and Pan Chengbiao, Editors

78 Braids, Joan S. Birman and Anatoly Libgober, Editors

79 Regular differential forms, Ernst Kunz and Rolf Waldi

80 Statistical inference from stochastic processes, N. U. Prabhu, Editor

81 Hamiltonian dynamical systems, Kenneth R. Meyer and Donald G. Saari, Editors

82 Classical groups and related topics, Alexander J. Hahn, Donald G. James, and Zhe-xian Wan, Editors

83 Algebraic K-theory and algebraic number theory, Michael R. Stein and R. Keith Dennis, Editors

84 Partition problems in topology, Stevo Todorcevic

85 Banach space theory, Bor-Luh Lin, Editor

86 Representation theory and number theory in connection with the local Langlands conjecture, J. Ritter, Editor

87 Abelian group theory, Laszlo Fuchs, Rüdiger Göbel, and Phillip Schultz, Editors

88 Invariant theory, R. Fossum, W. Haboush, M. Hochster, and V. Lakshmibai, Editors

89 Graphs and algorithms, R. Bruce Richter, Editor

90 Singularities, Richard Randell, Editor

91 Commutative harmonic analysis, David Colella, Editor

92 Categories in computer science and logic, John W. Gray and Andre Scedrov, Editors

93 Representation theory, group rings, and coding theory, M. Isaacs, A. Lichtman, D. Passman, S. Sehgal, N. J. A. Sloane, and H. Zassenhaus, Editors

94 Measure and measurable dynamics, R. Daniel Mauldin, R. M. Shortt, and Cesar E. Silva, Editors

95 Infinite algebraic extensions of finite fields, Joel V. Brawley and George E. Schnibben

96 Algebraic topology, Mark Mahowald and Stewart Priddy, Editors

97 Dynamics and control of multibody systems, J. E. Marsden, P. S. Krishnaprasad, and J. C. Simo, Editors

98 Every planar map is four colorable, Kenneth Appel and Wolfgang Haken

99 The connection between infinite dimensional and finite dimensional dynamical systems, Basil Nicolaenko, Ciprian Foias, and Roger Temam, Editors

100 Current progress in hyperbolic systems: Riemann problems and computations, W. Brent Lindquist, Editor

101 Recent developments in geometry, S.-Y. Cheng, H. Choi, and Robert E. Greene, Editors

102 Primes associated to an ideal, Stephen McAdam

103 Coloring theories, Steve Fisk

Titles in This Series

Volume

104 Accessible categories: The foundations of categorical model theory, Michael Makkai and Robert Paré

105 Geometric and topological invariants of elliptic operators, Jerome Kaminker, Editor

106 Logic and computation, Wilfried Sieg, Editor

107 Harmonic analysis and partial differential equations, Mario Milman and Tomas Schonbek, Editors

108 Mathematics of nonlinear science, Melvyn S. Berger, Editor

109 Combinatorial group theory, Benjamin Fine, Anthony Gaglione, and Francis C. Y. Tang, Editors

110 Lie algebras and related topics, Georgia Benkart and J. Marshall Osborn, Editors

111 Finite geometries and combinatorial designs, Earl S. Kramer and Spyros S. Magliveras, Editors

112 Statistical analysis of measurement error models and applications, Philip J. Brown and Wayne A. Fuller, Editors

113 Integral geometry and tomography, Eric Grinberg and Eric Todd Quinto, Editors

114 Mathematical developments arising from linear programming, Jeffrey C. Lagarias and Michael J. Todd, Editors

115 Statistical multiple integration, Nancy Flournoy and Robert K. Tsutakawa, Editors

116 Algebraic geometry: Sundance 1988, Brian Harbourne and Robert Speiser, Editors

117 Continuum theory and dynamical systems, Morton Brown, Editor

118 Probability theory and its applications in China, Yan Shi-Jian, Wang Jiagang, and Yang Chung-chun, Editors

119 Vision geometry, Robert A. Melter, Azriel Rosenfeld, and Prabir Bhattacharya, Editors

120 Selfadjoint and nonselfadjoint operator algebras and operator theory, Robert S. Doran, Editor

121 Spinor construction of vertex operator algebras, triality, and $E_8^{(1)}$, Alex J. Feingold, Igor B. Frenkel, and John F. X. Ries

Spinor Construction of Vertex Operator Algebras, Triality, and $E_8^{(1)}$

CONTEMPORARY MATHEMATICS

121

Spinor Construction of Vertex Operator Algebras, Triality, and $E_8^{(1)}$

Alex J. Feingold
Igor B. Frenkel
John F. X. Ries

American Mathematical Society
Providence, Rhode Island

The work of Alex J. Feingold was partially supported by National Science Foundation Grants DMS-840318 and DMS-8703156, and by the Institute for Advanced Study, Princeton, New Jersey. The work of Igor B. Frenkel was partially supported by the Sloan Foundation, by National Science Foundation Grant DMS-8602091, and by the Institute for Advanced Study, Princeton, New Jersey.

1991 *Mathematics Subject Classification.* Primary 17B65, 17B67, 81R10; Secondary 17A70, 17B25, 17B68, 81T40.

Library of Congress Cataloging-in-Publication Data

Feingold, Alex J., 1950–

Spinor construction of vertex operator algebras, triality, and $E_8^{(1)}$ / Alex J. Feingold, Igor B. Frenkel, John F. X. Ries.

p. cm. – (Contemporary mathematics; v. 121)

ISBN 0-8218-5128-4 (alk. paper)

1. Infinite dimensional Lie algebras. 2. Kac-Moody algebras. 3. Representations of algebras. I. Frenkel, Igor. II. Ries, John F. X., 1954– . III. Title. IV. Series: Contemporary mathematics (American Mathematical Society); v. 121.

QA252.3.F45   1991   91-24409

512$'$.55—dc20   CIP

Contents

Chapter 0	Introduction	1
Chapter 1	Summary	13
Chapter 2	Affine Algebras and Representations	34
Chapter 3	Spinor Construction of Vertex Operator Superalgebras	44
Chapter 4	Spinor Construction of the Chevalley Algebra and Triality for D_4	74
Chapter 5	Spinor Construction of Triality for $D_4^{(1)}$	91
Chapter 6	Spinor Construction of a Vertex Operator Para-algebra for $D_4^{(1)}$	106
Chapter 7	Spinor Construction of E_8	128
Chapter 8	Spinor Construction of Vertex Operator Algebras for $E_8^{(1)}$	132
References		144

CHAPTER 0

Introduction

Algebras range in diversity from such fundamental classes as associative and Lie algebras to rather exotic examples such as the Chevalley and Griess algebras. The Chevalley algebra is a commutative nonassociative algebra constructed on the direct sum of the three 8-dimensional representations of the Lie algebra D_4. The Griess algebra is a commutative nonassociative algebra constructed on the direct sum of the trivial representation and the minimal nontrivial representation of the Monster group. Both exceptional algebras have some similarity to the exceptional Lie algebra E_8, but one does not expect to obtain them as substructures of algebras in either general class of algebras. However, it does turn out that these exotic algebras are substructures of examples of a new class of algebras having remarkable similarities to the associative and Lie algebras.

The new class of algebras, called vertex operator algebras or chiral algebras, has been introduced independently in mathematics [**B,FLM**] and in physics (cf. [**BPZ**]). Vertex operator algebras can be thought of as an algebraic foundation of conformal field theory, which has become the crossroads of numerous mathematical and physical theories. The physical motivation for the study of conformal field theory and chiral algebras was the development of two-dimensional statistical mechanics and string theory, the modern approach to quantum gravity. In the most promising models of string theory the exceptional Lie algebra E_8 plays a central role. (See [**GSW**] for references to the physics literature.) An important mathematical motivation for the introduction of vertex operator algebras was the attempt to understand a natural realization of the Monster as a symmetry group, and to give an axiomatic approach to the Griess algebra [**G**]. After the axioms of the vertex operator algebras finally crystalized, it became clear that, in a deep sense, these algebraic structures are analogous to the associative and Lie algebras. Furthermore, one particular vertex operator algebra contains the Griess algebra as a (weight 2) substructure [**FLM**].

The purpose of this work is multifold. First we want to understand certain generalizations of vertex operator algebras. In one such algebra the Chevalley algebra can be found as a (weight $\frac{1}{2}$) substructure, showing the close analogy with the Griess algebra. The generalizations of vertex operator algebras which we study are natural and useful. The simplest generalizations, vertex operator superalgebras [**Go,T**], are involved in superconformal field theory. Beyond that, vertex operator para-algebras, introduced and developed independently in this paper and by Dong and Lepowsky [**DL1,DL2**], are related to one-dimensional representations of the braid group. Second we want to show that the "magic square" construction of E_8 is also quite natural from the vertex operator algebra point of view, and E_8 itself is a (weight 1) substructure in a certain example of a vertex operator algebra. This provides a unified approach to the Chevalley, Griess and E_8 algebras, and explains some of their similarities. A third purpose is to provide a purely spinor construction of the exceptional affine Lie algebra $E_8^{(1)}$, a natural continuation of our previous work on spinor and oscillator constructions of the classical affine Lie algebras [**FF**]. We expect that our constructions can rather easily be extended to include the rest of the exceptional affine Lie algebras. Our fourth objective is to develop an inductive technique of construction which could be applied to the Monster vertex operator algebra. Let us elaborate on this last point.

One may give definitions of untwisted and twisted representations of vertex operator algebras, superalgebras and para-algebras. It is possible that the sum of an untwisted and some twisted representations can yield a new representation, which is sometimes referred to as an "orbifold" construction. There is a close analogy between our construction and that of the Monster vertex operator algebra. In both cases one constructs the untwisted and some twisted representations, an action of the triality group, and finally, from the sum of these representations, an orbifold type vertex operator algebra which is closely connected to what is called supersymmetry in the physics literature [**GSW**]. The difference is that ours is a fermionic construction using Clifford algebras while the Monster construction [**FLM**] is bosonic, using the Weyl algebra. These two types of constructions have been studied in the physics literature, and when they yield isomorphic algebras one is said to have a "boson-fermion correspondence". There are many similar ideas in the fermionic and bosonic constructions, for example, the correction of vertex operators on the twisted sector by the exponential of a quadratic operator. Here we give an orbifold construction, which is completely different from the one in [**FLM**], of a vertex operator algebra associated with $E_8^{(1)}$. Our proof is inductive, uses the classical finite-dimensional construction of Chevalley as a base case, and lifts this structure to the infinite-dimensional vertex operator algebra. Although, in principle, we could have used techniques analogous to those in [**FLM**], this would have entailed unnecessary complications. On the other hand, our inductive method could be used to construct the Monster vertex operator algebra and the action of the Monster as automorphisms. The base case of the induction would require the finite-dimensional construction of the Griess algebra, and would be incomparably more complicated than our work on the Chevalley algebra.

We have also obtained our results from the untwisted bosonic picture by considering vertex constructions from rational lattices. But we want to have independent treatments of these two approaches and an isomorphism between them giving a boson-fermion correspondence. This will be included in a subsequent publication [**FR**].

We will now explain the main mathematical ideas behind our work, including definitions of vertex operator algebras (as defined in [**FLM**] and [**FHL**]), superalgebras and para-algebras. The main axiom of vertex operator algebras combines the Jacobi identity for Lie algebras and the Cauchy residue formula for a rational function on the Riemann sphere with at most three poles. Let $\mathbf{V}$ be a vector space over the complex numbers $\mathbb{C}$, and let

$$ad : \mathbf{V} \to \mathrm{End}(\mathbf{V}) \tag{0.1}$$

be a linear transformation satisfying the Jacobi identity

$$ad(u)ad(v) - ad(v)ad(u) = ad(ad(u)v) \tag{0.2}$$

for any $u, v \in \mathbf{V}$. With the bracket $[u, v] = ad(u)v$, $\mathbf{V}$ becomes a Lie algebra with trivial center if $\ker(ad)$ is trivial. The Cauchy residue formula for a rational function $f(z)$ with possible poles only at 0, z_0 and ∞ has little in common with the Jacobi identity besides the fact that it also has three terms

$$-\mathrm{Res}_{z=\infty} f(z) - \mathrm{Res}_{z=0} f(z) = \mathrm{Res}_{z=z_0} f(z), \tag{0.3}$$

but their combination gives the main axiom, which we call the *Jacobi Identity*, of the new theory of vertex operator algebras.

Let $\mathbf{V}$ be a vector space and let $\mathbf{V}[[z, z^{-1}]]$ denote the vector space of formal Laurent series with coefficients in $\mathbf{V}$. Let

$$ad_z : \mathbf{V} \to \mathrm{End}(\mathbf{V})[[z, z^{-1}]] \tag{0.4}$$

be a linear transformation, which may be viewed as a generating function

$$ad_z = \sum_{n \in \mathbb{Z}} ad_{(n)} z^{-n-1} \tag{0.5}$$

for the linear transformations

$$ad_{(n)} : \mathbf{V} \to \mathrm{End}(\mathbf{V}), \tag{0.6}$$

such that for $u, v \in \mathbf{V}$, we have $ad_{(n)}(u)v = 0$ for n sufficiently large. For any $z_0 \in \mathbb{C}^*$ assume that ad_z satisfies the *Jacobi Identity*

$$\begin{aligned} &- \mathrm{Res}_{z=\infty}(ad_z(u)ad_{z_0}(v)f(z)) - \mathrm{Res}_{z=0}(ad_{z_0}(v)ad_z(u)f(z)) \\ &= \mathrm{Res}_{z=z_0}(ad_{z_0}(ad_{z-z_0}(u)v)f(z)) \end{aligned} \tag{0.7}$$

for any $f(z) \in \mathbb{C}[z, z^{-1}, (z - z_0)^{-1}]$. Each residue in (0.7) is defined as the coefficient of the appropriate singular term in the Laurent expansion around the singular point. In fact, (0.7) is equivalent to infinitely many identities for the operators $ad_{(n)}$. For $f(z) = z^p(z - z_0)^r$, $p, r \in \mathbb{Z}$, (0.7) says that

$$\sum_{m,n\in\mathbb{Z}} \binom{r}{r+p-m}(-1)^{r+p-m}ad_{(m)}(u)ad_{(n)}(v)z_0^{r+p-m-n-1}$$

$$- \sum_{m,n\in\mathbb{Z}} \binom{r}{m-p}(-1)^{r-p+m}ad_{(n)}(v)ad_{(m)}(u)z_0^{r+p-m-n-1} \tag{0.8}$$

$$= \sum_{m,n\in\mathbb{Z}} \binom{p}{m-r}ad_{(n)}(ad_{(m)}(u)v)z_0^{r+p-m-n-1},$$

which is equivalent to the identities

$$\sum_{0\leq i\in\mathbb{Z}} \binom{r}{i}(-1)^i(ad_{(m+r-i)}(u)ad_{(n+i)}(v) - (-1)^r ad_{(n+r-i)}(v)ad_{(m+i)}(u))$$

$$= \sum_{0\leq k\in\mathbb{Z}} \binom{m}{k}ad_{(m+n-k)}(ad_{(r+k)}(u)v)$$

$$\tag{0.9}$$

for all $m, n, r \in \mathbb{Z}$. The finiteness of (0.9) when applied to any particular vector in $\mathbf{V}$ is guaranteed by our assumption that $ad_{(n)}(u)v = 0$ for n sufficiently large, so (0.7) is well-defined. In all later chapters we use the notation $Y(v, z)$ for $ad_z(v)$ and $\{v\}_n$ for $ad_{(n)}(v)$. As before, we also require that $\ker(ad_z)$ be trivial. Additional axioms for vertex operator algebras postulate the existence of two special vectors $\mathbf{1}, \omega \in \mathbf{V}$ such that

$$ad_z(\mathbf{1}) = \mathrm{Id}_{\mathbf{V}}, \tag{0.10}$$

$$ad_z(\omega) = \sum_{n\in\mathbb{Z}} D(n)z^{-n-2} \tag{0.11}$$

where $D(n) = ad_{(n+1)}(\omega)$, $n \in \mathbb{Z}$, generate the Virasoro algebra

$$[D(m), D(n)] = (n - m)D(m + n) + \frac{1}{12}(m^3 - m)\delta_{m,-n}c\,\mathrm{Id}_{\mathbf{V}}, \tag{0.12}$$

the scalar c is called the *rank* of $\mathbf{V}$, and

$$[D(-1), ad_z(v)] = -(d/dz)ad_z(v). \tag{0.13}$$

Note that (0.9), with $m = 0$, $r = 0$ and $u = \omega$, gives

$$[D(-1), ad_z(v)] = ad_z(D(-1)v). \tag{0.14}$$

The final axioms require that $D(0)$ defines a $\mathbb{Z}$-grading of $\mathbf{V}$,

$$\mathbf{V} = \coprod_{n\in\mathbb{Z}}(\mathbf{V})_n \tag{0.15}$$

where

$$(\mathbf{V})_n = \{v \in \mathbf{V} \mid D(0)v = -nv\} \tag{0.16}$$

is finite-dimensional for each $n \in \mathbb{Z}$ and $(\mathbf{V})_n = \{0\}$ for n sufficiently small. We define a vertex operator algebra to be a quadruple $(\mathbf{V}, ad_z, \mathbf{1}, \omega)$ satisfying the above axioms, but for brevity we sometimes just refer to $\mathbf{V}$ itself as a vertex operator algebra. The Virasoro operators $L_m = -D(m)$ are often used in the literature. We define the *weight* of a homogeneous vector by the rule

$$wt(v) = n \quad \text{if} \quad v \in (\mathbf{V})_n, \tag{0.17}$$

and note that (0.9), with $m = 0$, $r = 1$ and $u = \omega$, and (0.13) imply

$$wt(ad_{(n)}(v)w) = wt(v) + wt(w) - n - 1. \tag{0.18}$$

Note that with $r = 0$ in (0.9) we get the commutator formula

$$[ad_{(m)}(u), ad_{(n)}(v)] = \sum_{0 \leq k \in \mathbb{Z}} \binom{m}{k} ad_{(m+n-k)}(ad_{(k)}(u)v) \tag{0.19}$$

for all $m, n \in \mathbb{Z}$, in which the summation is actually finite. Letting

$$\hat{\mathbf{V}} = \mathbf{V} \otimes \mathbb{C}[t, t^{-1}] \tag{0.20}$$

and letting $D(-1)$ act on $\hat{\mathbf{V}}$ by

$$D(-1)(v \otimes t^n) = D(-1)v \otimes t^n \ - \ nv \otimes t^{n-1}, \tag{0.21}$$

it can be shown that $\hat{\mathbf{V}}/D(-1)\hat{\mathbf{V}}$ has the structure of a Lie algebra and is represented on $\mathbf{V}$ so that $ad_{(n)}(v)$ represents the coset of $v \otimes t^n$. For each $m \in \mathbb{Z}$, $ad_{(m)}(u)v$ defines a product on $\mathbf{V}$, so we have a family of (generally nonassociative) algebra structures on $\mathbf{V}$. With $m = n = 0$, (0.19) reduces to the classical Jacobi identity (0.2), and $\mathbf{V}/D(-1)\mathbf{V}$ is a Lie algebra under the bracket $[u, v] = ad_{(0)}(u)v$.

The definition of a module for a vertex operator algebra is analogous to the definition of a Lie algebra module. Let $\mathbf{V}$ be a Lie algebra with bracket $[u, v] = ad(u)v$ as in (0.1)-(0.2), and let $\mathbf{W}$ be a vector space. Then $(\mathbf{W}, \pi)$ is a $\mathbf{V}$-module if

$$\pi : \mathbf{V} \to \operatorname{End}(\mathbf{W}) \tag{0.22}$$

is a linear transformation satisfying

$$\pi(u)\pi(v) - \pi(v)\pi(u) = \pi(ad(u)v) \tag{0.23}$$

for any $u, v \in \mathbf{V}$. We say π is *faithful* if $\ker(\pi)$ is trivial. To define a vertex operator algebra module we combine (0.23) with the Cauchy identity (0.3). Let

$$\pi_z : \mathbf{V} \to \operatorname{End}(\mathbf{W})[[z, z^{-1}]] \tag{0.24}$$

be a linear transformation, which may be viewed as a generating function

$$\pi_z = \sum_{n \in \mathbb{Z}} \pi_{(n)} z^{-n-1} \tag{0.25}$$

for the linear transformations

$$\pi_{(n)} : \mathbf{V} \to \mathrm{End}(\mathbf{W}), \qquad (0.26)$$

such that for $v \in \mathbf{V}$, $w \in \mathbf{W}$, we have $\pi_{(n)}(v)w = 0$ for n sufficiently large. For any $z_0 \in \mathbb{C}^*$ assume that π_z satisfies the *Jacobi Identity*

$$\begin{aligned}
&- \mathrm{Res}_{z=\infty}(\pi_z(u)\pi_{z_0}(v)f(z)) - \mathrm{Res}_{z=0}(\pi_{z_0}(v)\pi_z(u)f(z)) \\
&= \mathrm{Res}_{z=z_0}(\pi_{z_0}(ad_{z-z_0}(u)v)f(z))
\end{aligned} \qquad (0.27)$$

for any $f(z) \in \mathbb{C}[z, z^{-1}, (z-z_0)^{-1}]$. As before, (0.27) is equivalent to the identities

$$\begin{aligned}
&\sum_{0 \le i \in \mathbb{Z}} \binom{r}{i} (-1)^i (\pi_{(m+r-i)}(u)\pi_{(n+i)}(v) - (-1)^r \pi_{(n+r-i)}(v)\pi_{(m+i)}(u)) \\
&= \sum_{0 \le k \in \mathbb{Z}} \binom{m}{k} \pi_{(m+n-k)}(ad_{(r+k)}(u)v)
\end{aligned} \qquad (0.28)$$

for all $m, n, r \in \mathbb{Z}$, which imply

$$[\pi_{(m)}(u), \pi_{(n)}(v)] = \sum_{0 \le k \in \mathbb{Z}} \binom{m}{k} \pi_{(m+n-k)}(ad_{(k)}(u)v). \qquad (0.29)$$

As additional axioms we require the analogs of (0.10)-(0.13) with ad_z replaced by π_z and $\mathrm{Id}_{\mathbf{V}}$ replaced by $\mathrm{Id}_{\mathbf{W}}$. The analog of (0.14) also follows. The only new feature is that $\mathbf{W}$ is not necessarily $\mathbb{Z}$-graded, but it is $\mathbb{Q}$-graded. In fact, if $\mathbf{W}$ is irreducible then it is graded by $\mathbb{Z} + \Delta_{\mathbf{W}}$ where $\Delta_{\mathbf{W}} \in \mathbb{Q}$, defined mod $\mathbb{Z}$, is an important characteristic of the module. Define a $\mathbf{V}$-module to be a pair $(\mathbf{W}, \pi_z)$ satisfying these axioms. From (0.29), $\mathbf{W}$ is a Lie algebra module for $\hat{\mathbf{V}}/D(-1)\hat{\mathbf{V}}$ with $\pi_{(n)}(v)$ representing the coset of $v \otimes t^n$.

The main task in constructing vertex operator algebras and their modules lies in the verification of the Jacobi Identity (0.7) and (0.27). One uses matrix coefficients which can be defined from either the dual spaces $(\mathbf{V})_n^*$ and $(\mathbf{W})_n^*$ or a nondegenerate Hermitian form on $\mathbf{V}$ and $\mathbf{W}$. It is sufficient to establish three analytic properties, *rationality*, *permutability* and *associativity*, of the matrix coefficients (correlation functions in the physics literature) made from products and compositions of the operators ad_z and π_z. If $(\cdot\, ,\, \cdot)$ denotes a suitable Hermitian form on $\mathbf{V}$, then by rationality we mean that for $u, v, w, w' \in \mathbf{V}$ the series $(ad_z(u)ad_{z_0}(v)w, w')$ converges absolutely in the domain $|z| > |z_0| > 0$ to a rational function $f(z, z_0)$ which may only have poles at $z = 0$, $z_0 = 0$ or $z = z_0$. By permutability we mean that the rational function, to which the series $(ad_{z_0}(v)ad_z(u)w, w')$ converges absolutely in the domain $|z_0| > |z| > 0$, is the same function $f(z, z_0)$. By associativity we mean that the series $(ad_{z_0}(ad_{z-z_0}(u)v)w, w')$ converges absolutely in the domain $|z_0| > |z - z_0| > 0$ to the same function $f(z, z_0)$. In the physics literature associativity is known as the associativity of the operator product expansion. Analogous definitions of these properties can be made on a module $\mathbf{W}$. One can also show, conversely, that the Jacobi Identity implies these three properties. For further details on equivalent axiomatic formulations of vertex operator algebras and modules we refer to [**FHL**].

Lie superalgebras are a natural generalization of Lie algebras. Let $\mathbf{V} = \mathbf{V}_0 \oplus \mathbf{V}_1$ be a $\mathbb{Z}_2$-graded vector space and let $|v| = 0$ if $v \in \mathbf{V}_0$, $|v| = 1$ if $v \in \mathbf{V}_1$. Let $ad : \mathbf{V} \to \mathrm{End}(\mathbf{V})$ be a linear transformation satisfying the super-Jacobi identity

$$ad(u)ad(v) - (-1)^{|u|\,|v|}ad(v)ad(u) = ad(ad(u)v) \tag{0.30}$$

for u and v homogeneous, and

$$|ad(u)v| = |u| + |v|. \tag{0.31}$$

With the super-bracket $[u, v] = ad(u)v$, $\mathbf{V}$ becomes a Lie superalgebra with trivial center if $\ker(ad)$ is trivial.

We may similarly generalize our definitions to define vertex operator super-algebras as follows (cf. [**T**]). First we allow $\mathbf{V} = \mathbf{V}_0 \oplus \mathbf{V}_1$ to be $\frac{1}{2}\mathbb{Z}$-graded in (0.15), where

$$\mathbf{V}_0 = \coprod_{n \in \mathbb{Z}} (\mathbf{V})_n, \quad \mathbf{V}_1 = \coprod_{n \in \mathbb{Z}+\frac{1}{2}} (\mathbf{V})_n, \tag{0.32}$$

and define $|v|$ as above. The only change to (0.7) is the additional factor $(-1)^{|u|\,|v|}$ in the $\mathrm{Res}_{z=0}$ term, which changes the corresponding term in (0.9). The left side of (0.19) becomes a super-bracket, and $\hat{\mathbf{V}}/D(-1)\hat{\mathbf{V}}$ has the structure of a Lie superalgebra. We also require the condition that ad_z respects the $\mathbb{Z}_2$-grading, that is,

$$|ad_{(n)}(u)v| = |u| + |v|. \tag{0.33}$$

The definition of a module $\mathbf{W}$ for a Lie superalgebra $\mathbf{V}$ requires a $\mathbb{Z}_2$-grading $\mathbf{W} = \mathbf{W}_0 \oplus \mathbf{W}_1$, the factor $(-1)^{|u|\,|v|}$ in the second term of (0.23), and $|\pi(v)w| = |v| + |w|$ for homogeneous $v \in \mathbf{V}$, $w \in \mathbf{W}$. A vertex operator superalgebra module $\mathbf{W} = \mathbf{W}_0 \oplus \mathbf{W}_1$ may be $\mathbb{Q}$-graded, but for $\mathbf{W}$ irreducible, it is graded by $\frac{1}{2}\mathbb{Z} + \Delta_{\mathbf{W}}$. The Jacobi Identity (0.27) is modified by the additional factor $(-1)^{|u|\,|v|}$ in the $\mathrm{Res}_{z=0}$ term, and we also require $|\pi_{(n)}(v)w| = |v| + |w|$. In the construction of vertex operator superalgebras and their modules, to establish the Jacobi Identity it is sufficient to prove the rationality, permutability and associativity properties of the matrix coefficients.

In the theory of affine Lie algebras one also considers their twisted construc-tions [**KP2,L**]. Analogous twisted constructions also apply to vertex operator algebras [**FLM2**] (see also Chapter 9 of [**FLM**], and [**DL2**]) and superalgebras. Let $(\mathbf{V}, ad_z, \mathbf{1}, \omega)$ be a vertex operator algebra and let ϑ be a vector space auto-morphism of $\mathbf{V}$ of order p such that

$$\vartheta ad_z(u)v = ad_z(\vartheta u)\vartheta v, \quad \vartheta \mathbf{1} = \mathbf{1}, \quad \vartheta \omega = \omega. \tag{0.34}$$

Let

$$\mathbf{V} = \bigoplus_{0 \le j \le p-1} \mathbf{V}^j \tag{0.35}$$

be the decomposition into the eigenspaces of ϑ, where $\mathbf{V}^j$ corresponds to the eigenvalue $e^{2\pi i j/p}$. We say that $(\mathbf{W}, \pi_z^\vartheta)$ is a ϑ-twisted $\mathbf{V}$-module (or a $\mathbf{V}^\vartheta$-module) if

$$\pi_z^\vartheta : \mathbf{V} \to \operatorname{End}(\mathbf{W})[[z^{1/p}, z^{-1/p}]] \tag{0.36}$$

is a linear transformation satisfying the axioms for a $\mathbf{V}$-module modified as follows. For $v \in \mathbf{V}^j$, we have

$$\pi_z^\vartheta(v) = \sum_{n \in \mathbb{Z}+j/p} ad_{(n)}(v) z^{-n-1}. \tag{0.37}$$

On $\mathbf{W}$ we have the twisted Jacobi Identity

$$\begin{aligned}
&- \operatorname{Res}_{z=\infty}(\pi_z^\vartheta(u)\pi_{z_0}^\vartheta(v)f(z)) - \operatorname{Res}_{z=0}(\pi_{z_0}^\vartheta(v)\pi_z^\vartheta(u)f(z)) \\
&= \operatorname{Res}_{z=z_0}(\pi_{z_0}^\vartheta(ad_{z-z_0}(u)v)f(z))
\end{aligned} \tag{0.38}$$

for $u \in \mathbf{V}^i$, $v \in \mathbf{V}^j$, $f(z) \in z^{i/p} z_0^{j/p} \mathbb{C}[z, z^{-1}, (z - z_0)^{-1}]$, which is equivalent to the identities (0.28) with $r \in \mathbb{Z}$, $m \in \mathbb{Z} + i/p$, $n \in \mathbb{Z} + j/p$. $\mathbf{W}$ is $\mathbb{Q}$-graded by $D(0)$, and if it is irreducible then it is graded by $(1/p)\mathbb{Z} + \Delta_{\mathbf{W}}$. Note that the twisted Jacobi Identity with $r = 0$ implies (0.29) with $m \in \mathbb{Z} + i/p$ and $n \in \mathbb{Z} + j/p$ if $u \in \mathbf{V}^i$ and $v \in \mathbf{V}^j$. Letting

$$\hat{\mathbf{V}}(\vartheta) = \bigoplus_{0 \le j \le p-1} \mathbf{V}^j \otimes t^{j/p} \mathbb{C}[t, t^{-1}], \tag{0.39}$$

it can be shown that $\hat{\mathbf{V}}(\vartheta)/D(-1)\hat{\mathbf{V}}(\vartheta)$ has the structure of a Lie algebra, and $\mathbf{W}$ is a module for that Lie algebra.

Let $(\mathbf{V}, ad_z, \mathbf{1}, \omega)$ be a vertex operator superalgebra and let ϑ be a vector space automorphism of $\mathbf{V}$ of order p satisfying (0.34). With notation as in (0.35)-(0.37), we say that $(\mathbf{W}, \pi_z^\vartheta)$ is a ϑ-twisted $\mathbf{V}$-module (or a $\mathbf{V}^\vartheta$-module) if the Jacobi Identity for a vertex operator superalgebra module is valid with π replaced by π^ϑ and the other axioms for a vertex operator superalgebra module are satisfied. There is a special role played by the order 2 automorphism of $\mathbf{V}$ which is the identity on $\mathbf{V}_0$ and minus the identity on $\mathbf{V}_1$. In that case, $\mathbf{W}$ is called a canonically $\mathbb{Z}_2$-twisted $\mathbf{V}$-module, and (0.39) becomes

$$\hat{\mathbf{V}}(\vartheta) = \mathbf{V}_0 \otimes \mathbb{C}[t, t^{-1}] \oplus \mathbf{V}_1 \otimes t^{1/2}\mathbb{C}[t, t^{-1}]. \tag{0.40}$$

The definitions of vertex operator algebras and superalgebras have a further generalization when $\mathbf{V}$ is graded by a finite abelian group Γ. Suppose that Γ has exponent g. For each $\gamma \in \Gamma$, suppose $\Delta_\gamma \in \mathbb{Q}$ and, for $\gamma_1, \gamma_2 \in \Gamma$, let

$$\Delta(\gamma_1, \gamma_2) = \Delta_{\gamma_1} + \Delta_{\gamma_2} - \Delta_{\gamma_1+\gamma_2} \in (1/g)\mathbb{Z} \tag{0.41}$$

be bilinear *mod* $\mathbb{Z}$. Assume that

$$\mathbf{V} = \bigoplus_{\gamma \in \Gamma} \mathbf{V}_\gamma \tag{0.42}$$

is a vector space and

$$ad_z : \mathbf{V} \to \mathrm{End}(\mathbf{V})[[z^{1/g}, z^{-1/g}]] \tag{0.43}$$

is a linear transformation with trivial kernel such that for $u \in \mathbf{V}_{\gamma_1}$, and $v \in \mathbf{V}_{\gamma_2}$, we have

$$ad_z(u)v = \sum_{n \in \mathbb{Z} + \Delta(\gamma_1, \gamma_2)} ad_{(n)}(u)v z^{-n-1} \tag{0.44}$$

and

$$ad_{(n)}(u)v \in \mathbf{V}_{\gamma_1 + \gamma_2} \tag{0.45}$$

is zero for n sufficiently large. Suppose that there are distinguished vectors $\mathbf{1}, \omega \in \mathbf{V}_0$ satisfying the usual axioms (0.10) - (0.13), that $D(0)$ defines a $\mathbb{Q}$-grading on $\mathbf{V}$ such that $\mathbf{V}_\gamma$ is graded by $\mathbb{Z} + \Delta_\gamma$, that each graded piece $(\mathbf{V}_\gamma)_n$ is finite-dimensional, and that $(\mathbf{V}_\gamma)_n = \{0\}$ for n sufficiently small. The main axiom is the Jacobi Identity

$$- \mathrm{Res}_{z=\infty}(ad_z(u)ad_{z_0}(v)wf(z, z_0)) - \eta\, \mathrm{Res}_{z=0}(ad_{z_0}(v)ad_z(u)wf(z, z_0))$$
$$= \mathrm{Res}_{z=z_0}(ad_{z_0}(ad_{z-z_0}(u)v)wf(z, z_0)) \tag{0.46}$$

for $u \in \mathbf{V}_{\gamma_1}$, $v \in \mathbf{V}_{\gamma_2}$, $w \in \mathbf{V}_{\gamma_3}$,

$$f(z, z_0) \in z^{\Delta(\gamma_1, \gamma_3)}\, z_0^{\Delta(\gamma_2, \gamma_3)}\, (z - z_0)^{\Delta(\gamma_1, \gamma_2)}\, \mathbb{C}[z, z^{-1}, z_0, z_0^{-1}, (z - z_0)^{-1}] \tag{0.47}$$

where $\eta = \eta(\gamma_1, \gamma_2) \in \mathbb{C}^*$ is a bilinear function on $\Gamma \times \Gamma$. Bilinearity implies that $\eta(\gamma_1, \gamma_2)$ is a root of unity and that η is a 2-cocycle representing a cohomology class in $H^2(\Gamma, \mathbb{Z}_g)$. This class is restricted by the further assumption that $\eta(\gamma_1, \gamma_2)\, e^{\pi i \Delta(\gamma_1, \gamma_2)}$ is skew-symmetric, that is, $\eta(\gamma_1, \gamma_2)\, e^{\pi i \Delta(\gamma_1, \gamma_2)} = \eta(\gamma_2, \gamma_1)^{-1}\, e^{-\pi i \Delta(\gamma_2, \gamma_1)}$. The reasons for these conditions on η will be explained below. We must be precise about how the three residues in (0.46) are to be computed. Let $p \in \mathbb{Z} + \Delta(\gamma_1, \gamma_2)$ and $m \in \mathbb{Z} + \Delta(\gamma_1, \gamma_3)$. In the first term of (0.46) we expand $(z - z_0)^p$ as

$$z^p (1 - z_0/z)^p = z^p \sum_{0 \le k \in \mathbb{Z}} \binom{p}{k}(-1)^k (z_0/z)^k \tag{0.48}$$

and find $-\mathrm{Res}_{z=\infty}$ to be the coefficient of z^{-1}. In the second term we expand $(z - z_0)^p$ as

$$e^{\pi i p} z_0^p (1 - z/z_0)^p = e^{\pi i p} z_0^p \sum_{0 \le k \in \mathbb{Z}} \binom{p}{k}(-1)^k (z/z_0)^k \tag{0.49}$$

and find $\mathrm{Res}_{z=0}$ to be the coefficient of z^{-1}. In the third term we let $\zeta = z - z_0$ and expand $z^m = (z_0 + \zeta)^m$ as

$$z_0^m (1 + \zeta/z_0)^m = z_0^m \sum_{0 \le k \in \mathbb{Z}} \binom{m}{k}(\zeta/z_0)^k \tag{0.50}$$

and find $\text{Res}_{z=z_0}$ to be the coefficient of ζ^{-1}. Note that the Jacobi Identity is equivalent to the identities

$$\sum_{0 \leq k \in \mathbb{Z}} \binom{p}{k} (-1)^k \big(ad_{(m+p-k)}(u) ad_{(n+k)}(v)$$
$$- \eta(\gamma_1, \gamma_2) e^{\pi i p} ad_{(n+p-k)}(v) ad_{(m+k)}(u) \big) w \qquad (0.51)$$
$$= \sum_{0 \leq k \in \mathbb{Z}} \binom{m}{k} ad_{(m+n-k)}(ad_{(p+k)}(u)v)w$$

for all $m \in \mathbb{Z} + \Delta(\gamma_1, \gamma_3)$, $n \in \mathbb{Z} + \Delta(\gamma_2, \gamma_3)$, $p \in \mathbb{Z} + \Delta(\gamma_1, \gamma_2)$. Under these assumptions we say that $(\mathbf{V}, ad_z, \mathbf{1}, \omega, \Gamma, \Delta, \eta)$ is a vertex operator para-algebra.

Let us explain further the meaning of the bilinearity of η and the skew-symmetry of $\eta e^{\pi i \Delta}$ in our definition. In the examples we consider, we prove the Jacobi Identity by proving rationality, permutability and associativity theorems. In fact, these three properties together can be shown to be equivalent to the Jacobi Identity. These properties apply to matrix coefficients multiplied by algebraic functions chosen so that the results are rational functions. When the permutability theorem is applied to do a transposition twice, the rational function should remain unchanged. Along with (0.49), it is this which requires the skew-symmetry condition on $\eta e^{\pi i \Delta}$. Applying the permutability and associativity theorems to rationalized matrix coefficients involving more than two vertex operators, the same considerations yield the bilinearity condition on η. In other approaches one considers the analytic continuations of the matrix coefficients, compares the branches in a common domain, and choices can lead to representations of the braid group.

Important examples of vertex operator para-algebras have been constructed in the following way. Let $(\mathbf{V}_0, ad_z, \mathbf{1}, \omega)$ be a vertex operator algebra, and let Γ be a finite abelian group acting on $\mathbf{V}_0$ such that

$$\gamma ad_z(u)v = ad_z(\gamma u)\gamma v, \quad \gamma \mathbf{1} = \mathbf{1}, \quad \gamma \omega = \omega \qquad (0.52)$$

for each $\gamma \in \Gamma$ and $u, v \in \mathbf{V}_0$. For each $\gamma \in \Gamma$ let $(\mathbf{V}_\gamma, \pi_z^\gamma)$ be an irreducible γ-twisted $\mathbf{V}_0$-module graded by $\mathbb{Z} + \Delta_\gamma$, and $\pi_z^0 = ad_z$. Then with $\mathbf{V}$ as in (0.42), it may be possible to define ad_z as in (0.43) so that $ad_z(u)v = \pi_z^\gamma(u)v$ for $u \in \mathbf{V}_0$, $v \in \mathbf{V}_\gamma$, and $(\mathbf{V}, ad_z, \mathbf{1}, \omega, \Gamma, \Delta, \eta)$ is a vertex operator para-algebra. A vertex operator superalgebra is then the special case of $\Gamma = \mathbb{Z}_2$ with $\Delta_0 = 0$, $\Delta_1 = \frac{1}{2}$ and $\eta(\gamma_1, \gamma_2) = -1$ if $\gamma_1 = \gamma_2 = 1$, 1 otherwise. One may generalize the above construction to the situation in which the pieces graded by Γ are a vertex operator superalgebra and modules for it, but this can be considered as a case of the above construction using the even part of the superalgebra as $\mathbf{V}_0$.

By analogy with the vertex operator algebra and superalgebra cases, letting

$$\hat{\mathbf{V}} = \mathbf{V} \otimes \mathbb{C}[t^{1/g}, t^{-1/g}] \qquad (0.53)$$

one might expect that some interesting new kind of algebraic structure exists on $\hat{\mathbf{V}}/D(-1)\hat{\mathbf{V}}$ and is represented on $\mathbf{V}$ so that the coset of $v \otimes t^n$ is represented by $ad_{(n)}(v)$. Looking at these operators for clues of what such a structure might be, one finds, for example, a generalized antisymmetry

$$\sum_{0 \leq k \in \mathbb{Z}} \binom{m}{k} ad_{(m+n-k)}(ad_{(p+k)}(u)v)w =$$
$$- \eta(\gamma_1,\gamma_2) e^{\pi i p} \sum_{0 \leq k \in \mathbb{Z}} \binom{n}{k} ad_{(m+n-k)}(ad_{(p+k)}(v)u)w \tag{0.54}$$

from (0.51) and the skew-symmetry of $\eta e^{\pi i \Delta}$. It is natural to introduce the following family of operations on $\hat{\mathbf{V}}$.

For $\gamma \in \Gamma$ let $\hat{\mathbf{V}}_\gamma = \mathbf{V}_\gamma \otimes \mathbb{C}[t^{1/g}, t^{-1/g}]$. For $i = 1,2$ let $v_i(p_i) = v_i \otimes t^{p_i} \in \hat{\mathbf{V}}_{\gamma_i}$, and for $r \in \frac{1}{g}\mathbb{Z}$ define the operation $\underset{r}{\circ}$ by

$$v_1(p_1) \underset{r}{\circ} v_2(p_2) = \sum_{0 \leq k \in \mathbb{Z}} \binom{p_1}{k} (ad_{(r+k)}(v_1)v_2)(p_1 + p_2 - k). \tag{0.55}$$

Note that each term on the right side of (0.55) will be zero unless $r \in \mathbb{Z} + \Delta(\gamma_1,\gamma_2)$. Although it can be shown that

$$v_1(p_1) \underset{r}{\circ} v_2(p_2) + \eta(\gamma_1,\gamma_2)\, e^{\pi i r}\, v_2(p_2) \underset{r}{\circ} v_1(p_1) \in D(-1)\hat{\mathbf{V}}, \tag{0.56}$$

there is a fundamental obstruction to the passage to the quotient by $D(-1)\hat{\mathbf{V}}$. It is easy to check that

$$D(-1)(v_1(p_1) \underset{r}{\circ} v_2(p_2)) = v_1(p_1) \underset{r}{\circ} (D(-1)\, v_2(p_2)) + r\, v_1(p_1) \underset{r-1}{\circ} v_2(p_2) \tag{0.57}$$

and

$$(D(-1)\, v_1(p_1)) \underset{r}{\circ} v_2(p_2) = r\, v_1(p_1) \underset{r-1}{\circ} v_2(p_2). \tag{0.58}$$

These show that the operation $\underset{r}{\circ}$ is not well-defined on the quotient unless $r = 0$. Before leaving this topic, one might mention that other relations among the operators $ad_{(n)}(v)$ follow from theorems about products and compositions of more than two vertex operators. For example, there is a generalization for vertex operator para-algebras of a "symmetric form of the Jacobi Identity", Proposition A.2.10 in [**FLM**]. We will investigate the implications of the axioms for vertex operator para-algebras in the future.

The axioms for vertex operator algebras, superalgebras, para-algebras and modules are quite restrictive, and a priori it is not at all clear that nontrivial examples exist. In trying to construct a vertex operator para-algebra one may find that the vertex operators satisfy the rationality theorem, but the permutability and associativity theorems have scalar factors depending on γ_1, γ_2 and γ_3. The factor in the associativity theorem should be a 3-cocycle representing a class in $H^3(\Gamma, \mathbb{Z}_{2g})$. By rescaling each vertex operator $Y(u,\zeta)v$ by a factor $\mu(\gamma_1,\gamma_2)$, we may attempt to remove the factor from the associativity theorem and make the permutability factor depend only on γ_1 and γ_2. This can only be done if the class represented by the factor in the associativity theorem is trivial. In [**FLM**] the vertex operator algebra associated with an even integral lattice was constructed, as well as an orbifold algebra. Here we give a spinor construction of a vertex operator superalgebra and its canonical $\mathbb{Z}_2$-twisted module from the four level 1 representations of the affine algebra $D_l^{(1)}$. In this case the basic and vector modules provide the vertex operator superalgebra, and the spinor modules provide the twisted module. When $l = 4$ we have triality and we can make the sum of these into a vertex operator para-algebra. In [**FR**] we will give the spinor para-algebra construction for $D_l^{(1)}$ with any rank l, and the lattice para-algebra construction for $A_l^{(1)}$, $D_l^{(1)}$ or $E_l^{(1)}$, and provide the boson-fermion correspondence between the two pictures for type $D_l^{(1)}$.

We gratefully acknowledge advice from J. Lepowsky, G. Seligman, and R. L. Wilson. We would also like to thank our patient and efficient technical typist, Marge Pratt.

CHAPTER 1

Summary

In Chapter 2 we begin with a brief review of facts about finite-dimensional simple Lie algebras $\mathbf{g}$ of type A, D, E, and their associated affine Kac-Moody Lie algebras $\hat{\mathbf{g}}$. The vertex construction $[\mathbf{FK}]$ of $\hat{\mathbf{g}}$ involves choosing a Cartan subalgebra $\mathbf{h}$ in $\mathbf{g}$, and is explicitly based on a choice of cohomology class $[\epsilon] \in H^2(P, \mathbb{C}^*)$ whose restriction to the root lattice Q from a finer lattice P is uniquely determined by

$$\epsilon(\alpha, \beta)\epsilon(\beta, \alpha)^{-1} = (-1)^{\langle \alpha, \beta \rangle}, \quad \alpha, \beta \in Q. \tag{1.1}$$

Finite-dimensional algebras $\mathbf{g}$ of type D_l (and B_l) can be constructed another way using Clifford algebras. Let $\mathbf{Cliff}$ be the Clifford algebra generated by $\mathbf{A} \cong \mathbb{C}^{2l}$ with a nondegenerate symmmetric bilinear form $\langle \, , \, \rangle$. A polarization $\mathbf{A} = \mathbf{A}^+ \oplus \mathbf{A}^-$ into maximally isotropic subspaces $\mathbf{A}^{\pm} \cong \mathbb{C}^l$ gives an irreducible left $\mathbf{Cliff}$-module $\mathbf{CM} = (\wedge \mathbf{A}^-) \cdot \mathbf{vac}$, where $0 \neq \mathbf{vac} \in \wedge^l \mathbf{A}^+$. Then the subspace of $\mathbf{Cliff}$ spanned by $\{\frac{1}{2}(ab - ba) \mid a, b \in \mathbf{A}\}$ acts by left multiplication on $\mathbf{CM}$, by (2.34) on $\mathbf{A}$ and by bracket (2.35) on itself. This provides the spinor, natural and adjoint representations of $\mathbf{g} = \mathbf{o}(2l)$ without fixing a choice of Cartan subalgebra. We review the spinor construction $[\mathbf{F1, KP1}]$ for $\hat{\mathbf{g}} = \hat{\mathbf{o}}(2l)$, providing four irreducible level 1 representations on which the Virasoro algebra $\mathbf{Vir}$ (2.26) is also represented. For $Z = \mathbb{Z}$ or $\mathbb{Z} + \frac{1}{2}$, let $\mathbf{Cliff}(Z)$ be the Clifford algebra generated by $\mathbf{A}(Z)$ given in (2.37) with the form $\langle \, , \, \rangle$ in (2.39). Let $\mathbf{A}(Z) = \mathbf{A}(Z)^+ \oplus \mathbf{A}(Z)^-$ be the polarization given in (2.40) and let $\mathfrak{J}(Z)$ be the left ideal in $\mathbf{Cliff}(Z)$ generated by $\mathbf{A}(Z)^+$. Then

$$\mathbf{CM}(Z) = \mathbf{Cliff}(Z)/\mathfrak{J}(Z) = (\wedge \mathbf{A}(Z)^-) \cdot \mathbf{vac}(Z) \tag{1.2}$$

is an irreducible left $\mathbf{Cliff}(Z)$-module, where $\mathbf{vac}(Z) = 1 + \mathfrak{J}(Z)$. Let

$$\mathbf{CM}(Z) = \mathbf{CM}(Z)^0 \oplus \mathbf{CM}(Z)^1 \tag{1.3}$$

be the decomposition into even and odd parity subspaces. For $a, b \in \mathbf{A}$, $k \in \mathbb{Z}$, the expressions in (2.51), (2.54) and (2.55) are well-defined operators on $\mathbf{CM}(Z)$. In Theorem 2.1 we see that, along with the identity transformation, these provide representations of the affine algebra $\hat{\mathbf{g}}$ and the Virasoro algebra $\mathbf{Vir}$ on $\mathbf{CM}(Z)$. For $Z = \mathbb{Z}$ and $\mathbb{Z} + \frac{1}{2}$, the decompositions (1.3) give the four irreducible level 1 $\hat{\mathbf{g}}$-modules, and the central element $z \in \mathbf{Vir}$ acts as the scalar l. Using a $\mathbb{C}$-antilinear involution ϑ on $\mathbf{A}$ satisfying (2.63)-(2.65), a positive Hermitian form

$(\ ,\ )$ is defined on $\mathbf{CM}(Z)$ in (2.66)-(2.68), and the adjoints of all the operators we have defined are given in Proposition 2.2. The eigenspaces of the Virasoro "energy" operator $D(0)$ provide $\mathbf{CM}(Z)$ with a grading in (2.69)-(2.71), and we define the weight of a homogeneous vector v in $\mathbf{CM}(Z)$ by $D(0)v = -\mathrm{wt}(v)v$. For $\mathbf{W} = \mathbf{CM}(Z)^{\alpha}$, $\alpha = 0, 1$, we define $\Delta_{\mathbf{W}}$ to be to be the minimal value of $\mathrm{wt}(w)$ for $w \in \mathbf{W}$, so that $\mathbf{W}$ is $\mathbb{Z} + \Delta_{\mathbf{W}}$-graded.

In Chapter 3, from the spinor constructions of type $D_l^{(1)}$, we construct a vertex operator superalgebra $(\mathbf{CM}(\mathbb{Z}+\tfrac{1}{2}), Y(\ , z), \mathbf{vac}(\mathbb{Z}+\tfrac{1}{2}), D(-2)\mathbf{vac}(\mathbb{Z}+\tfrac{1}{2}))$ and a representation on $\mathbf{CM}(\mathbb{Z})$ of the canonically $\mathbb{Z}_2$-twisted vertex operator superalgebra. In particular, we define a superalgebra structure on $\hat{\mathbf{V}}/D(-1)\hat{\mathbf{V}}$, where

$$\hat{\mathbf{V}} = \mathbf{CM}(\mathbb{Z} + \tfrac{1}{2}) \otimes \mathbb{C}[t, t^{-1}], \tag{1.4}$$

a representation of it on $\mathbf{CM}(\mathbb{Z}+\tfrac{1}{2})$, and we define a superalgebra structure on $\hat{\mathbf{V}}(\vartheta)/D(-1)\hat{\mathbf{V}}(\vartheta)$, where

$$\hat{\mathbf{V}}(\vartheta) = \mathbf{CM}(\mathbb{Z} + \tfrac{1}{2})^0 \otimes \mathbb{C}[t, t^{-1}] \oplus \mathbf{CM}(\mathbb{Z} + \tfrac{1}{2})^1 \otimes t^{1/2}\mathbb{C}[t, t^{-1}], \tag{1.5}$$

and a representation of it on $\mathbf{CM}(\mathbb{Z})$. This is done by defining a generalized vertex operator $Y(v, \zeta)$ on $\mathbf{CM}(\mathbb{Z} + \tfrac{1}{2})$ and on $\mathbf{CM}(\mathbb{Z})$ for any $v \in \mathbf{CM}(\mathbb{Z} + \tfrac{1}{2})$. Actually $Y(v, \zeta)$ is a generating function of operators

$$Y(v, \zeta) = \sum_{n \in \frac{1}{2}\mathbb{Z}} Y_{n+1-\mathrm{wt}(v)}(v)\zeta^{-n-1} = \sum_{n \in \frac{1}{2}\mathbb{Z}} \{v\}_n \zeta^{-n-1} \tag{1.6}$$

where

$$\mathrm{wt}(Y_m(v)w) = \mathrm{wt}(w) - m. \tag{1.7}$$

The coset of a vector $v \otimes t^n$ from (1.4) or (1.5) is represented by the operator $\{v\}_n$. The definition of $Y(v, \zeta)$ on $\mathbf{CM}(\mathbb{Z})$ is more complicated than it is on $\mathbf{CM}(\mathbb{Z} + \tfrac{1}{2})$, but there is a common part which we denote by $\bar{Y}(v, \zeta)$. With $\mathbf{vac} = \mathbf{vac}(\mathbb{Z} + \tfrac{1}{2})$, on $\mathbf{CM}(Z)$ define $\bar{Y}(\mathbf{vac}, \zeta) = 1$ (the identity operator), and for $a \in \mathbf{A}$, $0 \leq n \in \mathbb{Z}$ let

$$\bar{Y}(a(-n - \tfrac{1}{2})\mathbf{vac}, \zeta) = n!^{-1}(d/d\zeta)^n \sum_{m \in Z} a(m)\zeta^{-m-\frac{1}{2}}. \tag{1.8}$$

Using the fermionic normal ordering ${}^{\circ}_{\circ}a_1(n_1) \cdots a_r(n_r){}^{\circ}_{\circ}$ of a product of Clifford generators as defined in (3.16)-(3.20), for vectors of the form $v = a_1(-n_1 - \tfrac{1}{2}) \cdots a_r(-n_r - \tfrac{1}{2})\mathbf{vac} \in \mathbf{CM}(\mathbb{Z} + \tfrac{1}{2})$ we define

$$\bar{Y}(v, \zeta) = {}^{\circ}_{\circ}\bar{Y}(a_1(-n_1 - \tfrac{1}{2})\mathbf{vac}, \zeta) \cdots \bar{Y}(a_r(-n_r - \tfrac{1}{2})\mathbf{vac}, \zeta){}^{\circ}_{\circ} \tag{1.9}$$

and extend the definition to all $v \in \mathbf{CM}(\mathbb{Z} + \tfrac{1}{2})$ by linearity. With $n = 0$ in (1.8) we see that $\bar{Y}_m(a(-\tfrac{1}{2})\mathbf{vac}) = a(m)$ is a Clifford generator, and from (1.9) we have $\bar{Y}_m(a_1(-\tfrac{1}{2})a_2(-\tfrac{1}{2})\mathbf{vac}) = {}^{\circ}_{\circ}a_1(\zeta)a_2(\zeta){}^{\circ}_{\circ m}$ giving the operators (2.51) representing $\hat{\mathfrak{o}}(2l)$. Furthermore, with $\omega = D(-2)\mathbf{vac}$, we have

$$
\bar{Y}(\omega, \zeta) = \begin{cases} \zeta^{-2}D(\zeta) & \text{on } \mathbf{CM}(\mathbb{Z} + \tfrac{1}{2}) \\[2mm] \zeta^{-2}(D(\zeta) + l/8) & \text{on } \mathbf{CM}(\mathbb{Z}) \end{cases} \tag{1.10}
$$

where $D(\zeta) = \sum_{m \in \mathbb{Z}} D(m)\zeta^{-m}$ is the generating function of the Virasoro operators
in (2.54)-(2.55).

We prove algebraic properties of the components $Y_m(v)$ of the vertex operators $Y(v, \zeta)$ by proving analytic properties of the matrix coefficients (3.43), (3.103), such as $(Y(v_1, \zeta_1)Y(v_2, \zeta_2)w, w')$ and $(Y(Y(v_1, \zeta_1 - \zeta_2)v_2, \zeta_2)w, w')$. For $v_1, v_2 \in \mathbf{CM}(\mathbb{Z} + \tfrac{1}{2})$, $w, w' \in \mathbf{CM}(Z)$, these are series which converge absolutely in appropriate domains to algebraic functions of ζ_1 and ζ_2. The algebraic properties follow because, after multiplying by suitable algebraic functions, the resulting Laurent series $F_1(\zeta_2, \zeta_2)$ and $F_2(\zeta_1 - \zeta_2, \zeta_2)$ converge in some open domains in $\mathbb{C}^2$ to the same rational function in the ring $\mathbb{C}[\zeta_1, \zeta_1^{-1}, \zeta_2, \zeta_2^{-1}, (\zeta_1 - \zeta_2)^{-1}]$. We denote this relation by $F_1 \sim F_2$. The foundations of these convergence proofs are contained in Corollaries 3.3 and 3.5. We obtain these results by understanding how to commute exponentials of certain quadratic operators (3.48)-(3.50) on $\mathbb{Z}_2$-graded tensor products of the Clifford modules $\mathbf{CM}(Z_1) \otimes \cdots \otimes \mathbf{CM}(Z_r)$, where each Z_i is either $\mathbb{Z}$ or $\mathbb{Z} + \tfrac{1}{2}$. Theorem 3.6 and Corollary 3.7 show how to express $\bar{Y}(v_1, \zeta_1) \cdots \bar{Y}(v_r, \zeta_r)w$ in terms of these exponentials and some projection operators (3.47).

In order to define the operators $Y(v, \zeta)$ we need the additional quadratic operator $\Delta(\zeta)$ whose definition (3.57) involves the combinatorial coefficients

$$
C_{mn} = \frac{1}{2} \frac{m - n}{m + n + 1} \binom{-\tfrac{1}{2}}{m}\binom{-\tfrac{1}{2}}{n} \tag{1.11}
$$

which satisfy Lemma 3.9. For $v \in \mathbf{CM}(\mathbb{Z} + \tfrac{1}{2})$ we define

$$
Y(v, \zeta) = \begin{cases} \bar{Y}(v, \zeta) & \text{on } \mathbf{CM}(\mathbb{Z} + \tfrac{1}{2}) \\[2mm] \bar{Y}(\exp(\Delta(\zeta))v, \zeta) & \text{on } \mathbf{CM}(\mathbb{Z}). \end{cases} \tag{1.12}
$$

Note that $\exp(\Delta(\zeta))v$ is a finite sum of vectors whose weights are between 0 and $\mathrm{wt}(v)$. A series of technical results culminates in the following. For $v \in \mathbf{CM}(\mathbb{Z} + \tfrac{1}{2})$, $w \in \mathbf{CM}(Z)$, we have

$$
\begin{aligned}
\exp(\zeta D(-1))Y(v, z)\exp(-\zeta D(-1))w &= Y(\exp(\zeta D(-1))v, z)w \\
&= Y(v, z - \zeta)w
\end{aligned} \tag{1.13}
$$

where $Y(v, z - \zeta)w$ is expanded as a series in nonnegative powers of ζ. The first order terms of (1.13) yield

$$
[D(-1), Y(v, z)] = Y(D(-1)v, z) = -(d/dz)Y(v, z). \tag{1.14}
$$

Furthermore, one has

$$\exp(\zeta D(0))Y(v,z)\exp(-\zeta D(0))w = Y(\exp(\zeta D(0))v, ze^{-\zeta})w \qquad (1.15)$$

whose first order terms yield

$$[D(0), Y(v,z)] = -z(d/dz)Y(v,z) + Y(D(0)v,z). \qquad (1.16)$$

Another series of technical results about commutation relations among the exponentials of quadratic operators yields the main theorems of the chapter. Let $v_i \in \mathbf{CM}(\mathbb{Z}+\frac{1}{2})^{\alpha_i}$, $1 \leq i \leq n$, and let $w, w' \in \mathbf{CM}(Z)$. Define

$$\mathbf{s} = \mathbf{s}(v_1, \cdots, v_n; w) = (s_1, \cdots, s_n) \qquad (1.17)$$

where

$$s_i = \begin{cases} \frac{1}{2} & \text{if } \alpha_i = 1 \text{ and } Z = \mathbb{Z} \\[1em] 0 & \text{if } \alpha_i = 0 \text{ or } Z = \mathbb{Z} + \frac{1}{2}, \end{cases} \qquad (1.18)$$

and let $\zeta^{\mathbf{s}} = \displaystyle\prod_{1 \leq i \leq n} \zeta_i^{s_i}$.

RATIONALITY THEOREM. *The series $\zeta^{\mathbf{s}}(Y(v_1,\zeta_1)\cdots Y(v_n,\zeta_n)w, w')$ converges absolutely when $|\zeta_1| > \cdots > |\zeta_n| > 0$ to a rational function in $\mathbb{C}[\zeta_i, \zeta_i^{-1}, (\zeta_j - \zeta_k)^{-1}; 1 \leq i \leq n, 1 \leq j < k \leq n]$.*

Define the right action of a permutation $\sigma \in S_n$ on $\{0,1\}^n$ by $\mathbf{m}\sigma = (m_{\sigma 1}, \cdots, m_{\sigma n})$ where $\mathbf{m} = (m_1, \cdots, m_n) \in \{0,1\}^n$. We call a transposition $\tau \in S_n$ an adjacent transposition if it is of the form $\tau = (i, i+1)$ for some $1 \leq i \leq n-1$. For such $\mathbf{m}$ and τ, define the "fermionic sign"

$$fsgn(\mathbf{m}, \tau) = (-1)^{m_i m_{i+1}}. \qquad (1.19)$$

Writing an arbitrary $\sigma \in S_n$ as a product of adjacent transpositions, we define $fsgn(\mathbf{m}, \sigma)$ in (3.95), and note the validity of (3.96). For $v_1, \cdots, v_n, w, w'$ as above, let $\mathbf{m} = \mathbf{m}(v_1, \cdots, v_n) = (\alpha_1, \cdots, \alpha_n)$.

PERMUTABILITY THEOREM. *For any permutation $\sigma \in S_n$ we have*

$$\zeta^{\mathbf{s}}\, (Y(v_1,\zeta_1)\cdots Y(v_n,\zeta_n)w, w') \sim$$
$$fsgn(\mathbf{m},\sigma)\, \zeta^{\mathbf{s}}\, (Y(v_{\sigma 1},\zeta_{\sigma 1})\cdots Y(v_{\sigma n},\zeta_{\sigma n})w, w').$$

ASSOCIATIVITY THEOREM. *We have*

$$\zeta_1^{s_1}\zeta_2^{s_2}(Y(v_1,\zeta_1)Y(v_2,\zeta_2)w, w') \sim$$
$$\zeta_2^{s_1+s_2}(1+(\zeta_1-\zeta_2)/\zeta_2)^{s_1}(Y(Y(v_1,\zeta_1-\zeta_2)v_2,\zeta_2)w, w')$$

where the series on the right converges absolutely when $0 < |\zeta_1 - \zeta_2| < |\zeta_2|$.

Combining these three analytic theorems with the Cauchy integral theorem, we get the algebraic properties of vertex operators.

JACOBI IDENTITY. *Let $f(\zeta_1, \zeta_2) \in \zeta_1^{s_1} \zeta_2^{s_2} \mathbb{C}[\zeta_1, \zeta_1^{-1}, \zeta_2, \zeta_2^{-1}, (\zeta_1 - \zeta_2)^{-1}]$, let $C_i(r)$ be a circle in the ζ_i-plane of radius r centered at the origin, and let $C_1(\zeta_2, \epsilon)$ be a circle in the ζ_1-plane of radius ϵ centered at $\zeta_1 = \zeta_2$. For $0 < r < \rho < R$ and $0 < \epsilon < \min\{R - \rho, \rho - r\}$ we have*

$$\int_{C_2(\rho)} \int_{C_1(R)} (Y(v_1, \zeta_1)Y(v_2, \zeta_2)w, w')f(\zeta_1, \zeta_2)d\zeta_1 d\zeta_2$$

$$- \int_{C_2(\rho)} \int_{C_1(r)} (-1)^{\alpha_1 \alpha_2} (Y(v_2, \zeta_2)Y(v_1, \zeta_1)w, w')f(\zeta_1, \zeta_2)d\zeta_1 d\zeta_2$$

$$= \int_{C_2(\rho)} \int_{C_1(\zeta_2, \epsilon)} (Y(Y(v_1, \zeta_1 - \zeta_2)v_2, \zeta_2)w, w')f(\zeta_1, \zeta_2)d\zeta_1 d\zeta_2$$

where the function $f(\zeta_1, \zeta_2)$ is to be expanded in the three integrands as in (0.48)-(0.50), respectively, with $z = \zeta_1$ and $z_0 = \zeta_2$.

For operators L_1 and L_2 on $\mathbf{CM}(Z)$ let $[L_1, L_2]_{\pm} = L_1 L_2 \pm L_2 L_1$.

COROLLARY 1. *For $m \in \mathbb{Z} + s_1$, $n \in \mathbb{Z} + s_2$, on $\mathbf{CM}(Z)$, we have*

$$\sum_{0 \leq i \in \mathbb{Z}} (-1)^i \binom{r}{i} (\{v_1\}_{m+r-i}\{v_2\}_{n+i} - (-1)^{\alpha_1 \alpha_2 + r}\{v_2\}_{n+r-i}\{v_1\}_{m+i})$$

$$= \sum_{0 \leq k \in \mathbb{Z}} \binom{m}{k} \{\{v_1\}_{r+k} v_2\}_{m+n-k}.$$

The sum over k is finite, and $0 \leq k \leq wt(v_1) + wt(v_2) - r - 1$.

COROLLARY 2. *For $m \in \mathbb{Z} + s_1$, $n \in \mathbb{Z} + s_2$ on $\mathbf{CM}(Z)$, we have*

$$[\{v_1\}_m, \{v_2\}_n]_{\pm} = \sum_{0 \leq k \in \mathbb{Z}} \binom{m}{k} \{\{v_1\}_k v_2\}_{m+n-k}$$

where $\pm$ is $-(-1)^{\alpha_1 \alpha_2}$. This sum is finite, and $0 \leq k \leq wt(v_1) + wt(v_2) - 1$.

$D_l^{(1)}$ VERTEX OPERATOR SUPERALGEBRA THEOREM. *With $\mathbf{V} = \mathbf{CM}(\mathbb{Z} + \frac{1}{2})$, $\mathbf{1} = \mathbf{vac}(\mathbb{Z} + \frac{1}{2})$ and $\omega = D(-2)\mathbf{1}$, $(\mathbf{V}, Y(\ , z), \mathbf{1}, \omega)$ is a vertex operator superalgebra as defined in the Introduction. With $\mathbf{W} = \mathbf{CM}(\mathbb{Z})$, $(\mathbf{W}, Y(\ , z))$ is a canonically $\mathbb{Z}_2$-twisted $\mathbf{V}$- module.*

Corollary 2 and (1.14) show that, representing the coset of $v \otimes t^n$ by the operator $\{v\}_n$, we have superalgebra structures on $\hat{\mathbf{V}}/D(-1)\hat{\mathbf{V}}$ and $\hat{\mathbf{V}}(\vartheta)/D(-1)\hat{\mathbf{V}}(\vartheta)$ (recall (1.4) and (1.5)) represented, repectively, on $\mathbf{CM}(\mathbb{Z} + \frac{1}{2})$ and on $\mathbf{CM}(\mathbb{Z})$.

For $a_1, a_2 \in \mathbf{A}$, $x = {}^{\circ}_{\circ} a_1 a_2 {}^{\circ}_{\circ} \in \mathbf{o}(2l)$, $m \in \mathbb{Z}$, define the notation

$$x(m) = Y_m(a_1(-\tfrac{1}{2})a_2(-\tfrac{1}{2})\mathbf{vac}(\mathbb{Z} + \tfrac{1}{2})) = \{a_1(-\tfrac{1}{2})a_2(-\tfrac{1}{2})\mathbf{vac}(\mathbb{Z} + \tfrac{1}{2})\}_m \quad (1.20)$$

and note that

$$x(m) = \sum_{k \in \frac{1}{2}\mathbb{Z}} {}^{\circ}_{\circ} a_1(k)a_2(m - k) {}^{\circ}_{\circ}. \quad (1.21)$$

As was stated in Theorem 2.1, these are the operators on $\mathbf{CM}(Z)$ representing $\hat{\mathbf{o}}(2l)$.

COROLLARY 3. *For $v \in \mathbf{CM}(\mathbb{Z} + \frac{1}{2})$, $m \in \mathbb{Z}$, on $\mathbf{CM}(Z)$ we have*

(a) $\displaystyle [D(m), Y(v, z)] = \sum_{0 \leq k \in \mathbb{Z}} \binom{m+1}{k} z^{m+1-k} Y(D(k-1)v, z)$

and the sum is finite,

(b) *If $D(n)v = 0$ for all $n > 0$, then we have*

$$[D(m), Y(v, z)] = -z^{m+1}(d/dz)Y(v, z) + (m+1)z^m Y(D(0)v, z).$$

COROLLARY 4. *For $v \in \mathbf{CM}(\mathbb{Z} + \frac{1}{2})$, $m \in \mathbb{Z}$, on $\mathbf{CM}(Z)$ we have*

$$[x(m), Y(v, z)] = z^m \sum_{0 \leq k \in \mathbb{Z}} \binom{m}{k} z^{-k} Y(x(k)v, z).$$

COROLLARY 5. *For $v \in \mathbf{CM}(\mathbb{Z} + \frac{1}{2})$, $m \in \mathbb{Z}$, on $\mathbf{CM}(Z)$ we have*

$$Y(x(-m)v, z) =$$
$$\sum_{0 \leq i \in \mathbb{Z}} \binom{m+i-1}{i} [z^i x(-m-i) Y(v, z) - (-1)^m z^{-m-i} Y(v, z) x(i)],$$

$$Y(D(-m-1)v, z) =$$
$$\sum_{0 \leq i \in \mathbb{Z}} \binom{m+i-1}{i} [z^i D(-m-i-1) Y(v, z) - (-1)^m z^{-m-i} Y(v, z) D(i-1)].$$

Corollaries 4 and 5 will be used in an essential way in induction arguments in Chapters 5 and 6. Note that Corollary 3 generalizes (1.14) and (1.16).

The principle of triality for D_4 is more than just an S_3 acting as Lie algebra automorphisms on $\mathbf{o}(8)$. It also involves an action of S_3 on the direct sum of the three 8-dimensional $\mathbf{o}(8)$-modules intertwining the action of $\mathbf{o}(8)$ on the modules. The Chevalley algebra $[\mathbf{C}]$ is a commutative nonassociative algebra structure on the direct sum of the three 8-dimensional $\mathbf{o}(8)$-modules. In Chapter 4 we present the spinor construction of the Chevalley algebra and triality for D_4. Later we will see how the Chevalley algebra gives a spinor construction of E_8.

With $l = 4$, let $\mathbf{A}$, $\mathbf{Cliff}$, $\mathbf{CM}$, $\mathbf{vac}$ and ϑ be as described in Chapter 2. Let $\mathbf{C} = \mathbf{A} \oplus \mathbf{CM}$, and using the main antiautomorphism (4.6) of $\mathbf{Cliff}$, in (4.7) we extend the form $\langle \, , \, \rangle$ from $\mathbf{A}$ to $\mathbf{C}$. For $u \in \mathbf{C}$, with $u = a + b + c$, $a \in \mathbf{A}$, $b \in \mathbf{CM}^0$, $c \in \mathbf{CM}^1$, define the cubic form $F(u) = \langle ab, c \rangle$. By polarization, this gives a symmetric trilinear form Φ on $\mathbf{C}$, which allows us to define an operation $\circ$ on $\mathbf{C}$. For $u_1, u_2 \in \mathbf{C}$, let $u_1 \circ u_2$ be the unique element of $\mathbf{C}$ such that for all $u \in \mathbf{C}$, we have

$$\Phi(u_1, u_2, u) = \langle u_1 \circ u_2, u \rangle. \tag{1.22}$$

This operation is commutative and agrees with the Clifford action of $\mathbf{A}$ on $\mathbf{CM}$.

We introduce the notation

$$\mathbf{C}^{(1)} = \mathbf{A}, \quad \mathbf{C}^{(2)} = \mathbf{CM}^0, \quad \mathbf{C}^{(3)} = \mathbf{CM}^1, \tag{1.23}$$

and for $a \in \mathbf{C}^{(k)}$, $1 \le k \le 3$, such that $\langle a, a \rangle \neq 0$, let r_a be the reflection with respect to a on $\mathbf{C}^{(k)}$. For $e \in \mathbf{C}^{(k)}$ such that $\langle e, e \rangle = 2$, define $\rho(e) : \mathbf{C} \to \mathbf{C}$ by

$$\rho(e)u = \begin{cases} -r_e(u) & \text{for } u \in \mathbf{C}^{(k)} \\[2ex] e \circ u & \text{for } u \in \mathbf{C}^{(k')} \oplus \mathbf{C}^{(k'')} \end{cases} \tag{1.24}$$

where (k, k', k'') is a cyclic permutation of $(1, 2, 3)$. Then $\rho(e)$ is an order 2 automorphism of the algebra $(\mathbf{C}, \circ)$. Choosing $\mathbf{e}_1 \in \mathbf{C}^{(1)}$ and $\mathbf{e}_2 \in \mathbf{C}^{(2)}$, let

$$\sigma = \rho(\mathbf{e}_1)\rho(\mathbf{e}_2), \quad \tau = \rho(\mathbf{e}_1). \tag{1.25}$$

Then $\sigma^3 = 1 = \tau^2 = (\sigma\tau)^2$, so the group G generated by σ and τ is a group of automorphisms of the algebra $(\mathbf{C}, \circ)$ isomorphic to the symmetric group S_3. Furthermore, σ cyclically permutes $\mathbf{C}^{(1)}$, $\mathbf{C}^{(2)}$ and $\mathbf{C}^{(3)}$, while τ preserves $\mathbf{C}^{(1)}$ and switches $\mathbf{C}^{(2)}$ and $\mathbf{C}^{(3)}$. For $1 \le k \le 3$, let $\mathbf{Cliff}^{(k)}$ be the Clifford algebra generated by $\mathbf{C}^{(k)}$ and the form $\langle \, , \, \rangle$. Then

$$\mathbf{CM}^{(k)} = \mathbf{C}^{(k')} \oplus \mathbf{C}^{(k'')} \tag{1.26}$$

is an irreducible $\mathbf{Cliff}^{(k)}$-module under the action given by $\circ$, and the even and odd parity subspaces are

$$\mathbf{CM}^{(k)0} = \mathbf{C}^{(k')} \quad \text{and} \quad \mathbf{CM}^{(k)1} = \mathbf{C}^{(k'')}. \tag{1.27}$$

We may choose $\mathbf{vac}$ and a nonzero vector $\vartheta(\mathbf{vac}) \in (\wedge^4 \mathbf{A}^-) \cdot \mathbf{vac}$ such that $\langle \mathbf{vac}, \vartheta(\mathbf{vac}) \rangle = 1$ and the $\mathbb{C}$-antilinear involution ϑ on $\mathbf{A}$ can be extended to a $\mathbb{C}$-antilinear order 2 automorphism of the algebra $(\mathbf{C}, \circ)$ preserving each $\mathbf{C}^{(k)}$. If we choose $\mathbf{e}_1$ and $\mathbf{e}_2$ to be fixed by ϑ then ϑ commutes with G on $\mathbf{C}$.

For $1 \le k \le 3$, the span of $\{{}^{\circ}_{\circ}ab{}^{\circ}_{\circ} = \frac{1}{2}(ab - ba) \mid a, b \in \mathbf{C}^{(k)}\}$ in $\mathbf{Cliff}^{(k)}$ is a Lie algebra $\mathbf{g}^{(k)}$ of operators on $\mathbf{C}$. This is just a type D_4 spinor construction as given in Chapter 2. Under the action

$$g \cdot {}^{\circ}_{\circ}ab{}^{\circ}_{\circ} = {}^{\circ}_{\circ}(g \cdot a)(g \cdot b){}^{\circ}_{\circ}, \quad g \in G, \tag{1.28}$$

G permutes the three algebras $\mathbf{g}^{(k)}$, $1 \le k \le 3$, as it permutes the $\mathbf{g}^{(k)}$-modules $\mathbf{C}^{(k)}$, the Clifford algebras $\mathbf{Cliff}^{(k)}$ and the $\mathbf{Cliff}^{(k)}$-modules $\mathbf{CM}^{(k)}$. For $g \in G$, $x \in \mathbf{g}^{(k)}$, $u \in \mathbf{C}$, we have

$$g \cdot (x \cdot u) = (g \cdot x) \cdot (g \cdot u), \tag{1.29}$$

and $\mathbf{g}^{(k)}$ acts as derivations of $(\mathbf{C}, \circ)$. For $a_k \in \mathbf{C}^{(k)}$, $1 \le k \le 3$, as operators on $\mathbf{C}$ we prove that

$$\,{}^{\circ}_{\circ}a_1(a_2 \circ a_3){}^{\circ}_{\circ} + \,{}^{\circ}_{\circ}a_2(a_3 \circ a_1){}^{\circ}_{\circ} + \,{}^{\circ}_{\circ}a_3(a_1 \circ a_2){}^{\circ}_{\circ} = 0. \tag{1.30}$$

This implies that as Lie algebras of operators on $\mathbf{C}$, $\mathbf{g}^{(1)} = \mathbf{g}^{(2)} = \mathbf{g}^{(3)}$, so that we have one Lie algebra, $\mathbf{o}(8)$, of operators on $\mathbf{C}$, constructed in three ways. From (1.28) and (1.29), G acts as Lie algebra automorphisms of $\mathbf{o}(8)$ intertwining the action of $\mathbf{o}(8)$ on $\mathbf{C}$. With the additional restriction $\langle \mathbf{e}_2, \mathbf{vac} \rangle = 0$ on the choice of $\mathbf{e}_2$, we can choose a Cartan subalgebra $\mathbf{h}$ in $\mathbf{o}(8)$ such that the action of G permutes the weight spaces of $\mathbf{C}$ with respect to $\mathbf{h}$. This will be of use in later proofs.

We can define a positive Hermitian form on $\mathbf{C}$ by $(a,b) = \langle a, \vartheta b \rangle$ for $a, b \in \mathbf{C}$. Then G acts as unitary transformations of $\mathbf{C}$ and the adjoint with respect to $(\ ,\ )$ of $x = {}^{\circ}_{\circ} ab {}^{\circ}_{\circ} \in \mathbf{o}(8)$ is $x^* = {}^{\circ}_{\circ}(\vartheta b)(\vartheta a){}^{\circ}_{\circ}$.

We also show how the product $*$ on $\mathbf{C}^{(1)}$ defined by

$$a * b = \sigma\tau(a) \circ \sigma^2\tau(b) \tag{1.31}$$

gives $\mathbf{C}^{(1)}$ the structure of the complex octonians with $\mathbf{e}_1$ the unit element and τ the conjugation. (See also [**GNORS**].) The principle of local triality is then a restatement of the fact that $\mathbf{o}(8)$ acts as derivations of $(\mathbf{C}, \circ)$. We conclude this chapter with a listing of certain tensor product decompositions which are used in later proofs.

In Chapter 5 we lift the action of the triality group $G \cong S_3$ from $\mathbf{o}(8)$ and $\mathbf{C}$ to the affine Lie algebra $\hat{\mathbf{o}}(8)$ and the direct sum of its level 1 modules. Let $\mathbf{CM}(Z)$ be the $\mathbf{Cliff}(Z)$-module for $l = 4$ constructed in Chapter 2. Let

$$\mathbf{U}_0 = \mathbf{CM}(\mathbb{Z}+\tfrac{1}{2})^0, \ \mathbf{U}_1 = \mathbf{CM}(\mathbb{Z}+\tfrac{1}{2})^1, \ \mathbf{U}_2 = \mathbf{CM}(\mathbb{Z})^0, \ \mathbf{U}_3 = \mathbf{CM}(\mathbb{Z})^1 \tag{1.32}$$

and let

$$\mathbf{U} = \mathbf{U}_0 \oplus \mathbf{U}_1 \oplus \mathbf{U}_2 \oplus \mathbf{U}_3. \tag{1.33}$$

The lifting $\hat{\sigma}$ of σ to $\mathbf{U}$ allows the definition of vertex operators $Y(u, \zeta)$ for all $u \in \mathbf{U}$. For any subspace $\mathbf{S}$ of $\mathbf{U}$ and any $n \in \tfrac{1}{2}\mathbb{Z}$, $(\mathbf{S})_n$ is the set of all vectors in $\mathbf{S}$ of weight n. Define a unitary isomorphism $\Upsilon : \mathbf{C} \to (\mathbf{U})_{1/2}$ by

$$\begin{aligned}
\Upsilon(a) &= a(-\tfrac{1}{2})\mathbf{vac}(\mathbb{Z}+\tfrac{1}{2}) \ \text{ for } \ a \in \mathbf{A}, \\
\Upsilon(\mathbf{vac}) &= \mathbf{vac}(\mathbb{Z}), \ \ \Upsilon(a \circ v) = a(0)\Upsilon(v) \ \text{ for } \ a \in \mathbf{A}, \ v \in \mathbf{CM}.
\end{aligned} \tag{1.34}$$

Then we have

$$\Upsilon(x \cdot u) = x(0)\Upsilon(u) \ \text{ for } \ x \in \mathbf{o}(8), \ u \in \mathbf{C}. \tag{1.35}$$

We use Υ to transport the action of G from $\mathbf{C}$ to $(\mathbf{U})_{1/2}$,

$$g\Upsilon(u) = \Upsilon(gu) \ \text{ for } \ g \in G, \ u \in \mathbf{C}. \tag{1.36}$$

We lift the action of G from $\mathbf{o}(8)$ to $\hat{\mathbf{o}}(8)$ by

$$\hat{g}(x(n)) = (gx)(n), \ \ \hat{g}(c) = c, \ \ \hat{g}(d) = d. \tag{1.37}$$

From (1.35) and (1.36) we see that $g : (\mathbf{U})_{1/2} \to (\mathbf{U})_{1/2}$ satisfies

$$gx(0)g^{-1} = (gx)(0). \tag{1.38}$$

LIFTING THEOREM. *The triality group G, acting on $\mathbf{C}$ so as to commute with ϑ, lifts uniquely to a group of unitary automorphisms $\hat{g} : \mathbf{U} \to \mathbf{U}$ such that $\hat{g}$ restricts to the identity operator on $(\mathbf{U})_0$, $\hat{g}$ restricts to g on $(\mathbf{U})_{1/2}$, and $\hat{g}x(n)\hat{g}^{-1} = (gx)(n)$ on $\mathbf{U}$ for $x(n) \in \hat{\mathbf{o}}(8)$. Furthermore, G commutes with the adjoint $*$ on $\hat{\mathbf{o}}(8)$, and G commutes with the Virasoro operators $D(n)$, $n \in \mathbb{Z}$, on $\mathbf{U}$.*

For $u \in \mathbf{U}_1$ define the vertex operators

$$Y(\hat{\sigma}u, \zeta) = \hat{\sigma}Y(u, \zeta)\hat{\sigma}^{-1}, \quad Y(\hat{\sigma}^2 u, \zeta) = \hat{\sigma}^2 Y(u, \zeta)\hat{\sigma}^{-2}. \tag{1.39}$$

For all homogeneous vectors $u \in \mathbf{U}$ this defines

$$Y(u, \zeta) = \sum_{n \in \frac{1}{2}\mathbb{Z}} Y_{n+1-\mathrm{wt}(u)}(u)\zeta^{-n-1} = \sum_{n \in \frac{1}{2}\mathbb{Z}} \{u\}_n \zeta^{-n-1}. \tag{1.40}$$

INTERTWINING LEMMA. *For $g \in G$, $u \in \mathbf{U}_0$, on $\mathbf{U}$ we have*

$$\hat{g}Y(u, \zeta)\hat{g}^{-1} = Y(\hat{g}u, \zeta).$$

From (1.39) and this lemma we see that for all $u \in \mathbf{U}$,

$$Y(\hat{\sigma}u, \zeta) = \hat{\sigma}Y(u, \zeta)\hat{\sigma}^{-1}. \tag{1.41}$$

Let $\mathbf{Cliff}^{(k)}(Z)$ be the Clifford algebra generated by

$$\mathbf{C}^{(k)}(Z) = \{Y_m(\Upsilon(a)) \mid a \in \mathbf{C}^{(k)}, m \in Z\} \tag{1.42}$$

and the form

$$\langle Y_m(\Upsilon(a)), Y_n(\Upsilon(b)) \rangle = \langle a, b \rangle \delta_{m,-n}. \tag{1.43}$$

Conjugation of the Clifford relations in $\mathbf{Cliff}(Z)$ by $\hat{\sigma}$ shows that we have an irreducible representation of $\mathbf{Cliff}^{(k)}(Z)$ on

$$\mathbf{CM}^{(k)}(Z) = \begin{cases} \mathbf{U}_0 \oplus \mathbf{U}_k & \text{if } Z = \mathbb{Z} + \frac{1}{2} \\[2ex] \mathbf{U}_{k'} \oplus \mathbf{U}_{k''} & \text{if } Z = \mathbb{Z} \end{cases} \tag{1.44}$$

and these are the decompositions into even and odd parity subspaces. We thus have three spinor constructions of a type $D_4^{(1)}$ affine algebra and of the Virasoro algebra on $\mathbf{U}$. Combining the Lifting Theorem and the Intertwining Lemma, we see that these three constructions coincide, providing three ways of constructing one $\hat{\mathbf{o}}(8)$ and one $\mathbf{Vir}$ on $\mathbf{U}$. The bracket formulas

$$[x(m), Y_n(\Upsilon(a))] = Y_{m+n}(\Upsilon(x \cdot a)) \tag{1.45}$$

and

$$[D(m), Y_n(\Upsilon(a))] = (n + \tfrac{1}{2}m)Y_{m+n}(\Upsilon(a)) \tag{1.46}$$

for $x \in \mathbf{o}(8)$, $a \in \mathbf{C}$, $m \in \mathbb{Z}$, $n \in \frac{1}{2}\mathbb{Z}$, play an important role in later proofs. Also note that for $u \in \mathbf{U}_i$, $0 \le i, j \le 3$, $n \in \frac{1}{2}\mathbb{Z}$, we have

$$Y_n(u) : \mathbf{U}_j \to \mathbf{U}_{i+j} \tag{1.47}$$

where we have put the additive $\mathbb{Z}_2 \times \mathbb{Z}_2$ group structure on $\Gamma = \{0, 1, 2, 3\}$. The group $G \cong \mathrm{Aut}(\mathbb{Z}_2 \times \mathbb{Z}_2)$ also acts on Γ such that $\hat{g}(\mathbf{U}_i) = \mathbf{U}_{g(i)}$.

All the results of Chapter 3 can also be applied from any of the three points of view, $1 \leq k \leq 3$, provided by the three Clifford constructions of $\mathbf{U}$. We have a vertex operator superalgebra $(\mathbf{U}_0 \oplus \mathbf{U}_k, Y(\ ,\zeta), \mathbf{vac}(\mathbb{Z} + \frac{1}{2}), D(-2)\mathbf{vac}(\mathbb{Z} + \frac{1}{2}))$ and its canonically $\mathbb{Z}_2$-twisted representation on $\mathbf{U}_{k'} \oplus \mathbf{U}_{k''}$. In particular, from Corollary 2 we have a superalgebra structure on $(\hat{\mathbf{U}}_0 \oplus \hat{\mathbf{U}}_k)/D(-1)(\hat{\mathbf{U}}_0 \oplus \hat{\mathbf{U}}_k)$, where

$$(\hat{\mathbf{U}}_0 \oplus \hat{\mathbf{U}}_k) = (\mathbf{U}_0 \oplus \mathbf{U}_k) \otimes \mathbb{C}[t, t^{-1}], \tag{1.48}$$

represented on $\mathbf{U}_0 \oplus \mathbf{U}_k$, and a superalgebra structure on $(\hat{\mathbf{U}}_0 \oplus \hat{\mathbf{U}}'_k)/D(-1)(\hat{\mathbf{U}}_0 \oplus \hat{\mathbf{U}}'_k)$, where

$$(\hat{\mathbf{U}}_0 \oplus \hat{\mathbf{U}}'_k) = \mathbf{U}_0 \otimes \mathbb{C}[t, t^{-1}] \oplus \mathbf{U}_k \otimes t^{1/2}\mathbb{C}[t, t^{-1}], \tag{1.49}$$

represented on $\mathbf{U}_{k'} \oplus \mathbf{U}_{k''}$. Furthermore, we can apply Corollaries 4 and 5 for all $v \in \mathbf{U}$. Along with certain tensor product decompositions given at the end of Chapter 4, these allow us to prove the last theorem of Chapter 5.

INTERTWINING THEOREM. *For $g \in G$, $u \in \mathbf{U}$, on $\mathbf{U}$ we have*

$$\hat{g}Y(u, \zeta)\hat{g}^{-1} = Y(\hat{g}u, \zeta). \tag{1.50}$$

In Chapter 6 we extend the Rationality, Permutability and Associativity Theorems to apply to operators $Y(u_1, \zeta_1)$ and $Y(u_2, \zeta_2)$ for any $u_1, u_2 \in \mathbf{U}$. They play a vital role in the spinor construction of $E_8^{(1)}$ in Chapter 8. They also imply the Jacobi Identity (0.46), showing that for appropriate Δ and η, $(\mathbf{U}, Y(\ ,\zeta), \mathbf{vac}(\mathbb{Z} + \frac{1}{2}), D(-2)\mathbf{vac}(\mathbb{Z} + \frac{1}{2}), \Gamma, \Delta, \eta)$ is a vertex operator paraalgebra as defined in the Introduction.

In Propositions 6.2 and 6.5 we prove identities which serve as the base cases for inductive proofs of the Rationality, Permutability and Associativity Theorems. We now set up notation for the following theorems. For $0 \leq m, n \leq 3$ define

$$s(m, n) = \begin{cases} \frac{1}{2} & \text{if } 0 \neq m \neq n \neq 0 \\ \\ 0 & \text{otherwise.} \end{cases} \tag{1.51}$$

Let $u_1 \in \mathbf{U}_{n_1}$, $u_2 \in \mathbf{U}_{n_2}$, $u \in \mathbf{U}_n$, $u' \in \mathbf{U}$ and let $s_1 = s(n_1, n)$, $s_2 = s(n_2, n)$, $s_{12} = s(n_1, n_2)$.

RATIONALITY THEOREM. *The series*

$$\zeta_1^{s_1} \zeta_2^{s_2} \zeta_1^{s_{12}} (1 - \zeta_2/\zeta_1)^{s_{12}} (Y(u_1, \zeta_1)Y(u_2, \zeta_2)u, u')$$

converges absolutely when $|\zeta_1| > |\zeta_2| > 0$ to a rational function in $\mathbb{C}[\zeta_1, \zeta_1^{-1}, \zeta_2, \zeta_2^{-1}, (\zeta_1 - \zeta_2)^{-1}]$.

It is obvious how to modify the statement of this theorem to assert the rationality of matrix coefficients for a product of more than two vertex operators, and we prove Theorem 3.28 in that generality. But since we only prove Theorem

6.6 for two vertex operators, we can only prove this theorem for two operators. While the inductive steps of the proof of Theorem 6.6 seem to work just as well for more than two operators, the base cases seem to become more and more complex as the number of operators increases. In principle our techniques could establish these base cases, but in order to prove the Jacobi Identity we only need the above theorem, so we will not pursue the matter here.

PERMUTABILITY THEOREM. *We have*

$$\zeta_1^{s_1}\zeta_2^{s_2}\zeta_1^{s_{12}}(1-\zeta_2/\zeta_1)^{s_{12}}(Y(u_1,\zeta_1)Y(u_2,\zeta_2)u,u') \sim$$
$$\kappa\,\zeta_1^{s_1}\zeta_2^{s_2}\zeta_2^{s_{12}}(1-\zeta_1/\zeta_2)^{s_{12}}(Y(u_2,\zeta_2)Y(u_1,\zeta_1)u,u')$$

where

$$\kappa = \begin{cases} 1 & \text{if } n_1 = 0 \text{ or } n_2 = 0 \\ -1 & \text{if } n_1 = n_2 \neq 0 \\ 1 & \text{if } 0 \neq n_1 \neq n_2 \neq 0 \text{ and } (n = 0 \text{ or } n = n_1 + n_2) \\ -1 & \text{if } 0 \neq n_1 \neq n_2 \neq 0 \text{ and } (n = n_1 \text{ or } n = n_2). \end{cases}$$

ASSOCIATIVITY THEOREM. *We have*

$$\zeta_1^{s_1}\zeta_2^{s_1}\zeta_1^{s_{12}}(1-\zeta_2/\zeta_1)^{s_{12}}(Y(u_1,\zeta_1)Y(u_2,\zeta_2)u,u') \sim$$
$$\kappa'\zeta_2^{s_1+s_2}(\zeta_1-\zeta_2)^{s_{12}}(1+(\zeta_1-\zeta_2)/\zeta_2)^{s_1}(Y(Y(u_1,\zeta_1-\zeta_2)u_2,\zeta_2)u,u')$$

where

$$\kappa' = \begin{cases} 1 & \text{if } n_1 = 0 \text{ or } n_2 = 0 \text{ or } n_1 = n_2 \\ 1 & \text{if } 0 \neq n_1 \neq n_2 \neq 0 \text{ and } (n = 0, n_2, n_1 + n_2) \\ -1 & \text{if } 0 \neq n_1 \neq n_2 \neq 0 \text{ and } n = n_1 \end{cases}$$

and the second series converges absolutely when $0 < |\zeta_1 - \zeta_2| < |\zeta_2|$.

As in Chapter 3, the Rationality, Permutability and Associativity Theorems above imply a modified version of the Jacobi Identity in which $fsgn$ is replaced by κ, and κ' appears on the right side. This does not give the Jacobi Identity axiom (0.46) for vertex operator para-algebras because κ depends on n_1, n_2 and n, and κ' is not identically 1. To solve this problem we modify these factors by the rescaling

$$Y^\mu(u_1,\zeta)u = \mu(n_1,n)Y(u_1,\zeta)u \tag{1.52}$$

for $u_1 \in \mathbf{U}_{n_1}$, $u \in \mathbf{U}_n$ and $\mu(n_1,n) \in \mathbb{C}^*$. As a 2-cochain, μ has coboundary $\delta\mu(n_1,n_2,n) = \mu(n_2,n)\mu(n_1,n_2+n)\mu(n_1,n_2)^{-1}\mu(n_1+n_2,n)^{-1}$. With Y replaced by Y^μ, the Rationality Theorem still holds, the Permutability Theorem holds with κ replaced by

$$\kappa\frac{\mu(n_2,n)\mu(n_1,n_2+n)}{\mu(n_1,n)\mu(n_2,n_1+n)} = \kappa\frac{\mu(n_1,n_2)}{\mu(n_2,n_1)}\frac{\delta\mu(n_1,n_2,n)}{\delta\mu(n_2,n_1,n)} \tag{1.53}$$

and the Associativity Theorem still holds with κ' replaced by

$$\kappa'\frac{\mu(n_2,n)\mu(n_1,n_2+n)}{\mu(n_1,n_2)\mu(n_1+n_2,n)} = \kappa'\delta\mu(n_1,n_2,n). \tag{1.54}$$

We want to choose μ such that (1.54) equals 1 for all $0 \le n, n_1, n_2 \le 3$, and (1.53) equals a function $\beta(n_1,n_2)$ which is independent of n, and $\beta(n_1,n_2)$ equals $f\,sgn$ when $\{n_1,n_2\} \subseteq \{0,k\}$ for some $1 \le k \le 3$. We can achieve these goals by choosing $\mu(1,2)$, $\mu(2,3)$ and $\mu(3,1)$ arbitrarily,

$$\mu(0,k) = \mu(k,0) = \mu(0,0) = 1,$$
$$\mu(k',k) = \mathbf{i}^{-\mathbf{a}}\mu(k,k'), \tag{1.55}$$
$$\mu(k,k) = \mathbf{i}^{-\mathbf{a}}\mu(k,k')\mu(k'',k)$$

for $1 \le k \le 3$, where $\mathbf{a} = \pm 1$ is fixed independent of k. Then we get

$$\beta(n_1,n_2) = \begin{cases} 1 & \text{if } n_1 = 0 \text{ or } n_2 = 0 \\[2ex] -1 & \text{if } n_1 = n_2 \ne 0 \\[2ex] \mathbf{i}^{\mathbf{a}} & \text{if } n_1 \ne 0 \text{ and } n_2 = n_1' \\[2ex] \mathbf{i}^{-\mathbf{a}} & \text{if } n_1 \ne 0 \text{ and } n_2 = n_1''. \end{cases} \tag{1.56}$$

Before we prove the Jacobi Identity (0.46) for the operators Y^μ, we need a few more definitions. For $0 \le k \le 3$ define Δ_k to be the minimal value of $\mathrm{wt}(u)$ for $u \in \mathbf{U}_k$. From (2.71) or (3.6) with $l = 4$, we see that $\Delta_0 = 0$ and $\Delta_k = \frac{1}{2}$ for $1 \le k \le 3$. Let $\Delta(n_1,n_2)$ be defined as in (0.41) and check that $\Delta(n_1,n_2)$ is bilinear $mod\ \mathbb{Z}$. Also note that $\Delta(n_1,n_2) = s(n_1,n_2)$ unless $n_1 = n_2 \ne 0$, in which case $\Delta(n_1,n_2) = 1$ but $s(n_1,n_2) = 0$. For $n_1, n_2 \in \Gamma$ define

$$\eta(n_1,n_2) = \beta(n_1,n_2)\, e^{-\pi i s_{12}}$$

$$= \begin{cases} 1 & \text{if } n_1 = 0 \text{ or } n_2 = 0 \\[2ex] -1 & \text{if } n_1 = n_2 \ne 0 \\[2ex] \mathbf{a} & \text{if } n_1 \ne 0 \text{ and } n_2 = n_1' \\[2ex] -\mathbf{a} & \text{if } n_1 \ne 0 \text{ and } n_2 = n_1''. \end{cases} \tag{1.57}$$

and check that η is bilinear and that $\eta(n_1, n_2)\, e^{\pi i \Delta(n_1, n_2)}$ is skew-symmetric.

JACOBI IDENTITY. *With notations as above, let*

$$f(\zeta_1, \zeta_2) \in \zeta_1^{\Delta(n_1, n)} \, \zeta_2^{\Delta(n_2, n)} \, (\zeta_1 - \zeta_2)^{\Delta(n_1, n_2)} \, \mathbb{C}[\zeta_1, \zeta_1^{-1}, \zeta_2, \zeta_2^{-1}, (\zeta_1 - \zeta_2)^{-1}],$$

let $C_i(r)$ be a circle in the ζ_i-plane of radius r centered at the origin, and let $C_1(\zeta_2, \epsilon)$ be a circle in the ζ_1-plane of radius ϵ centered at $\zeta_1 = \zeta_2$. Then for $0 < r < \rho < R$ and for $0 < \epsilon < \min(R - \rho, \rho - r)$, we have

$$\int_{C_2(\rho)} \int_{C_1(R)} (Y^\mu(u_1, \zeta_1) Y^\mu(u_2, \zeta_2) u, u') f(\zeta_1, \zeta_2) d\zeta_1 d\zeta_2$$

$$- \int_{C_2(\rho)} \int_{C_1(r)} \eta(n_1, n_2)(Y^\mu(u_2, \zeta_2) Y^\mu(u_1, \zeta_1) u, u') f(\zeta_1, \zeta_2) d\zeta_1 d\zeta_2$$

$$= \int_{C_2(\rho)} \int_{C_1(\zeta_2, \epsilon)} (Y^\mu(Y^\mu(u_1, \zeta_1 - \zeta_2) u_2, \zeta_2) u, u') f(\zeta_1, \zeta_2) d\zeta_1 d\zeta_2$$

where the function $f(\zeta_1, \zeta_2)$ is to be expanded in the three integrands as in (0.48)-(0.50), respectively, with $z = \zeta_1$ and $z_0 = \zeta_2$.

We modify the notation of (1.40) to define the components of the vertex operator

$$Y^\mu(u, \zeta) = \sum_{n \in \frac{1}{2}\mathbb{Z}} Y^\mu_{n+1-\mathrm{wt}(u)}(u)\zeta^{-n-1} = \sum_{n \in \frac{1}{2}\mathbb{Z}} \{u; \mu\}_n \zeta^{-n-1}. \tag{1.58}$$

Note that if $u \in \mathbf{U}_0$ then $Y^\mu(u, \zeta) = Y(u, \zeta)$ on $\mathbf{U}$, so the components of the vertex operators $Y^\mu(u, \zeta)$ for $\mathrm{wt}(u) \in \{0, 1\}$ represent the affine algebra $\hat{\mathfrak{o}}(8)$, and if $\omega = D(-2)\mathbf{vac}$ then the components of $Y^\mu(\omega, \zeta)$ represent the Virasoro operators $D(n)$, $n \in \mathbb{Z}$. From the above Jacobi Identity we get the following modified version of Corollary 1.

COROLLARY 6. *With notations as above, for $p \in \mathbb{Z} + \Delta(n_1, n)$, $q \in \mathbb{Z} + \Delta(n_2, n)$, $r \in \mathbb{Z} + \Delta(n_1, n_2)$, we have*

$$\sum_{0 \leq k \in \mathbb{Z}} \binom{r}{i} (-1)^k (\{u_1; \mu\}_{p+r-k} \{u_2; \mu\}_{q+k} - \eta(n_1, n_2) e^{\pi i r} \{u_2; \mu\}_{q+r-k} \{u_1; \mu\}_{p+k}) u$$

$$= \sum_{0 \leq k \in \mathbb{Z}} \binom{p}{k} \{\{u_1; \mu\}_{r+k} u_2; \mu\}_{p+q-k} u. \tag{1.59}$$

The number of nonzero terms in these sums is finite. In the sum of first terms on the left side, $0 \leq k \leq \mathrm{wt}(u) + \mathrm{wt}(u_2) - q - 1$. In the sum of second terms on the left side, $0 \leq k \leq \mathrm{wt}(u) + \mathrm{wt}(u_1) - p - 1$. In the sum on the right side, $0 \leq k \leq \mathrm{wt}(u_1) + \mathrm{wt}(u_2) - r - 1$.

Note that when $\mathrm{wt}(u_1) = \mathrm{wt}(u_2) = \frac{1}{2}$, $r = -\frac{1}{2}$ and $n_1 \neq n_2$, this corollary gives a relation between Clifford generators from different Clifford algebras in which the right side reduces to a single term involving the Chevalley operation

$\circ$. This type of relation has occurred in the work of Lepowsky and Wilson [**LW1,LW2,LW3**], where it is called a "Z-algebra" relation. With $\mathrm{wt}(u_1) = \mathrm{wt}(u_2) = \frac{1}{2}$, $r = 0$ and $n_1 = n_2 \neq 0$, the corollary shows that the rescaling of Y by μ is equivalent to rescaling the form on $\mathbf{C}$ to $\mu(n_1, n_2)\langle a_1, a_2 \rangle$ and the operation $\circ$ on $\mathbf{C}$ to the noncommutative operation $\mu(n_1, n_2)a_1 \circ a_2$, where $a_1 \in \mathbf{C}^{(n_1)}$, $a_2 \in \mathbf{C}^{(n_2)}$.

The rescaling of Y by μ has the following effect on the Intertwining Theorem. For $u_1 \in \mathbf{U}_{n_1}$, $u \in \mathbf{U}_{n_2}$, $g \in G$, we have

$$\hat{g}Y^\mu(u_1, \zeta)u = \mu(n_1, n)\mu(gn_1, gn)^{-1}Y^\mu(\hat{g}u_1, \zeta)\hat{g}u. \tag{1.60}$$

If we choose $\mu(1,2) = \mu(2,3) = \mu(3,1)$ then the extra factor in (1.60) equals 1 for all $0 \leq n_1, n \leq 3$ when $g = \sigma$. The factor is 1 when $n = 0$ or $n_1 = 0$ for all $g \in G$, but no choice of μ can make the factor 1 for all $0 \leq n_1, n \leq 3$ when $g = \tau$.

$D_4^{(1)}$ VERTEX OPERATOR PARA-ALGEBRA THEOREM. *Let* $\Gamma = \mathbb{Z}_2 \times \mathbb{Z}_2 = \{0, 1, 2, 3\}$ *and let* $\Delta_0 = 0$, $\Delta_k = \frac{1}{2}$ *for* $1 \leq k \leq 3$. *With* η *as given in (1.57),* $\mathbf{1} = \mathbf{vac}(\mathbb{Z} + \frac{1}{2})$ *and* $\omega = D(-2)\mathbf{1}$, $(\mathbf{U}, Y(\ , z), \mathbf{1}, \omega, \Gamma, \Delta, \eta)$ *is a vertex operator para-algebra as defined in the Introduction.*

In Chapter 7 we give a spinor construction of the finite-dimensional Lie algebra $\mathbf{g}$ of type E_8. Let $(\mathbf{C}_1, \circ)$ and $(\mathbf{C}_2, \circ)$ be two copies of the Chevalley algebra, and let $\mathbf{o}_1(8)$ and $\mathbf{o}_2(8)$ be corresponding copies of the type D_4 Lie algebra, each constructed from three points of view as in Chapter 4. Let $\mathbf{k}_0 = \mathbf{o}_1(8) \oplus \mathbf{o}_2(8)$, $\mathbf{p}_k = \mathbf{C}_1^{(k)} \otimes \mathbf{C}_2^{(k)}$, $1 \leq k \leq 3$, and $\mathbf{g} = \mathbf{k}_0 \oplus \mathbf{p}_1 \oplus \mathbf{p}_2 \oplus \mathbf{p}_3$. Then $\mathbf{g}$ is a type E_8 Lie algebra under the brackets

$$\begin{aligned} [x_1, a \otimes b] &= (x_1 \cdot a) \otimes b, \quad [x_2, a \otimes b] = a \otimes (x_2 \cdot b), \\ [a \otimes b, c \otimes d] &= -\langle a, c \rangle_\circ^\circ bd_\circ^\circ - \langle b, d \rangle_\circ^\circ ac_\circ^\circ, \end{aligned} \tag{1.61}$$

for $x_1 \in \mathbf{o}_1(8)$, $x_2 \in \mathbf{o}_2(8)$, $a \otimes b, c \otimes d \in \mathbf{p}_k$, $1 \leq k \leq 3$, and

$$[a \otimes b, c \otimes d] = \lambda(a \circ c) \otimes (b \circ d) \tag{1.62}$$

where

$$\lambda = \begin{cases} 1 & \text{if } a \otimes b \in \mathbf{p}_k, \, c \otimes d \in \mathbf{p}_{k'} \\[2ex] -1 & \text{if } a \otimes b \in \mathbf{p}_k, \, c \otimes d \in \mathbf{p}_{k''}. \end{cases} \tag{1.63}$$

Furthermore, $\mathbf{g}$ has three subalgebras $\mathbf{o}_k(16) = \mathbf{k}_0 \oplus \mathbf{p}_k$ of type D_8 whose intersection is $\mathbf{k}_0$. For $i = 1, 2$, we extend the $\mathbb{C}$-antilinear involution ϑ_i from $\mathbf{C}_i$ and $\mathbf{o}_i(8)$ to an involution ϑ on $\mathbf{g}$, we extend the form $\langle \ , \ \rangle$ to a nondegenerate invariant symmetric bilinear form on $\mathbf{g}$, and we define a positive Hermitian form on $\mathbf{g}$ by $(x, y) = \langle x, \vartheta y \rangle$. We also have an action of the triality group G as Lie algebra automorphisms of $\mathbf{g}$.

In Chapter 8 we give a spinor construction of a vertex operator algebra of type $E_8^{(1)}$ and a vertex operator para-algebra containing that algebra and various superalgebras of type $D_8^{(1)}$. The Rationality, Permutability and Associativity

Theorems for the type $E_8^{(1)}$ vertex operator algebra are proved easily from those theorems for type $D_4^{(1)}$.

Let $\mathbf{Cliff}_i^{(k)}(Z)$ be the Clifford algebra generated by $\mathbf{C}_i^{(k)}(Z)$ (see (1.42)) and the form (1.43), with irreducible module $\mathbf{CM}_i^{(k)}(Z)$ having vacuum vector $\mathbf{vac}_i^{(k)}(Z)$. We use the notation $\mathbf{vac}_i = \mathbf{vac}_i^{(k)}(\mathbb{Z} + \frac{1}{2})$, which is independent of k. We have seen how to identify as one space

$$\mathbf{U} = \mathbf{CM}_1^{(k)}(\mathbb{Z} + \tfrac{1}{2}) \oplus \mathbf{CM}_1^{(k)}(\mathbb{Z}) = \mathbf{U}_0 \oplus \mathbf{U}_1 \oplus \mathbf{U}_2 \oplus \mathbf{U}_3 \qquad (1.64)$$

where for $1 \leq k \leq 3$,

$$\begin{aligned}
\mathbf{U}_0 &= \mathbf{CM}_1^{(k)}(\mathbb{Z} + \tfrac{1}{2})^0, & \mathbf{U}_k &= \mathbf{CM}_1^{(k)}(\mathbb{Z} + \tfrac{1}{2})^1, \\
\mathbf{U}_{k'} &= \mathbf{CM}_1^{(k)}(\mathbb{Z})^0, & \mathbf{U}_{k''} &= \mathbf{CM}_1^{(k)}(\mathbb{Z})^1,
\end{aligned} \qquad (1.65)$$

are the four irreducible level 1 $\hat{\mathbf{o}}_1(8)$-modules and (k, k', k'') is a cyclic permutation of $(1, 2, 3)$. Similarly we have

$$\mathbf{W} = \mathbf{CM}_2^{(k)}(\mathbb{Z} + \tfrac{1}{2}) \oplus \mathbf{CM}_2^{(k)}(\mathbb{Z}) = \mathbf{W}_0 \oplus \mathbf{W}_1 \oplus \mathbf{W}_2 \oplus \mathbf{W}_3 \qquad (1.66)$$

where for $1 \leq k \leq 3$,

$$\begin{aligned}
\mathbf{W}_0 &= \mathbf{CM}_2^{(k)}(\mathbb{Z} + \tfrac{1}{2})^0, & \mathbf{W}_k &= \mathbf{CM}_2^{(k)}(\mathbb{Z} + \tfrac{1}{2})^1, \\
\mathbf{W}_{k'} &= \mathbf{CM}_2^{(k)}(\mathbb{Z})^0, & \mathbf{W}_{k''} &= \mathbf{CM}_2^{(k)}(\mathbb{Z})^1,
\end{aligned} \qquad (1.67)$$

are the four irreducible level 1 $\hat{\mathbf{o}}_2(8)$-modules. Let $\mathbf{Vir}_1$ (resp., $\mathbf{Vir}_2$) be the Virasoro algebra represented on $\mathbf{U}$ (resp., $\mathbf{W}$) by the operators $D_1(n)$ (resp., $D_2(n)$), $n \in \mathbb{Z}$. Let $(\ ,\ )$ denote the Hermitian form on $\mathbf{U}$ or on $\mathbf{W}$. We have unitary isomorphisms (1.34)

$$\Upsilon_1 : \mathbf{C}_1 \to (\mathbf{U})_{1/2}, \quad \Upsilon_2 : \mathbf{C}_2 \to (\mathbf{W})_{1/2} \qquad (1.68)$$

used to transport the actions of the triality groups G_1 and G_2. The Lifting Theorem gives actions of G_1 on $\mathbf{U}$ and G_2 on $\mathbf{W}$ such that

$$\hat{g}_i x_i(n) \hat{g}_i^{-1} = (g_i x_i)(n) \qquad (1.69)$$

for $g_i \in G_i$, $x_i(n) \in \hat{\mathbf{o}}_i(8)$. For $u \in \mathbf{U}$, $w \in \mathbf{W}$, we have constructed vertex operators $Y(u, \zeta)$ on $\mathbf{U}$, and $Y(w, \zeta)$ on $\mathbf{W}$, satisfying the theorems in Chapters 3, 5 and 6. Let $\Gamma_1 = \Gamma_2 = \{0, 1, 2, 3\}$ be given the additive group structure $\mathbb{Z}_2 \times \mathbb{Z}_2$, and note that G_i acts on Γ_i (see (5.45)). Define Δ on $\Gamma_1 = \Gamma_2$ as in the $D_4^{(1)}$ Vertex Operator Para-algebra Theorem. Let μ_1 and μ_2 be functions from Γ_1 and Γ_2 to $\mathbb{C}^*$ satisfying (1.55) with $\mathbf{a}_1$ and $\mathbf{a}_2$, respectively, in place of $\mathbf{a}$. Let β_1 and β_2 be the corresponding functions (1.56), and let η_1 and η_2 be similarly obtained from (1.57). Write $Y^1(u, \zeta) = Y^{\mu_1}(u, \zeta)$ and $Y^2(w, \zeta) = Y^{\mu_2}(w, \zeta)$ for the rescaled vertex operators (1.52). Then these operators satisfy the Rationality,

Permutability and Associativity Theorems, the Jacobi Identity and its corollaries with the notations

$$Y^1(u,\zeta) = \sum_{n\in\frac{1}{2}\mathbb{Z}} Y^1_{n+1-\mathrm{wt}(u)}(u)\zeta^{-n-1} = \sum_{n\in\frac{1}{2}\mathbb{Z}} \{u;\mu_1\}_n\zeta^{-n-1}, \tag{1.70}$$

$$Y^2(w,\zeta) = \sum_{n\in\frac{1}{2}\mathbb{Z}} Y^2_{n+1-\mathrm{wt}(w)}(w)\zeta^{-n-1} = \sum_{n\in\frac{1}{2}\mathbb{Z}} \{w;\mu_2\}_n\zeta^{-n-1}. \tag{1.71}$$

If $\omega_i = D_i(-2)\mathbf{vac}_i$ then the components of $Y^i(\omega_i,\zeta)$ represent the Virasoro operators $D_i(n)$, $n \in \mathbb{Z}$.

Define $\Gamma = \Gamma_1 \times \Gamma_2$ and Δ on Γ as follows. For $\gamma = (m,n) \in \Gamma$, let

$$\Delta_\gamma = \Delta_m + \Delta_n \tag{1.72}$$

and for $\gamma_1 = (m_1,n_1), \gamma_2 = (m_2,n_2) \in \Gamma$, as in (0.41), let

$$\Delta(\gamma_1,\gamma_2) = \Delta_{\gamma_1} + \Delta_{\gamma_2} - \Delta_{\gamma_1+\gamma_2} = \Delta(m_1,m_2) + \Delta(n_1,n_2) \tag{1.73}$$

which is bilinear *mod* $\mathbb{Z}$. Define

$$\mathbf{V} = \mathbf{U} \otimes \mathbf{W} \tag{1.74}$$

and give $\mathbf{V}$ the product Hermitian form

$$(u_1 \otimes w_1, u_2 \otimes w_2) = (u_1,u_2)(w_1,w_2). \tag{1.75}$$

For homogeneous vectors $u \in \mathbf{U}$, $w \in \mathbf{W}$, define

$$\mathrm{wt}(u \otimes w) = \mathrm{wt}(u) + \mathrm{wt}(w). \tag{1.76}$$

For $\gamma = (m,n) \in \Gamma$ let

$$\mathbf{V}_\gamma = \mathbf{U}_m \otimes \mathbf{W}_n, \tag{1.77}$$

and note that Δ_γ is the minimal weight of $\mathbf{V}_\gamma$. Define four subspaces of $\mathbf{V}$ by

$$\begin{aligned}
\mathbf{V}_0 &= (\mathbf{U}_0 \otimes \mathbf{W}_0) \oplus (\mathbf{U}_1 \otimes \mathbf{W}_1) \oplus (\mathbf{U}_2 \otimes \mathbf{W}_2) \oplus (\mathbf{U}_3 \otimes \mathbf{W}_3), \\
\mathbf{V}_k &= (\mathbf{U}_0 \otimes \mathbf{W}_k) \oplus (\mathbf{U}_k \otimes \mathbf{W}_0) \oplus (\mathbf{U}_{k'} \otimes \mathbf{W}_{k''}) \oplus (\mathbf{U}_{k''} \otimes \mathbf{W}_{k'})
\end{aligned} \tag{1.78}$$

for $1 \leq k \leq 3$, so that

$$\mathbf{V} = \mathbf{V}_0 \oplus \mathbf{V}_1 \oplus \mathbf{V}_2 \oplus \mathbf{V}_3. \tag{1.79}$$

Note that (1.76) provides a grading on each of the summands in (1.79).

We now define vertex operators $Y(v,\zeta)$ on $\mathbf{V}$ for $v \in \mathbf{V}$. For $u_1,u \in \mathbf{U}$, $w_1,w \in \mathbf{W}$, define

$$Y^1(u_1,\zeta)(u \otimes w) = (Y^1(u_1,\zeta)u) \otimes w, \tag{1.80}$$

and, for $u \in \mathbf{U}_m$, $w_1 \in \mathbf{W}_{n_1}$, define

$$Y^2(w_1,\zeta)(u \otimes w) = \epsilon\, u \otimes Y^2(w_1,\zeta)w \tag{1.81}$$

where

$$\epsilon = \epsilon(m, n_1) = \begin{cases} -1 & \text{if } m \neq 0,\ n_1 \neq 0 \text{ and } m \neq n_1' \\[2mm] 1 & \text{otherwise.} \end{cases} \tag{1.82}$$

Note that ϵ is bilinear (see (8.27)). For $u_1 \in \mathbf{U}_{m_1}$, $w_1 \in \mathbf{W}_{n_1}$ define

$$Y(u_1 \otimes w_1, \zeta) = Y^1(u_1, \zeta)Y^2(w_1, \zeta) = \epsilon(m_1, n_1)Y^2(w_1, \zeta)Y^1(u_1, \zeta) \tag{1.83}$$

on $\mathbf{V}$, and extend this definition to all $u_1 \in \mathbf{U}$, $w_1 \in \mathbf{W}$ by linearity. Then with the notation

$$Y(v_1, \zeta) = \sum_{n \in \frac{1}{2}\mathbb{Z}} \{v_1\}_n \zeta^{-n-1} \tag{1.84}$$

we have

$$\{u_1 \otimes w_1\}_n = \sum_{k \in \frac{1}{2}\mathbb{Z}} \{u_1; \mu_1\}_k \{w_1; \mu_2\}_{n-k-1}. \tag{1.85}$$

Let $\omega = \omega_1 \otimes \mathbf{vac}_2 + \mathbf{vac}_1 \otimes \omega_2$ and define the operators $D(n) = \{\omega\}_{n+1}$ by

$$Y(\omega, \zeta) = \sum_{n \in \mathbb{Z}} D(n)\zeta^{-n-2}. \tag{1.86}$$

It follows from (1.83) that $Y(\omega, \zeta) = Y^1(\omega_1, \zeta) + Y^2(\omega_2, \zeta)$ so $D(n) = D_1(n) + D_2(n)$ for $n \in \mathbb{Z}$ are Virasoro operators on $\mathbf{V}$ satisfying (2.62) with $l = 8$.

We derive the Rationality, Permutability, Associativity and Jacobi Identity theorems for the operators (1.84) on $\mathbf{V}$ from those theorems on $\mathbf{U}$ and $\mathbf{W}$.

RATIONALITY THEOREM. *For $v_i \in \mathbf{U}_{m_i} \otimes \mathbf{W}_{n_i}$, $v \in \mathbf{U}_m \otimes \mathbf{W}_n$, $v' \in \mathbf{V}$, $1 \leq i \leq 2$, let $\mathbf{s} = (s_1, s_2)$ and $\mathbf{t} = (t_1, t_2)$ be defined by $s_i = s(m_i, m)$, $t_i = s(n_i, n)$, let $s_{12} = s(m_1, m_2)$, $t_{12} = s(n_1, n_2)$ and $\zeta^{\mathbf{s}+\mathbf{t}} = \zeta_1^{s_1+t_1}\zeta_2^{s_2+t_2}$. Then the series*

$$\zeta^{\mathbf{s}+\mathbf{t}}\zeta_1^{s_{12}+t_{12}}(1 - \zeta_2/\zeta_1)^{s_{12}+t_{12}}(Y(v_1, \zeta_1)Y(v_2, \zeta_2)v, v')$$

converges absolutely when $|\zeta_1| > |\zeta_2| > 0$ to a rational function in the ring $\mathbb{C}[\zeta_1, \zeta_1^{-1}, \zeta_2, \zeta_2^{-1}, (\zeta_1 - \zeta_2)^{-1}]$.

PERMUTABILITY THEOREM. *With notation as above, we have*

$$\zeta^{\mathbf{s}+\mathbf{t}}\zeta_1^{s_{12}+t_{12}}(1 - \zeta_2/\zeta_1)^{s_{12}+t_{12}}(Y(v_1, \zeta_1)Y(v_2, \zeta_2)v, v') \sim$$
$$\kappa\, \zeta^{\mathbf{s}+\mathbf{t}}\zeta_2^{s_{12}+t_{12}}(1 - \zeta_1/\zeta_2)^{s_{12}+t_{12}}(Y(v_2, \zeta_2)Y(v_1, \zeta_1)v, v')$$

where

$$\kappa = \epsilon(m_1, n_2)\epsilon(m_2, n_1)^{-1}\beta_1(m_1, m_2)\beta_2(n_1, n_2).$$

ASSOCIATIVITY THEOREM. *With notation as above, we have*

$$\zeta^{\mathbf{s}+\mathbf{t}}\zeta_1^{s_{12}+t_{12}}(1 - \zeta_2/\zeta_1)^{s_{12}+t_{12}}(Y(v_1, \zeta_1)Y(v_2, \zeta_2)v, v') \sim$$
$$\zeta_2^{s_1+s_2+t_1+t_2}(\zeta_1 - \zeta_2)^{s_{12}+t_{12}}(1 + (\zeta_1 - \zeta_2)/\zeta_2)^{s_1+t_1}(Y(Y(v_1, \zeta_1 - \zeta_2)v_2, \zeta_2)v, v')$$

and the second series converges absolutely when $0 < |\zeta_1 - \zeta_2| < |\zeta_2|$.

For $i = 1, 2$ let $\gamma_i = (m_i, n_i) \in \Gamma$ and let $\kappa(\gamma_1, \gamma_2)$ be the factor defined in the Permutability Theorem. Define

$$\begin{aligned}
\eta(\gamma_1, \gamma_2) &= \kappa(\gamma_1, \gamma_2)\, e^{\pi i (s_{12} + t_{12})} \\
&= \epsilon(m_1, n_2)\, \epsilon(m_2, n_1)^{-1}\, \eta_1(m_1, m_2)\, \eta_2(n_1, n_2)
\end{aligned} \tag{1.87}$$

and note that η is bilinear since ϵ, η_1 and η_2 are bilinear. Furthermore, from (1.73) we see that $\eta(\gamma_1, \gamma_2)\, e^{\pi i \Delta(\gamma_1, \gamma_2)}$ is skew-symmetric.

JACOBI IDENTITY. *Let $v_i \in \mathbf{V}_{\gamma_i}$ for $i = 1, 2$, $v \in \mathbf{V}_\gamma$, $v' \in \mathbf{V}$. Let*

$$f(\zeta_1, \zeta_2) \in \zeta_1^{\Delta(\gamma_1, \gamma)}\, \zeta_2^{\Delta(\gamma_2, \gamma)}\, (\zeta_1 - \zeta_2)^{\Delta(\gamma_1, \gamma_2)}\, \mathbb{C}[\zeta_1, \zeta_1^{-1}, \zeta_2, \zeta_2^{-1}(\zeta_1 - \zeta_2)^{-1}],$$

let $C_i(r)$ be a circle in the ζ_i-plane of radius r centered at the origin, and let $C_1(\zeta_2, \epsilon)$ be a circle in the ζ_1-plane of radius ϵ centered at $\zeta_1 = \zeta_2$. Then for $0 < r < \rho < R$ and for $0 < \epsilon < \min(R - \rho, \rho - r)$ we have

$$\int_{C_2(\rho)} \int_{C_1(R)} (Y(v_1, \zeta_1) Y(v_2, \zeta_2) v, v') f(\zeta_1, \zeta_2) d\zeta_1 d\zeta_2$$

$$- \int_{C_2(\rho)} \int_{C_1(r)} \eta(\gamma_1, \gamma_2)\, (Y(v_2, \zeta_2) Y(v_1, \zeta_1) v, v') f(\zeta_1, \zeta_2) d\zeta_1 d\zeta_2$$

$$= \int_{C_2(\rho)} \int_{C_1(\zeta_2, \epsilon)} (Y(Y(v_1, \zeta_1 - \zeta_2) v_2, \zeta_2) v, v') f(\zeta_1, \zeta_2) d\zeta_1 d\zeta_2$$

where the function $f(\zeta_1, \zeta_2)$ is to be expanded in the three integrands as in (0.48)-(0.50), respectively, with $z = \zeta_1$ and $z_0 = \zeta_2$.

COROLLARY 7. *With notation as above, for $p \in \mathbb{Z} + \Delta(\gamma_1, \gamma)$, $q \in \mathbb{Z} + \Delta(\gamma_2, \gamma)$, $r \in \mathbb{Z} + \Delta(\gamma_1, \gamma_2)$, we have*

$$\sum_{0 \le k \in \mathbb{Z}} \binom{r}{k} (-1)^k [\{v_1\}_{p+r-k} \{v_2\}_{q+k} - \eta(\gamma_1, \gamma_2)\, e^{\pi i r}\, \{v_2\}_{q+r-k} \{v_1\}_{p+k}] v$$

$$= \sum_{0 \le k \in \mathbb{Z}} \binom{p}{k} \{\{v_1\}_{r+k} v_2\}_{p+q-k} v. \tag{1.88}$$

COROLLARY 8. *With notation as above, let $m_1 = n_1$ and $m_2 = n_2$ so that $v_1, v_2 \in \mathbf{V}_0$. If $\mathbf{a}_1 \mathbf{a}_2 = -1$ then we have*

$$[\{v_1\}_p, \{v_2\}_q] v = \sum_{0 \le k \in \mathbb{Z}} \binom{p}{k} \{\{v_1\}_k v_2\}_{p+q-k} v$$

where $p \in \mathbb{Z} + \Delta(\gamma_1, \gamma)$, $q \in \mathbb{Z} + \Delta(\gamma_2, \gamma)$. Let $\hat{\mathbf{V}}_0 = \mathbf{V}_0 \otimes \mathbb{C}[t, t^{-1}]$ and for $1 \le k \le 3$, let

$$\hat{\mathbf{V}}_k = (\mathbf{U}_0 \otimes \mathbf{W}_0\ \oplus\ \mathbf{U}_k \otimes \mathbf{W}_k) \otimes \mathbb{C}[t, t^{-1}]\ \oplus$$

$$(\mathbf{U}_{k'} \otimes \mathbf{W}_{k'}\ \oplus\ \mathbf{U}_{k''} \otimes \mathbf{W}_{k''}) \otimes t^{1/2} \mathbb{C}[t, t^{-1}].$$

Then $\hat{\mathbf{V}}_0/D(-1)\hat{\mathbf{V}}_0$ is a Lie algebra represented on $\mathbf{V}_0$ and for $1 \leq k \leq 3$, $\hat{\mathbf{V}}_k/D(-1)\hat{\mathbf{V}}_k$ is a Lie algebra represented on $\mathbf{V}_k$. In these representations the coset of $v \otimes t^n$ is represented by the operator $\{v\}_n$.

We identify $\mathbf{V}_0$ as the homogeneously graded basic representation of the affine algebra $E_8^{(1)}$, and $\mathbf{V}_1$, $\mathbf{V}_2$ and $\mathbf{V}_3$ as "k-p" graded basic representations of $E_8^{(1)}$. The homogeneously graded affine algebra $\hat{\mathbf{g}}$ associated with a finite-dimensional Lie algebra $\mathbf{g}$ is defined in (2.17)-(2.18). When $\mathbf{g} = \mathbf{k} \oplus \mathbf{p}$ is a decomposition such that $[\mathbf{k},\mathbf{k}] \subseteq \mathbf{k}$, $[\mathbf{k},\mathbf{p}] \subseteq \mathbf{p}$, $[\mathbf{p},\mathbf{p}] \subseteq \mathbf{k}$, then the associated "k-p" graded affine algebra is defined to be

$$\hat{\mathbf{g}}' = \mathbf{k} \otimes \mathbb{C}[t,t^{-1}] \oplus \mathbf{p} \otimes t^{1/2}\mathbb{C}[t,t^{-1}] \oplus \mathbb{C}c \oplus \mathbb{C}d \tag{1.89}$$

with brackets as in (2.18). From Theorem 7.1, $\mathbf{g}$ of type E_8 has three such decompositions, $\mathbf{g} = \mathbf{k}^{(k)} \oplus \mathbf{p}^{(k)}$, where

$$\mathbf{k}^{(k)} = \mathbf{k}_0 \oplus \mathbf{p}_k = \mathbf{o}_k(16), \quad \mathbf{p}^{(k)} = \mathbf{p}_{k'} \oplus \mathbf{p}_{k''}. \tag{1.90}$$

Let the three corresponding "k-p" graded affine algebras be denoted $\hat{\mathbf{g}}^{(k)}$, $1 \leq k \leq 3$, and write $\hat{\mathbf{g}}^{(0)}$ for the homogeneously graded affine algebra $\hat{\mathbf{g}}$.

We can use the isomorphisms in (1.68) to define an isomorphism Υ from $\mathbf{g}$ to the weight 1 vectors in $\mathbf{V}_0$. For $x \in \mathbf{g}$, $n \in \frac{1}{2}\mathbb{Z}$, define the notation

$$x(n) = Y_n(\Upsilon(x)) = \{\Upsilon(x)\}_n. \tag{1.91}$$

The following theorem gives spinor constructions of four basic $E_8^{(1)}$ representations when $\mu_1 = \mu_2^{-1}$, an assumption we make for the rest of this chapter.

SPINOR CONSTRUCTION OF AFFINE ALGEBRA $E_8^{(1)}$. *Let $\mu_1(i,j) = \mu_2(i,j)^{-1}$ for $0 \leq i,j \leq 3$. The homogeneous affine algebra $\hat{\mathbf{g}}^{(0)}$ of type $E_8^{(1)}$ is represented on $\mathbf{V}_0$ by the identity operator, $D(0)$ and the operators $x(q)$, $x \in \mathbf{g}$, $q \in \mathbb{Z}$. For $1 \leq k \leq 3$, the "k-p" graded affine algebra $\hat{\mathbf{g}}^{(k)}$ of type $E_8^{(1)}$ is represented on $\mathbf{V}_k$ by the identity operator, $D(0)$ and the operators $x(q)$, where $q \in \mathbb{Z}$ if $x \in \mathbf{k}^{(k)} = \mathbf{o}_k(16)$, and $q \in \mathbb{Z} + \frac{1}{2}$ if $x \in \mathbf{p}^{(k)}$.*

It is possible to prove this theorem using only Proposition 6.2, Corollary 6.3 and an induction argument. This method of proof is more elementary in that it uses the theory of the affine algebra $D_4^{(1)}$ and triality (as in Chapters 2 and 5), and avoids the vertex operator superalgebra constructions in Chapter 3. It's disadvantage is its length, there being many cases to check.

For $g_1 \in G_1$, $g_2 \in G_2$, let $g = (g_1,g_2) \in G_1 \times G_2$, and for $u \otimes w \in \mathbf{U} \otimes \mathbf{W}$ define $\hat{g} : \mathbf{V} \to \mathbf{V}$ by $\hat{g}(u \otimes w) = \hat{g}_1 u \otimes \hat{g}_2 w$.

INTERTWINING THEOREM. *For $v_1 \in \mathbf{U}_{m_1} \otimes \acute{\mathbf{W}}_{n_1}$, $v \in \mathbf{U}_m \otimes \mathbf{W}_n$, $g = (g_1,g_2) \in G_1 \times G_2$, we have*

$$\hat{g}Y(v_1,\zeta)v = \lambda Y(\hat{g}v_1,\zeta)\hat{g}v$$

where

$$\lambda = \lambda(m_1, n_1; m, n; g_1, g_2) = \frac{\epsilon(m, n_1)\mu_1(m_1, m)\mu_2(n_1, n)}{\epsilon(g_1 m, g_2 n_1)\mu_1(g_1 m_1, g_1 m)\mu_2(g_2 n_1, g_2 n)}.$$

If $m_1 = n_1 = 0$, then $\lambda = 1$ for all $g_1 \in G_1$, $g_2 \in G_2$, $0 \le m, n \le 3$. If $g_1 = g_2 = \sigma$ and $\mu_1(1, 2) = \mu_1(2, 3) = \mu_1(3, 1)$ (which implies that $\mu_1(\sigma i, \sigma j) = \mu_1(i, j)$ for all $0 \le i, j \le 3$), then $\lambda = 1$ for all $0 \le m_1, n_1, m, n \le 3$.

For each k, $1 \le k \le 3$, and each choice of Z_1 and Z_2, the set of operators

$$\{Y_m^1(u), Y_n^2(w) \mid u \in \mathbf{U}_k, w \in \mathbf{W}_k, \mathrm{wt}(u) = \mathrm{wt}(w) = \tfrac{1}{2}, m \in Z_1, n \in Z_2\} \quad (1.92)$$

represents on

$$\mathbf{CM}^{(k)}(Z_1, Z_2) = \mathbf{CM}_1^{(k)}(Z_1) \otimes \mathbf{CM}_2^{(k)}(Z_2), \quad (1.93)$$

the generators of a Clifford algebra

$$\mathbf{Cliff}^{(k)}(Z_1, Z_2) = \mathbf{Cliff}_1^{(k)}(Z_1; \mu_1) \otimes \mathbf{Cliff}_2^{(k)}(Z_2; \mu_2). \quad (1.94)$$

These are $\mathbb{Z}_2$-graded tensor products (see (8.85)-(8.86)). This is equivalent to the usual construction from the space $A^{(k)}(Z_1, Z_2) = \mathbf{C}_1^{(k)}(Z_1) \oplus \mathbf{C}_2^{(k)}(Z_2)$. The module (1.93) is an irreducible $\mathbf{Cliff}^{(k)}(Z_1, Z_2)$-module. We see that for $1 \le k \le 3$,

$$\begin{aligned}
\mathbf{V}_0 \oplus \mathbf{V}_k &= \mathbf{CM}^{(k)}(\mathbb{Z} + \tfrac{1}{2}, \mathbb{Z} + \tfrac{1}{2}) \oplus \mathbf{CM}^{(k)}(\mathbb{Z}, \mathbb{Z}) \\
&= (\mathbf{U}_0 \oplus \mathbf{U}_k) \otimes (\mathbf{W}_0 \oplus \mathbf{W}_k) \oplus (\mathbf{U}_{k'} \oplus \mathbf{U}_{k''}) \otimes (\mathbf{W}_{k'} \oplus \mathbf{W}_{k''})
\end{aligned} \quad (1.95)$$

and

$$\begin{aligned}
\mathbf{V}_{k'} \oplus \mathbf{V}_{k''} &= \mathbf{CM}^{(k)}(\mathbb{Z} + \tfrac{1}{2}, \mathbb{Z}) \oplus \mathbf{CM}^{(k)}(\mathbb{Z}, \mathbb{Z} + \tfrac{1}{2}) \\
&= (\mathbf{U}_0 \oplus \mathbf{U}_k) \otimes (\mathbf{W}_{k'} \oplus \mathbf{W}_{k''}) \oplus (\mathbf{U}_{k'} \oplus \mathbf{U}_{k''}) \otimes (\mathbf{W}_0 \oplus \mathbf{W}_k).
\end{aligned} \quad (1.96)$$

From Chapter 2 we see that we have a homogeneous representation of the affine algebra $\hat{\mathbf{o}}_k(16)$ of type $D_8^{(1)}$ on $\mathbf{CM}^{(k)}(\mathbb{Z} + \tfrac{1}{2}, \mathbb{Z} + \tfrac{1}{2})$ and on $\mathbf{CM}^{(k)}(\mathbb{Z}, \mathbb{Z})$, and the even and odd parity subspaces of each one are irreducible, giving the four level 1 $\hat{\mathbf{o}}_k(16)$-modules. Similarly, from $\mathbf{CM}^{(k)}(\mathbb{Z} + \tfrac{1}{2}, \mathbb{Z})$ and $\mathbf{CM}^{(k)}(\mathbb{Z}, \mathbb{Z} + \tfrac{1}{2})$ we get four level 1 "k-p" graded representations of $\hat{\mathbf{o}}'_k(16)$. The next theorem follows from Corollary 7.

THEOREM. *For $1 \le k \le 3$, we have*

(a) $\mathbf{CM}^{(k)}(\mathbb{Z} + \tfrac{1}{2}, \mathbb{Z} + \tfrac{1}{2})$ *is a vertex operator superalgebra,*

(b) $\mathbf{CM}^{(k)}(\mathbb{Z}, \mathbb{Z})$ *is a $\mathbb{Z}_2$-twisted $\mathbf{CM}^{(k)}(\mathbb{Z} + \tfrac{1}{2}, \mathbb{Z} + \tfrac{1}{2})$-module, with the twisting given by the order 2 automorphism of $\mathbf{CM}^{(k)}(\mathbb{Z} + \tfrac{1}{2}, \mathbb{Z} + \tfrac{1}{2})$ which is 1 on $(\mathbf{U}_0 \otimes \mathbf{W}_0) \oplus (\mathbf{U}_k \otimes \mathbf{W}_k)$ and -1 on $(\mathbf{U}_0 \otimes \mathbf{W}_k) \oplus (\mathbf{U}_k \otimes \mathbf{W}_0)$,*

(c) $\mathbf{CM}^{(k)}(\mathbb{Z} + \tfrac{1}{2}, \mathbb{Z})$ *is a $\mathbb{Z}_2$-twisted $\mathbf{CM}^{(k)}(\mathbb{Z} + \tfrac{1}{2}, \mathbb{Z} + \tfrac{1}{2})$-module, with the twisting given by the order 2 automorphism of $\mathbf{CM}^{(k)}(\mathbb{Z} + \tfrac{1}{2}, \mathbb{Z} + \tfrac{1}{2})$ which is 1 on $(\mathbf{U}_0 \otimes \mathbf{W}_0) \oplus (\mathbf{U}_k \otimes \mathbf{W}_0)$ and -1 on $(\mathbf{U}_0 \otimes \mathbf{W}_k) \oplus (\mathbf{U}_k \otimes \mathbf{W}_k)$,*

(d) $\mathbf{CM}^{(k)}(\mathbb{Z}, \mathbb{Z} + \frac{1}{2})$ *is a* $\mathbb{Z}_2$*-twisted* $\mathbf{CM}^{(k)}(\mathbb{Z} + \frac{1}{2}, \mathbb{Z} + \frac{1}{2})$*-module, with the twisting given by the order 2 automorphism of* $\mathbf{CM}^{(k)}(\mathbb{Z} + \frac{1}{2}, \mathbb{Z} + \frac{1}{2})$ *which is 1 on* $(\mathbf{U}_0 \otimes \mathbf{W}_0) \oplus (\mathbf{U}_0 \otimes \mathbf{W}_k)$ *and -1 on* $(\mathbf{U}_k \otimes \mathbf{W}_0) \oplus (\mathbf{U}_k \otimes \mathbf{W}_k)$.

$E_8^{(1)}$ VERTEX OPERATOR ALGEBRA AND PARA-ALGEBRA THEOREM. *Let* $\Gamma = \mathbb{Z}_2^2 \times \mathbb{Z}_2^2$, $\Delta_{(i,j)}$ *be the minimal weight of* $\mathbf{U}_i \otimes \mathbf{W}_j$, η *be given in (1.87) and* $\mathbf{1} = \mathbf{vac}_1 \otimes \mathbf{vac}_2$. *Then, as defined in the Introduction,* $(\mathbf{V}_0, Y(\ , z), \mathbf{1}, \omega)$ *is a vertex operator algebra,* $(\mathbf{V}_k, Y(\ , z))$ *is a* $\mathbf{V}_0$*-module for* $1 \leq k \leq 3$, *and* $(\mathbf{V}, Y(\ , z), \mathbf{1}, \omega, \Gamma, \Delta, \eta)$ *is a vertex operator para-algebra.*

CHAPTER 2

Affine Algebras and Representations

Let $\mathbf{g}$ be a rank l simple Lie algebra over $\mathbb{C}$ of type A, D, or E. Then $\mathbf{g}$ is determined up to isomorphism by an associated $l \times l$ Cartan matrix $A = (a_{ij})$, or by an associated l-vertex graph, a Dynkin diagram. Let $\mathbf{h}$ be a Cartan subalgebra of $\mathbf{g}$, and let $\mathbf{h}^*$ be the dual space of $\mathbf{h}$. With respect to $\mathbf{h}$ we have a root-space decomposition

$$\mathbf{g} = \mathbf{h} \oplus \sum_{\alpha \in \Delta} \mathbf{g}_\alpha \tag{2.1}$$

where for $\alpha \in \mathbf{h}^*$,

$$\mathbf{g}_\alpha = \{x \in \mathbf{g} \mid [h, x] = \alpha(h)x, \ \forall h \in \mathbf{h}\} \tag{2.2}$$

is the α root-space of $\mathbf{g}$, and

$$\Delta = \{\alpha \in \mathbf{h}^* \mid \mathbf{g}_\alpha \neq 0, \ \alpha \neq 0\} \tag{2.3}$$

is the set of roots of $\mathbf{g}$. Let $\langle \, , \, \rangle$ denote the nondegenerate symmetric invariant bilinear form on $\mathbf{g}$, normalized so that the induced form on $\mathbf{h}^*$ satisfies $\langle \alpha, \alpha \rangle = 2$ for $\alpha \in \Delta$. Let $\alpha_1, \ldots, \alpha_l \in \Delta$ be simple roots, so that $A = (\langle \alpha_i, \alpha_j \rangle)$, and the root lattice

$$Q = \sum_{1 \leq i \leq l} \mathbb{Z}\alpha_i \tag{2.4}$$

is the $\mathbb{Z}$-span of Δ. For $\mathbf{g}$ of type A, D, or E, and for $i \neq j$, $\langle \alpha_i, \alpha_j \rangle = 0$ or -1. The ith and j^{th} vertices of the Dynkin diagram are directly connected by an edge when $\langle \alpha_i, \alpha_j \rangle = -1$. Since

$$\Delta = \{\alpha \in Q \mid \langle \alpha, \alpha \rangle = 2\}, \tag{2.5}$$

for $\alpha, \beta \in \Delta$ we clearly have $\alpha + \beta \in \Delta$ iff $\langle \alpha, \beta \rangle = -1$, and $\alpha + \beta = 0$ iff $\langle \alpha, \beta \rangle = -2$. For $\alpha, \beta \in Q$ we have $\langle \alpha, \beta \rangle \in \mathbb{Z}$ and $\langle \alpha, \alpha \rangle \in 2\mathbb{Z}$.

The restriction of the invariant form $\langle \, , \, \rangle$ to $\mathbf{h}$ is nondegenerate, allowing the identification of $\mathbf{h}$ with its dual space $\mathbf{h}^*$. The weight lattice

$$P = \{\beta \in \mathbf{h}^* \mid \langle \alpha, \beta \rangle \in \mathbb{Z}, \ \forall \alpha \in Q\} \tag{2.6}$$

34

contains Q with index

$$[P, Q] = \det(A) = \begin{cases} l+1 & \text{if } \mathbf{g} \text{ is type } A_l, \\ 4 & \text{if } \mathbf{g} \text{ is type } D_l, \\ 3, 2, 1 & \text{if } \mathbf{g} \text{ is type } E_6, \ E_7, \ E_8, \text{ respectively.} \end{cases} \tag{2.7}$$

Let $\omega_1, \ldots, \omega_l \in P$ be the fundamental weights of $\mathbf{g}$ such that $\langle \omega_i, \alpha_j \rangle = \delta_{ij}$. Then

$$P = \sum_{1 \le i \le l} \mathbb{Z} \omega_i \tag{2.8}$$

and

$$A^{-1} = (\langle \omega_i, \omega_j \rangle) \tag{2.9}$$

determines the structure of the quotient P/Q. The Weyl group W of $\mathbf{g}$ is the group of orthogonal transformations of $\mathbf{h}$ generated by the reflections with respect to the simple roots.

We can give a simple construction [**FK**] of $\mathbf{g}$ based on the unique cohomology class $[\epsilon] \in H^2(Q, \{\pm 1\})$ such that

$$\epsilon(\alpha, \beta)\epsilon(\beta, \alpha)^{-1} = (-1)^{\langle \alpha, \beta \rangle} \quad \text{for} \quad \alpha, \beta \in Q, \tag{2.10}$$

where ϵ is any 2-cocycle representing $[\epsilon]$. Take a vector space with basis $\{\alpha_i, x_\alpha \mid 1 \le i \le l, \ \alpha \in \Delta\}$ and define the Lie brackets to be

$$[\alpha_i, \alpha_j] = 0, \quad [\alpha_i, x_\alpha] = \langle \alpha_i, \alpha \rangle x_\alpha,$$

$$[x_\alpha, x_\beta] = \begin{cases} 0 & \text{if } \langle \alpha, \beta \rangle \ge 0 \\ \epsilon(\alpha, \beta) x_{\alpha+\beta} & \text{if } \langle \alpha, \beta \rangle = -1 \\ \epsilon(\alpha, -\alpha)\alpha & \text{if } \langle \alpha, \beta \rangle = -2. \end{cases} \tag{2.11}$$

Changing each basis vector x_α by a scalar multiple corresponds to changing the cocycle by a coboundary. It is routine to verify that (2.11) defines a simple Lie algebra whose root system is Δ with respect to the Cartan subalgebra $\mathbf{h}$ spanned by $\alpha_1, \ldots, \alpha_l$. The invariance of the form on $\mathbf{g}$,

$$\langle x, [y, z] \rangle = \langle [x, y], z \rangle \quad \text{for all } x, y, z \in \mathbf{g}, \tag{2.12}$$

requires that for $\alpha, \beta \in \Delta$,

$$\langle x_\alpha, x_\beta \rangle = \begin{cases} 0 & \text{if } \alpha + \beta \ne 0 \\ \epsilon(\alpha, -\alpha) & \text{if } \alpha + \beta = 0. \end{cases} \tag{2.13}$$

We can also give a simple construction of some representations of $\mathbf{g}$. Let Γ be a coset of Q in P, and let $[\epsilon] \in H^2(P, \mathbb{C}^*)$ be a class whose restriction to Q is the unique class satisfying (2.10). Let

$$\Gamma_1 = \{\gamma \in \Gamma \mid \langle \gamma, \gamma \rangle \text{ is minimal}\}. \tag{2.14}$$

Define a vector space V with basis $\{v_\gamma \mid \gamma \in \Gamma_1\}$. For $\alpha \in \Delta$, $\gamma \in \Gamma_1$, we have $\langle \alpha, \gamma \rangle = -1$ iff $\alpha + \gamma \in \Gamma_1$. Define an action of $\mathbf{g}$ on V by

$$h \cdot v_\gamma = \langle \gamma, h \rangle v_\gamma$$

$$x_\alpha \cdot v_\gamma = \begin{cases} \epsilon(\alpha, \gamma) v_{\gamma + \alpha} & \text{if } \langle \alpha, \gamma \rangle = -1 \\ \\ 0 & \text{otherwise,} \end{cases} \tag{2.15}$$

for $h \in \mathbf{h}$, $\alpha \in \Delta$, $\gamma \in \Gamma_1$. Using (2.10) one checks directly that

$$[x, y] \cdot v_\gamma = x \cdot (y \cdot v_\gamma) - y \cdot (x \cdot v_\gamma) \tag{2.16}$$

for all $x, y \in \mathbf{g}$, $\gamma \in \Gamma_1$, so V is a $\mathbf{g}$-module. V is an irreducible $\mathbf{g}$-module corresponding to the one Weyl group orbit in Γ_1, thus providing the miniscule representations of $\mathbf{g}$. (See [**LP**], Theorem 5.6.)

The homogeneously graded affine Lie algebra $\hat{\mathbf{g}}$ associated with $\mathbf{g}$ is defined to be

$$\hat{\mathbf{g}} = \mathbf{g} \otimes \mathbb{C}[t, t^{-1}] \oplus \mathbb{C}c \oplus \mathbb{C}d \tag{2.17}$$

where c is central and $d = t(d/dt)$. For $x \in \mathbf{g}$, $n \in \mathbb{Z}$, we write $x(n) = x \otimes t^n$. Then the brackets in $\hat{\mathbf{g}}$ are

$$\begin{aligned} [x(m), y(n)] &= [x, y](m + n) + m\delta_{m,-n}\langle x, y \rangle c, \\ [d, x(m)] &= mx(m). \end{aligned} \tag{2.18}$$

We say that $\hat{\mathbf{g}}$ is of type $A_l^{(1)}$, $D_l^{(1)}$, or $E_l^{(1)}$. We identify $\mathbf{g} \subset \hat{\mathbf{g}}$ by identifying x with $x(0)$. A Cartan subalgebra of $\hat{\mathbf{g}}$ is

$$\mathbf{H} = \mathbf{h} \oplus \mathbb{C}c \oplus \mathbb{C}d, \tag{2.19}$$

and the invariant form $\langle \, , \, \rangle$ on $\mathbf{g}$ is extended to an invariant form on $\hat{\mathbf{g}}$ by

$$\langle x(m) + r_1 c + s_1 d, \ y(n) + r_2 c + s_2 d \rangle = \langle x, y \rangle \delta_{m,-n} + r_1 s_2 + s_1 r_2. \tag{2.20}$$

Since c is central in $\hat{\mathbf{g}}$, the root system

$$\Delta(\hat{\mathbf{g}}) = \{\alpha \in \mathbf{H}^* \mid \hat{\mathbf{g}}_\alpha \neq 0, \ \alpha \neq 0\} \tag{2.21}$$

is contained in the codimension 1 subspace $\{\alpha \in \mathbf{H}^* \mid \alpha(c) = 0\}$. Define $c^*, d^* \in \mathbf{H}^*$ by $c^*(c) = d^*(d) = 1$, $c^*(d) = d^*(c) = 0$ and $c^*(h) = d^*(h) = 0$ for all $h \in \mathbf{h}$. Let $\alpha^+ \in \Delta$ be the highest root of $\mathbf{g}$. Then simple roots for $\hat{\mathbf{g}}$ can be chosen to be

$$\alpha_0 = d^* - \alpha^+, \ \alpha_1, \alpha_2, \ldots, \alpha_l, \tag{2.22}$$

and the root system of $\hat{\mathbf{g}}$ is

$$\Delta(\hat{\mathbf{g}}) = \{nd^* + \alpha,\ md^* \mid \alpha \in \Delta,\ n \in \mathbb{Z},\ 0 \neq m \in \mathbb{Z}\}. \tag{2.23}$$

The Weyl group $\hat{W}$ of $\hat{\mathbf{g}}$ is the group of orthogonal transformations of $\mathbf{H}$ generated by the reflections with respect to the simple roots (2.22). The $(l+1) \times (l+1)$ Cartan matrix of $\hat{\mathbf{g}}$ is $\hat{A} = (\langle \alpha_i, \alpha_j \rangle)$, $0 \leq i, j \leq l$, and corresponds to an $(l+1)$-vertex Dynkin diagram which contains the Dynkin diagram of $\mathbf{g}$. The extra vertex corresponding to the new simple root α_0 is connected by new edges to those vertices $1 \leq i \leq l$ such that $\langle \alpha^+, \alpha_i \rangle = 1$. The fundamental weights of $\hat{\mathbf{g}}$, $\hat{\omega}_0, \hat{\omega}_1, \ldots, \hat{\omega}_l$, are determined up to multiples of d^* by the conditions

$$\langle \hat{\omega}_i, \alpha_j \rangle = \delta_{ij}, \quad 0 \leq i, j \leq l. \tag{2.24}$$

One finds that

$$\hat{\omega}_0 = c^*, \ \hat{\omega}_i = n_i c^* + \omega_i, \ 1 \leq i \leq l, \tag{2.25}$$

where $n_i = \langle \omega_i, \alpha^+ \rangle$ so $\alpha^+ = \sum_{1 \leq i \leq l} n_i \alpha_i$.

The Virasoro algebra is the abstract Lie algebra $\mathbf{Vir}$ with basis $\{z, L(m) \mid m \in \mathbb{Z}\}$ such that z is central and

$$[L(m), L(n)] = (n - m)L(m + n) + \frac{1}{12}(m^3 - m)\delta_{m,-n} z. \tag{2.26}$$

We may form the semidirect product of $\mathbf{Vir}$ with the subalgebra of $\hat{\mathbf{g}}$,

$$\tilde{\mathbf{g}} = \mathbf{g} \otimes \mathbb{C}[t, t^{-1}] \oplus \mathbb{C}c \tag{2.27}$$

where c and z are central and we have the brackets

$$[L(m), x(n)] = nx(m + n). \tag{2.28}$$

We identify $L(0) = d$ so that this semidirect product contains $\hat{\mathbf{g}}$.

An affine algebra $\hat{\mathbf{g}}_\tau$ can be associated with $\mathbf{g}$ and a finite order automorphism τ of $\mathbf{g}$ [**KP2,L**]. We will be concerned with the homogeneously graded affine algebras where τ is the identity automorphism, and with "$\mathbf{k} - \mathbf{p}$" graded affine algebras associated with order two automorphisms. Suppose we have a decomposition

$$\mathbf{g} = \mathbf{k} \oplus \mathbf{p} \tag{2.29}$$

such that

$$[\mathbf{k}, \mathbf{k}] \subseteq \mathbf{k}, \quad [\mathbf{k}, \mathbf{p}] \subseteq \mathbf{p}, \quad [\mathbf{p}, \mathbf{p}] \subseteq \mathbf{k}. \tag{2.30}$$

This is equivalent to having an order two autormorphism τ on $\mathbf{g}$ with ± 1-eigenspace decomposition (2.29). The "$\mathbf{k} - \mathbf{p}$" graded affine algebra associated with τ (or (2.29)) is defined to be $\hat{\mathbf{g}}_\tau = \tilde{\mathbf{g}}_\tau \oplus \mathbb{C}d$ with the brackets (2.18), where

$$\tilde{\mathbf{g}}_\tau = \mathbf{k} \otimes \mathbb{C}[t, t^{-1}] \oplus \mathbf{p} \otimes t^{1/2}\mathbb{C}[t, t^{-1}] \oplus \mathbb{C}c. \tag{2.31}$$

If τ is an inner automorphism of $\mathbf{g}$ then $\tilde{\mathbf{g}}_\tau$ is isomorphic to $\tilde{\mathbf{g}}$, and these affine algebras have the same Dynkin diagrams. But if τ is an outer automorphism of $\mathbf{g}$ then it can be associated with a Dynkin diagram automorphism of $\mathbf{g}$, and

$\tilde{\mathbf{g}}_\tau$ and $\tilde{\mathbf{g}}$ are not isomorphic, with distinct Dynkin diagrams. We will only be concerned with "$\mathbf{k} - \mathbf{p}$" gradings associated with inner automorphisms.

The representation theory of $\hat{\mathbf{g}}$ is remarkably similar to that of $\mathbf{g}$. The "standard" modules of $\hat{\mathbf{g}}$ are irreducible highest weight modules whose highest weights are dominant integral (of the form $\sum_{0 \le i \le l} k_i \hat{\omega}_i$ for $0 \le k_i \in \mathbb{Z}$). The Weyl group of $\hat{\mathbf{g}}$ preserves the weights of each standard module, and one has a theory of characters of standard modules in which the Weyl-Kac character and denominator formulas generalize the classical results. The "fundamental" modules are those with highest weights $\hat{\omega}_i$, $0 \le i \le l$, and the one with highest weight $\hat{\omega}_0$ is further distinguished as the "basic" module. The unique coefficient of c^* in the weights of an irreducible $\hat{\mathbf{g}}$-module is called the "level" of the module. The central element c acts by that scalar on the module. The level 1 standard modules of $\hat{\mathbf{g}}$ for $\mathbf{g}$ of type A, D, or E can be constructed using vertex operators [**FK**]. If $\mathbf{g}$ is of type B or D then there is another method [**F1,KP1**] of constructing the level 1 standard $\hat{\mathbf{g}}$-modules, analogous to the spinor representations of $\mathbf{g}$. For $\mathbf{g}$ of type D, the isomorphism of the two constructions of $\hat{\mathbf{g}}$-modules is called a "boson-fermion correspondence" in physics [**F2**]. A remarkable feature of the representation theory of $\hat{\mathbf{g}}$ is the existence of an additional infinite-dimensional Lie algebra, the Virasoro algebra, whose semidirect product with $\tilde{\mathbf{g}}$ is represented on each standard $\hat{\mathbf{g}}$-module.

We wish to provide a "spinor" construction for $\hat{\mathbf{g}}$ of type $E_8^{(1)}$ by using spinor constructions of type $D_4^{(1)}$ and triality. In [**GOS**] vertex operators were used to define Clifford algebras and Clifford modules from which $E_8^{(1)}$ representations were constructed. But the key point, that the operators constructed close under brackets so as to represent $E_8^{(1)}$, then relied on vertex operator calculations. We would not call that a purely spinor construction, but rather a part of the vertex picture necessary for carrying out the boson-fermion correspondence. In this paper we will give an independent spinor construction of $E_8^{(1)}$ and a proof of closure which uses only properties of the Clifford algebras and modules. Later [**FR**] we will give the boson-fermion correspondence between the two constructions. In this paper we will also construct superalgebra representations from the level 1 $D_l^{(1)}$-modules in the spinor picture. First we review the spinor construction of those level 1 $\hat{\mathbf{g}}$-modules.

Recall the spinor construction of the finite-dimensional algebra $\mathbf{o}(2l)$. Begin with a nondegenerate symmetric bilinear form $\langle \, , \, \rangle$ on a vector space $\mathbf{A} \cong \mathbb{C}^{2l}$ such that $\mathbf{A} = \mathbf{A}^+ \oplus \mathbf{A}^-$ is a polarization into maximally isotropic subspaces $\mathbf{A}^\pm \cong \mathbb{C}^l$. Define an associative algebra **Cliff** with unit element 1, generators from $\mathbf{A}$ and relations

$$ab + ba = \langle a, b \rangle 1 \quad \text{for} \quad a, b \in \mathbf{A}. \tag{2.32}$$

Letting

$$\,_\circ^\circ ab \,_\circ^\circ = \tfrac{1}{2}(ab - ba) \quad \text{for} \quad a, b \in \mathbf{A}, \tag{2.33}$$

the span of all such elements in the Clifford algebra **Cliff** is closed under brackets, providing the adjoint representation of $\mathbf{o}(2l)$. Three more irreducible representations of $\mathbf{o}(2l)$ are constructed as follows. On $\mathbf{A}$, $\mathbf{o}(2l)$ acts by brackets

$$[{}^{\circ}_{\circ}ab{}^{\circ}_{\circ}, c] = \langle b, c\rangle a - \langle a, c\rangle b, \tag{2.34}$$

providing the natural representation. In the exterior algebra $\wedge\mathbf{A}^+$ choose a nonzero vector $\mathbf{vac} \in \wedge^l\mathbf{A}^+$. Then $\mathbf{A}^+\cdot\mathbf{vac} = 0$ and $\mathbf{CM} = \mathbf{Cliff}\cdot\mathbf{vac} = (\wedge\mathbf{A}^-)\cdot\mathbf{vac}$ is an irreducible left **Cliff**-module. **CM** decomposes into two irreducible $\mathbf{o}(2l)$-modules $\mathbf{CM}^0 = (\wedge^{even}\mathbf{A}^-)\cdot\mathbf{vac}$ and $\mathbf{CM}^1 = (\wedge^{odd}\mathbf{A}^-)\cdot\mathbf{vac}$, called semispinors, each of dimension 2^{l-1}.

If $x = {}^{\circ}_{\circ}r_1r_2{}^{\circ}_{\circ}$, $y = {}^{\circ}_{\circ}s_1s_2{}^{\circ}_{\circ} \in \mathbf{o}(2l)$ then

$$[x, y] = \langle r_1, s_2\rangle{}^{\circ}_{\circ}r_2s_1{}^{\circ}_{\circ} + \langle r_2, s_1\rangle{}^{\circ}_{\circ}r_1s_2{}^{\circ}_{\circ} - \langle r_1, s_1\rangle{}^{\circ}_{\circ}r_2s_2{}^{\circ}_{\circ} - \langle r_2, s_2\rangle{}^{\circ}_{\circ}r_1s_1{}^{\circ}_{\circ} \tag{2.35}$$

and the invariant form is

$$\langle x, y\rangle = \langle r_1, s_2\rangle\langle r_2, s_1\rangle - \langle r_1, s_1\rangle\langle r_2, s_2\rangle. \tag{2.36}$$

Let $\{a_1, \ldots, a_l\}$ be a basis of $\mathbf{A}^+$ and let $\{a_1^*, \ldots, a_l^*\}$ be a basis of $\mathbf{A}^-$ such that $\langle a_i, a_j^*\rangle = \delta_{ij}$. We call $\{a_1, \ldots, a_l, a_1^*, \ldots, a_l^*\}$ a "canonical" basis of $\mathbf{A}$. Then $\{h_i = {}^{\circ}_{\circ}a_ia_i^*{}^{\circ}_{\circ} \mid 1 \leq i \leq l\}$ is an orthonormal basis of a Cartan subalgebra of $\mathbf{o}(2l)$.

The spinor construction of $\hat{\mathbf{g}}$ for $\mathbf{g}$ of type D_l begins as above with a non-degenerate symmetric bilinear form $\langle\ ,\ \rangle$ on a vector space $\mathbf{A} \cong \mathbb{C}^{2l}$ such that $\mathbf{A} = \mathbf{A}^+ \oplus \mathbf{A}^-$ is a polarization into maximally isotropic subspaces $\mathbf{A}^{\pm} \cong \mathbb{C}^l$. Let

$$\mathbf{A}(\mathbb{Z}) = \mathbf{A} \otimes \mathbb{C}[t, t^{-1}] \quad \text{and} \quad \mathbf{A}(\mathbb{Z} + \tfrac{1}{2}) = \mathbf{A} \otimes t^{1/2}\mathbb{C}[t, t^{-1}]. \tag{2.37}$$

For $Z = \mathbb{Z}$ or $\mathbb{Z} + \tfrac{1}{2}$ denote $a \otimes t^n$ by $a(n)$ for $a \in \mathbf{A}$, $n \in Z$, and let

$$\iota = \begin{cases} +1 & \text{if } Z = \mathbb{Z} \\[2mm] -1 & Z = \mathbb{Z} + \tfrac{1}{2}. \end{cases} \tag{2.38}$$

Extend the form to $\mathbf{A}(Z)$ by setting

$$\langle a(m), b(n)\rangle = \langle a, b\rangle\delta_{m,-n}. \tag{2.39}$$

Then $\mathbf{A}(Z) = \mathbf{A}(Z)^+ \oplus \mathbf{A}(Z)^-$ is a polarization with respect to the extended form, where

$$\begin{aligned} \mathbf{A}(\mathbb{Z})^{\pm} &= \mathbf{A}^{\pm} \oplus (\mathbf{A} \otimes t^{\pm 1}\mathbb{C}[t^{\pm 1}]), \\ \mathbf{A}(\mathbb{Z} + \tfrac{1}{2})^{\pm} &= \mathbf{A} \otimes t^{\pm 1/2}\mathbb{C}[t^{\pm 1}]. \end{aligned} \tag{2.40}$$

Define an associative algebra $\mathbf{Cliff}(Z)$ with unit element 1, generators from $\mathbf{A}(Z)$ and relations

$$a(m)b(n) + b(n)a(m) = \langle a(m), b(n)\rangle 1 = \langle a, b\rangle\delta_{m,-n}1. \tag{2.41}$$

Let $\mathfrak{I}(Z)$ be the left ideal in $\mathbf{Cliff}(Z)$ generated by $\mathbf{A}(Z)^+$. Then

$$\mathbf{CM}(Z) = \mathbf{Cliff}(Z)/\mathfrak{I}(Z) \tag{2.42}$$

is an irreducible left $\mathbf{Cliff}(Z)$-module having a "vacuum vector"

$$\mathbf{vac}(Z) = 1 + \mathfrak{I}(Z) \tag{2.43}$$

satisfying $\mathbf{A}(Z)^+ \cdot \mathbf{vac}(Z) = 0$ and $1 \cdot \mathbf{vac}(Z) = \mathbf{vac}(Z)$. The subalgebra of $\mathbf{Cliff}(Z)$ generated by 1 and $\mathbf{A}(Z)^-$ is the exterior algebra $\wedge \mathbf{A}(Z)^-$, and

$$\mathbf{CM}(Z) = (\wedge \mathbf{A}(Z)^-) \cdot \mathbf{vac}(Z). \tag{2.44}$$

Therefore, $\mathbf{CM}(Z)$ is spanned by vectors of the form

$$a_1(-m_1)\dots a_r(-m_r)\mathbf{vac}(Z) \tag{2.45}$$

for $a_i(-m_i) \in \mathbf{A}(Z)^-$, $1 \leq i \leq r$. We have the decomposition

$$\mathbf{CM}(Z) = \mathbf{CM}(Z)^0 \oplus \mathbf{CM}(Z)^1 \tag{2.46}$$

into the even and odd parity subspaces,

$$\mathbf{CM}(Z)^0 = (\wedge^{even}\mathbf{A}(Z)^-) \cdot \mathbf{vac}(Z), \quad \mathbf{CM}(Z)^1 = (\wedge^{odd}\mathbf{A}(Z)^-) \cdot \mathbf{vac}(Z). \tag{2.47}$$

Let ζ be a formal complex variable and for $a \in \mathbf{A}$ define the generating function

$$a(\zeta) = \sum_{n \in Z} a(n)\zeta^{-n}. \tag{2.48}$$

For $a(m), b(n) \in \mathbf{A}(Z)$ define the "fermionic normal ordering"

$$\substack{\circ\\\circ}a(m)b(n)\substack{\circ\\\circ} = \begin{cases} a(m)b(n) & \text{if } n > m \\[2mm] \frac{1}{2}(a(m)b(n) - b(n)a(m)) & \text{if } n = m \\[2mm] -b(n)a(m) & \text{if } n < m. \end{cases} \tag{2.49}$$

Then

$$\substack{\circ\\\circ}a(\zeta)b(\zeta)\substack{\circ\\\circ} = \sum_{k \in \mathbb{Z}}\left(\sum_{n \in Z}\substack{\circ\\\circ}a(n)b(k-n)\substack{\circ\\\circ}\right)\zeta^{-k} \tag{2.50}$$

is the generating function defining the homogeneous components

$$\substack{\circ\\\circ}a(\zeta)b(\zeta)\substack{\circ\\\circ}_k = \sum_{n \in Z}\substack{\circ\\\circ}a(n)b(k-n)\substack{\circ\\\circ} \quad \text{for } k \in \mathbb{Z}. \tag{2.51}$$

These are well-defined operators on $\mathbf{CM}(Z)$. For $1 \leq i \leq l$ define the generating functions

$$D_i^Z(\zeta) = -\frac{1+\iota}{16} + \sum_{k \in \mathbb{Z}}\sum_{n \in Z}(n - \tfrac{1}{2}k)\,\substack{\circ\\\circ}a_i(n)a_i^*(k-n)\substack{\circ\\\circ}\zeta^{-k} \tag{2.52}$$

and

$$D^Z(\zeta) = \sum_{1 \le i \le l} D_i^Z(\zeta). \tag{2.53}$$

Then for $k \in \mathbb{Z}$ we have their homogeneous components $D_i^Z(k)$ and $D^Z(k)$, so

$$D^Z(k) = \sum_{1 \le i \le l} \sum_{n \in Z} (n - \tfrac{1}{2}k) \, {}^\circ_\circ a_i(n) a_i^*(k-n) {}^\circ_\circ \quad \text{for } k \ne 0, \tag{2.54}$$

and

$$D^Z(0) = -\frac{l(1+\iota)}{16} + \sum_{1 \le i \le l} \sum_{n \in Z} n \, {}^\circ_\circ a_i(n) a_i^*(-n) {}^\circ_\circ \tag{2.55}$$

These are well-defined operators on $\mathbf{CM}(Z)$, independent of the choice of canonical basis $\{a_1, \ldots, a_l, a_1^*, \ldots, a_l^*\}$. Let $Z' = \tfrac{1}{2}\mathbb{Z} - Z$ and for $a \in \mathbf{A}$, $n \in Z$, define $a(n)$ to act on $\mathbf{CM}(Z')$ as zero. Then the operators

$$D(m) = D^Z(m) + D^{Z+\frac{1}{2}}(m), \quad m \in \mathbb{Z}, \tag{2.56}$$

are well-defined on $\mathbf{CM}(\mathbb{Z}) \oplus \mathbf{CM}(\mathbb{Z}+\tfrac{1}{2})$ and $D(m) = D^Z(m)$ on $\mathbf{CM}(Z)$. Using the Wick theorem [**W**] one can prove the following (see [**FF,F2**]).

THEOREM 2.1. *For $Z = \mathbb{Z}$ or $\mathbb{Z} + \tfrac{1}{2}$ the operators (2.51) along with the operator (2.55) and the identity operator represent the affine algebra $\hat{\mathfrak{o}}(2l)$ on $\mathbf{CM}(Z)$. $\mathbf{CM}(Z) = \mathbf{CM}(Z)^0 \oplus \mathbf{CM}(Z)^1$ is the decomposition of $\mathbf{CM}(Z)$ into two irreducible $\hat{\mathfrak{o}}(2l)$-modules. The operators (2.56) and the identity operator represent the Virasoro algebra on $\mathbf{CM}(\mathbb{Z}) \oplus \mathbf{CM}(\mathbb{Z} + \tfrac{1}{2})$.*

More precisely, we have the following. Let $a, a_1, a_2, b_1, b_2 \in \mathbf{A}$, $x = {}^\circ_\circ a_1 a_2 {}^\circ_\circ, y = {}^\circ_\circ b_1 b_2 {}^\circ_\circ \in \mathfrak{o}(2l)$, $m, n \in \mathbb{Z}$, $k \in Z$, and let $a(k) \in \mathbf{Cliff}(Z)$. Then on $\mathbf{CM}(Z)$

$$x(m) = {}^\circ_\circ a_1(\zeta) a_2(\zeta) {}^\circ_{\circ m}, \, y(n) = {}^\circ_\circ b_1(\zeta) b_2(\zeta) {}^\circ_{\circ n}, \, c = 1, \, d = D(0) \tag{2.57}$$

represent $\hat{\mathfrak{o}}(2l)$, $L(m) = D(m)$ and $z = l$ represent $\mathbf{Vir}$, and we have the brackets

$$[x(m), a(k)] = (x \cdot a)(m+k), \tag{2.58}$$

$$[x(m), y(n)] = [x, y](m+n) + m\delta_{m,-n}\langle x, y \rangle 1, \tag{2.59}$$

$$[D(n), a(k)] = (k + \tfrac{1}{2}n)a(k+n), \tag{2.60}$$

$$[D(n), x(m)] = mx(n+m), \tag{2.61}$$

$$[D(m), D(n)] = (n-m)D(m+n) + \tfrac{1}{12}(m^3 - m)\delta_{m,-n}l, \tag{2.62}$$

where $[x, y]$ and $\langle x, y \rangle$ are given by (2.35) and (2.36).

We can define a positive Hermitian form on the representation space $\mathbf{CM}(Z)$ of $\hat{\mathfrak{o}}(2l)$, and we can determine the adjoints of the operators acting on $\mathbf{CM}(Z)$. We may choose a $\mathbb{C}$-antilinear involution $\vartheta : \mathbf{A} \to \mathbf{A}$ such that for $a, b \in \mathbf{A}$ we have

$$\langle \vartheta a, \vartheta b \rangle = \overline{\langle a, b \rangle}, \tag{2.63}$$

$$0 < \langle a, \vartheta a \rangle \in \mathbb{R} \quad \text{if } a \ne 0, \tag{2.64}$$

$$\mathbf{A}^- = \vartheta(\mathbf{A}^+). \tag{2.65}$$

Note that given any $\mathbb{C}$-antilinear involution $\vartheta : \mathbf{A} \to \mathbf{A}$ satisfying (2.63) and (2.64), for any choice of maximal isotropic subspace $\mathbf{A}^+$, $\vartheta(\mathbf{A}^+)$ is a maximal isotropic subspace and $\mathbf{A} = \mathbf{A}^+ \oplus \vartheta(\mathbf{A}^+)$ is a polarization of $\mathbf{A}$. For $a, b \in \mathbf{A}$ we define a form

$$(a, b) = \langle a, \vartheta b \rangle \tag{2.66}$$

and note that it is positive Hermitian because of (2.63) and (2.64). We extend this form to $\mathbf{A}(Z)^-$ by

$$(a(m), b(n)) = \delta_{m,n}(a, b), \tag{2.67}$$

and we further extend it to $\mathbf{CM}(Z)$ (2.44)-(2.45) by

$$\begin{aligned}
(a_1(m_1) &\ldots a_r(m_r)\mathbf{vac}(Z), \; b_1(n_1)\ldots b_s(n_s)\mathbf{vac}(Z)) \\
&= \delta_{r,s}\det[(a_i(m_i), b_j(n_j))] \\
&= \delta_{r,s}\sum_{\sigma \in S_r} sgn(\sigma) \prod_{1 \le k \le r} \left(a_k(m_k), b_{\sigma(k)}(n_{\sigma(k)})\right),
\end{aligned} \tag{2.68}$$

giving a positive Hermitian form. If L is any linear operator on $\mathbf{CM}(Z)$, denote by L^* the adjoint of L with respect to this form.

PROPOSITION 2.2. *Let* $(\,,\,)$ *denote the positive Hermitian form on* $\mathbf{CM}(Z)$ *defined by (2.66)-(2.68). Then we have*

(a) $a(m)^* = (\vartheta a)(-m)$ *for* $a(m) \in \mathbf{A}(Z)$,
(b) $({}^\circ_\circ a(\zeta)b(\zeta){}^\circ_\circ{}_k)^* = {}^\circ_\circ(\vartheta b)(\zeta)(\vartheta a)(\zeta){}^\circ_\circ{}_{-k}$ *for* $a, b \in \mathbf{A}$, $k \in \mathbb{Z}$,
(c) $D^Z(m)^* = D^Z(-m)$ *for* $m \in \mathbb{Z}$.

PROOF. (a) follows from the definitions (2.66)-(2.68). (b) From (2.49) we have $({}^\circ_\circ a(m)b(n){}^\circ_\circ)^* = {}^\circ_\circ(\vartheta b)(-n)(\vartheta a)(-m){}^\circ_\circ$, giving (b). Finally, since ϑ takes one canonical basis to another canonical basis, and the operators $D^Z(k)$, $k \in \mathbb{Z}$, are independent of the choice of such a basis, we get (c). ∎

The eigenspaces of $D(0)$ provide $\mathbf{CM}(Z)$ with a grading as follows. Let

$$(\mathbf{CM}(Z))_n = \{u \in \mathbf{CM}(Z) \mid D(0)u = -nu\} \tag{2.69}$$

be the subspace of vectors of weight n, and write $wt(u)$ for the weight of u. From (2.55) it is clear that

$$D(0)\mathbf{vac}(Z) = -\frac{l(1 + \iota)}{16}\mathbf{vac}(Z), \tag{2.70}$$

so using (2.60), for $u = a_1(-m_1)\ldots a_r(-m_r)\mathbf{vac}(Z)$ as in (2.45), we have

$$D(0)u = -\left(m_1 + \ldots + m_r + \frac{l(1 + \iota)}{16}\right)u. \tag{2.71}$$

Then $\mathbf{CM}(Z)$ is the direct sum of the eigenspaces (2.69), and the $\frac{1}{2}\mathbb{Z}$-grading of $\mathbf{CM}(\mathbb{Z} + \frac{1}{2})$ begins at 0 and the $\mathbb{Z}$-grading of $\mathbf{CM}(\mathbb{Z})$ begins at $l/8$. For $\mathbf{W} = \mathbf{CM}(Z)^\alpha$, $\alpha = 0, 1$, we define $\Delta_{\mathbf{W}}$ to be to be the minimal value of $wt(w)$ for $w \in \mathbf{W}$, so that $\mathbf{W}$ is $\mathbb{Z} + \Delta_{\mathbf{W}}$-graded. Since $D(0)$ is self-adjoint, the eigenspaces for distinct eigenvalues are orthogonal with respect to the Hermitian form on $\mathbf{CM}(Z)$. If $A(m)$ is any of the operators $a(m) \in \mathbf{Cliff}(Z)$, $x(m) \in \hat{\mathfrak{o}}(2l)$ or $D(m) \in \mathbf{Vir}$, then from (2.60)- (2.62) we see that

$$A(m) : (\mathbf{CM}(Z))_n \to (\mathbf{CM}(Z))_{n-m}. \qquad (2.72)$$

CHAPTER 3

Spinor Construction of Vertex Operator Superalgebras

We will now describe vertex operator superalgebras constructed from certain representation spaces in the spinor construction. First recall from Chapter 2 the spinor construction of the $\mathbf{Cliff}(Z)$-module

$$\mathbf{CM}(Z) = \mathbf{Cliff}(Z) \cdot \mathbf{vac}(Z) = (\wedge \mathbf{A}(Z)^-) \cdot \mathbf{vac}(Z), \tag{3.1}$$

from a nondegenerate symmetric bilinear form $\langle\ ,\ \rangle$ on $\mathbf{A} \approx \mathbb{C}^{2l}$, with $Z = \mathbb{Z}$ or $\mathbb{Z} + \frac{1}{2}$, and $\mathbf{A}(Z)^-$ defined in (2.40). $\mathbf{CM}(Z)$ has the decomposition (2.46) into even and odd parity subspaces, each of which is preserved by the Virasoro operators $D(m)$, $m \in \mathbb{Z}$, defined in (2.54)-(2.56) using a canonical basis of $\mathbf{A}$. Using a $\mathbb{C}$-antilinear involution ϑ on $\mathbf{A}$ satisfying (2.63)-(2.65), $\mathbf{CM}(Z)$ has a positive Hermitian form $(\ ,\ )$ defined by (2.66)-(2.68). Proposition 2.2 gives the adjoints of the operators $a(m) \in \mathbf{A}(Z)$, $x(k) \in \hat{\mathbf{o}}(2l)$, and $D(m)$ with respect to $(\ ,\ )$. With the notation as in Theorem 2.1, we have the formulas (2.58)-(2.62).

Let

$$\mathbf{V}_0 = \mathbf{CM}(\mathbb{Z}+\tfrac{1}{2})^0, \ \ \mathbf{V}_1 = \mathbf{CM}(\mathbb{Z}+\tfrac{1}{2})^1, \ \ \mathbf{V}_2 = \mathbf{CM}(\mathbb{Z})^0, \ \ \mathbf{V}_3 = \mathbf{CM}(\mathbb{Z})^1, \tag{3.2}$$

$$\mathbf{V}(\tfrac{1}{2}\mathbb{Z}) = \mathbf{CM}(\mathbb{Z} + \tfrac{1}{2}) \oplus \mathbf{CM}(\mathbb{Z}) = \mathbf{V}_0 \oplus \mathbf{V}_1 \oplus \mathbf{V}_2 \oplus \mathbf{V}_3, \tag{3.3}$$

$$\mathbf{V} = \mathbf{V}_0 \oplus \mathbf{V}_1, \tag{3.4}$$

and extend the Hermitian form $(\ ,\ )$ to $\mathbf{V}(\tfrac{1}{2}\mathbb{Z})$ so that $\mathbf{CM}(\mathbb{Z} + \tfrac{1}{2})$ and $\mathbf{CM}(\mathbb{Z})$ are orthogonal. For $a \in \mathbf{A}$, $n \in \tfrac{1}{2}\mathbb{Z}$, let $a(n)$ act on $\mathbf{V}(\tfrac{1}{2}\mathbb{Z})$ as usual on $\mathbf{CM}(Z)$ if $n \in Z$, but as zero if $n \in Z' = \tfrac{1}{2}\mathbb{Z} - Z$. From (2.60) and (2.69) we see that $D(0)$ provides a grading of $\mathbf{V}_i$, $0 \le i \le 3$, into mutually orthogonal eigenspaces

$$\mathbf{V}_i = \bigoplus_{0 \le k \in \mathbb{Z}+\Delta_i} (\mathbf{V}_i)_k \tag{3.5}$$

where we define

$$\Delta_i = \Delta_{\mathbf{V}_i} = \begin{cases} 0 & \text{if } i = 0 \\ \tfrac{1}{2} & \text{if } i = 1 \\ l/8 & \text{if } i = 2, 3. \end{cases} \tag{3.6}$$

44

Our objective is to construct a vertex operator superalgebra $(\mathbf{CM}(\mathbb{Z}+\frac{1}{2})$, $Y(\ ,\zeta), \mathbf{vac}(\mathbb{Z}+\frac{1}{2}), D(-2)\mathbf{vac}(\mathbb{Z}+\frac{1}{2}))$ and the canonically $\mathbb{Z}_2$-twisted representation of it on $\mathbf{CM}(\mathbb{Z})$. In particular, we will define superalgebra structures on $\hat{\mathbf{V}}/D(-1)\hat{\mathbf{V}}$, where

$$\hat{\mathbf{V}} = \mathbf{V} \otimes \mathbb{C}[t, t^{-1}], \tag{3.7}$$

and on $\hat{\mathbf{V}}'/D(-1)\hat{\mathbf{V}}'$, where

$$\hat{\mathbf{V}}' = \mathbf{V}_0 \otimes \mathbb{C}[t, t^{-1}] \oplus \mathbf{V}_1 \otimes t^{1/2}\mathbb{C}[t, t^{-1}] \tag{3.8}$$

so that $\hat{\mathbf{V}}/D(-1)\hat{\mathbf{V}}$ is represented on $\mathbf{V}$ and $\hat{\mathbf{V}}'/D(-1)\hat{\mathbf{V}}'$ is represented on $\mathbf{V}_2 \oplus \mathbf{V}_3$. The first step is to define generalized vertex operators $Y(v, \zeta)$ on $\mathbf{CM}(\mathbb{Z}+\frac{1}{2})$ and on $\mathbf{CM}(\mathbb{Z})$ for any $v \in \mathbf{V}$. The definition of $Y(v, \zeta)$ on $\mathbf{CM}(\mathbb{Z})$ is more complicated than it is on $\mathbf{CM}(\mathbb{Z}+\frac{1}{2})$, but there is a common part which we denote by $\bar{Y}(v, \zeta)$. With $\mathbf{vac} = \mathbf{vac}(\mathbb{Z}+\frac{1}{2})$ define $\bar{Y}(\mathbf{vac}, \zeta) = 1$ (the identity operator) on $\mathbf{V}(\frac{1}{2}\mathbb{Z})$, and for $a \in \mathbf{A}$, $0 \le n \in \mathbb{Z}$, on $\mathbf{CM}(Z)$ let

$$\bar{Y}(a(-n-\tfrac{1}{2})\mathbf{vac}, \zeta) = n!^{-1}(d/d\zeta)^n(\zeta^{-1/2}a(\zeta)) \tag{3.9}$$

where

$$a(\zeta) = \sum_{m \in Z} a(m)\zeta^{-m}. \tag{3.10}$$

Define the notations $a^{(n)}(\zeta)$, $a^{(n)}(\zeta)^{\pm}$ and $a^{(n)}(\zeta)^0$ by

$$a^{(n)}(\zeta) = \bar{Y}(a(-n-\tfrac{1}{2})\mathbf{vac}, \zeta) = \sum_{m \in Z} \binom{-m-\frac{1}{2}}{n} a(m)\zeta^{-m-n-\frac{1}{2}}, \tag{3.11}$$

$$a^{(n)}(\zeta)^+ = \sum_{0 < m \in Z} \binom{-m-\frac{1}{2}}{n} a(m)\zeta^{-m-n-\frac{1}{2}}, \tag{3.12}$$

$$a^{(n)}(\zeta)^- = \sum_{0 > m \in Z} \binom{-m-\frac{1}{2}}{n} a(m)\zeta^{-m-n-\frac{1}{2}}, \tag{3.13}$$

$$a^{(n)}(\zeta)^0 = \binom{-\frac{1}{2}}{n} a(0)\zeta^{-n-\frac{1}{2}} \quad \text{if} \ \ Z = \mathbb{Z}, \tag{3.14}$$

so that

$$a^{(n)}(\zeta) = \begin{cases} a^{(n)}(\zeta)^+ + a^{(n)}(\zeta)^- & \text{on } \mathbf{CM}(\mathbb{Z}+\tfrac{1}{2}) \\[2mm] a^{(n)}(\zeta)^+ + a^{(n)}(\zeta)^0 + a^{(n)}(\zeta)^- & \text{on } \mathbf{CM}(\mathbb{Z}). \end{cases} \tag{3.15}$$

For $a_1, \ldots, a_r \in \mathbf{A}$, $n_1, \ldots, n_r \in \mathbb{Z}+\frac{1}{2}$ extend the definition of fermionic normal ordering (2.49) to any product of Clifford generators

$$^\circ_\circ a_1(n_1) \ldots a_r(n_r)^\circ_\circ = sgn(\sigma)a_{\sigma 1}(n_{\sigma 1}) \ldots a_{\sigma r}(n_{\sigma r}), \tag{3.16}$$

where $\sigma \in S_r$ is any permutation such that $n_{\sigma 1} \le \ldots \le n_{\sigma r}$. The Clifford relations (2.41) show that this is well-defined and agrees with (2.49) for $r = 2$. For $n_1, \ldots, n_r \in \mathbb{Z}$, we have to modify the definition of the normally ordered product ${}^\circ_\circ a_1(n_1)\ldots a_r(n_r){}^\circ_\circ$ in (3.16) because of the possibility that some $n_i = 0$. We wish to make the definition so that for any permutation $\sigma \in S_r$ we have

$$\,{}^\circ_\circ a_1(n_1)\ldots a_r(n_r){}^\circ_\circ = sgn(\sigma)\,{}^\circ_\circ a_{\sigma 1}(n_{\sigma 1})\ldots a_{\sigma r}(n_{\sigma r}){}^\circ_\circ. \tag{3.17}$$

It is therefore enough to make the definition when $n_1 \leq \ldots \leq n_r$ and then define the other cases by (3.17). If $n_1 \leq \ldots \leq n_r$ and each $0 \neq n_i \in \mathbb{Z}$, let

$$\,{}^\circ_\circ a_1(n_1)\ldots a_r(n_r){}^\circ_\circ = a_1(n_1)\ldots a_r(n_r), \tag{3.18}$$

but if $n_{i-1} < 0 = n_i = n_{i+1} = \ldots = n_{i+s-1} < n_{i+s}$ for some $s \geq 1$, then define

$$\begin{aligned}
&{}^\circ_\circ a_1(n_1)\ldots a_r(n_r){}^\circ_\circ \\
&\quad = a_1(n_1)\ldots a_{i-1}(n_{i-1})({}^\circ_\circ a_i(0)\ldots a_{i+s-1}(0){}^\circ_\circ)a_{i+s}(n_{i+s})\ldots a_r(n_r),
\end{aligned} \tag{3.19}$$

where for any $a_1,\ldots,a_k \in \mathbf{A}$,

$$\,{}^\circ_\circ a_1(0)\ldots a_k(0){}^\circ_\circ = k!^{-1}\sum_{\sigma \in S_k} sgn(\sigma)a_{\sigma 1}(0)\ldots a_{\sigma k}(0). \tag{3.20}$$

For $0 \leq n_1,\ldots,n_r \in \mathbb{Z}$, and for elements of the form

$$v = a_1(-n_1 - \tfrac{1}{2})\ldots a_r(-n_r - \tfrac{1}{2})\mathbf{vac} \in (\mathbf{V})_n, \tag{3.21}$$

$n = \mathrm{wt}(v) = n_1 + \ldots + n_r + \tfrac{1}{2}r$, on $\mathbf{CM}(Z)$ we define

$$\begin{aligned}
\bar{Y}(v,\zeta) &= {}^\circ_\circ \bar{Y}(a_1(-n_1 - \tfrac{1}{2})\mathbf{vac},\zeta)\ldots \bar{Y}(a_r(-n_r - \tfrac{1}{2})\mathbf{vac},\zeta){}^\circ_\circ \\
&= {}^\circ_\circ a_1^{(n_1)}(\zeta)\ldots a_r^{(n_r)}(\zeta){}^\circ_\circ.
\end{aligned} \tag{3.22}$$

Then on $\mathbf{CM}(\mathbb{Z} + \tfrac{1}{2})$ we have

$$\bar{Y}(v,\zeta) = \sum sgn(\sigma)a_{\sigma 1}^{(n_{\sigma 1})}(\zeta)^{\pm}\ldots a_{\sigma r}^{(n_{\sigma r})}(\zeta)^{\pm} \tag{3.23}$$

where the summation is over the 2^r terms obtained by splitting each factor as in the first line of (3.15) and in each term permuting the resulting factors by some $\sigma \in S_r$ so that all $^-$ factors are to the left of all $^+$ factors. Note that any two $^-$ factors anticommute, and any two $^+$ factors anticommute. On $\mathbf{CM}(\mathbb{Z})$ we have

$$\bar{Y}(v,\zeta) = \sum sgn(\sigma)a_{\sigma 1}^{(n_{\sigma 1})}(\zeta)^{-}\ldots({}^\circ_\circ a_{\sigma i}^{(n_{\sigma i})}(\zeta)^{0}\ldots a_{\sigma j}^{(n_{\sigma j})}(\zeta)^{0}{}^\circ_\circ)\ldots a_{\sigma r}^{(n_{\sigma r})}(\zeta)^{+} \tag{3.24}$$

where the summation is over the 3^r terms obtained by splitting each factor as in the second line of (3.15) and in each term permuting the resulting factors by some $\sigma \in S_r$ so that all $^-$ factors are to the left and all $^+$ factors are to the right. Note that any two $^-$ factors anticommute, and any two $^+$ factors anticommute. Extend the definitions of $\bar{Y}(v,\zeta)$ to any $v \in \mathbf{CM}(\mathbb{Z} + \tfrac{1}{2})$ by linearity. For $v \in (\mathbf{V})_k$ define the homogeneous components $\bar{Y}_m(v)$ of $\bar{Y}(v,\zeta)$ and the notation $\{\bar{v}\}_n$ by

$$\bar{Y}(v,\zeta) = \sum_{n \in \frac{1}{2}\mathbb{Z}} \bar{Y}_{n+1-\mathrm{wt}(v)}(v)\zeta^{-n-1} = \sum_{n \in \frac{1}{2}\mathbb{Z}} \{\bar{v}\}_n \zeta^{-n-1}. \tag{3.25}$$

On $\mathbf{CM}(\mathbb{Z}+\frac{1}{2})$ the coefficient of ζ^{-n-1} is zero unless $n \in \mathbb{Z}$. On $\mathbf{CM}(\mathbb{Z})$ the coefficient of ζ^{-n-1} is zero unless $n \in \mathbb{Z}$ for $v \in \mathbf{V}_0$, but for $v \in \mathbf{V}_1$, it is zero unless $n \in \mathbb{Z}+\frac{1}{2}$. It is easy to see that the components $\bar{Y}_m(v)$ are well-defined operators on $\mathbf{V}(\frac{1}{2}\mathbb{Z})$ such that

$$\mathrm{wt}(\bar{Y}_m(v)w) = \mathrm{wt}(w) - m. \tag{3.26}$$

PROPOSITION 3.1. *For $v \in \mathbf{V}$ we have*

(a) $\lim\limits_{\zeta \to 0} \bar{Y}(v,\zeta)\mathbf{vac} = v$,

(b) $\bar{Y}(D(-1)v,\zeta) = -(d/d\zeta)\bar{Y}(v,\zeta)$ *on* $\mathbf{V}(\frac{1}{2}\mathbb{Z})$,

(c) $[D(0),\bar{Y}(v,\zeta)] = -\zeta(d/d\zeta)\bar{Y}(v,\zeta) + \bar{Y}(D(0)v,\zeta)$ *on* $\mathbf{V}(\frac{1}{2}\mathbb{Z})$.

PROOF. It is enough to check these for $v = a_1(-n_1 - \frac{1}{2})\ldots a_r(-n_r - \frac{1}{2})\mathbf{vac} \in \mathbf{V}$.

(a) From (3.23) we have

$$\bar{Y}(v,\zeta)\mathbf{vac} = a_1^{(n_1)}(\zeta)^- \ldots a_r^{(n_r)}(\zeta)^- \mathbf{vac} \tag{3.27}$$

in which only terms with nonnegative integral powers of ζ occur. From (3.13) with $Z = \mathbb{Z}+\frac{1}{2}$, the ζ^0 term is $a_1(-n-\frac{1}{2})\ldots a_r(-n_r-\frac{1}{2})\mathbf{vac} = v$.

(b) It is clear from (2.54) and (2.60) that

$$D(-1)\mathbf{vac} = 0 \tag{3.28}$$

and

$$[D(-1), a(-n-\tfrac{1}{2})] = (-n-1)a(-n-\tfrac{3}{2}) \tag{3.29}$$

so that

$$D(-1)v = \sum_{1\le i\le r} a_1(-n_1-\tfrac{1}{2})\ldots[D(-1),a_i(-n_i-\tfrac{1}{2})]\ldots a_r(-n_r-\tfrac{1}{2})\mathbf{vac}$$

$$= -\sum_{1\le i\le r}(n_i+1)a_1(-n_1-\tfrac{1}{2})\ldots a_i(-n_i-\tfrac{3}{2})\ldots a_r(-n_r-\tfrac{1}{2})\mathbf{vac}. \tag{3.30}$$

Therefore,

$$\bar{Y}(D(-1)v,\zeta) = -\sum_{1\le i\le r}(n_i+1){}^\circ_\circ a_1^{(n_1)}(\zeta)\ldots a_i^{(n_i+1)}(\zeta)\ldots a_r^{(n_r)}(\zeta){}^\circ_\circ$$

$$= -(d/d\zeta){}^\circ_\circ a_1^{(n_1)}(\zeta)\ldots a_r^{(n_r)}(\zeta){}^\circ_\circ \tag{3.31}$$

$$= -(d/d\zeta)\bar{Y}(v,\zeta).$$

(c) With v as above, we have

$$D(0)v = -\sum_{1\le i\le r}(n_i+\tfrac{1}{2})v, \tag{3.32}$$

$$[D(0),a^{(n)}(\zeta)] = \sum_{m\in\mathbb{Z}}\binom{-m-\frac{1}{2}}{n}ma(m)\zeta^{-m-n-\frac{1}{2}} \tag{3.33}$$

$$= -\zeta(d/d\zeta)a^{(n)}(\zeta) - (n+\tfrac{1}{2})a^{(n)}(\zeta)$$

and so

$$[D(0), \bar{Y}(v,\zeta)] = \sum_{1 \leq i \leq r} {}^{\circ}_{\circ} a_1^{(n_1)}(\zeta) \ldots [D(0), a_i^{(n_i)}(\zeta)] \ldots a_r^{(n_r)}(\zeta) {}^{\circ}_{\circ}$$

$$= \sum_{1 \leq i \leq r} {}^{\circ}_{\circ} a_1^{(n_1)}(\zeta) \ldots ((-\zeta(d/d\zeta) - (n_i + \tfrac{1}{2})) a_i^{(n_i)}(\zeta)) \ldots a_r^{(n_r)}(\zeta) {}^{\circ}_{\circ}$$

$$= -\zeta(d/d\zeta)\bar{Y}(v,\zeta) + \bar{Y}(D(0)v,\zeta).$$

$$(3.34)$$

Note that we have used the fact that

$$[D(0), {}^{\circ}_{\circ} a_1(m_1) \ldots a_r(m_r) {}^{\circ}_{\circ}]$$

$$= \sum_{1 \leq i \leq r} {}^{\circ}_{\circ} a_1(m_1) \ldots [D(0), a_i(m_i)] \ldots a_r(m_r) {}^{\circ}_{\circ}, \qquad (3.35)$$

which would not be true with $D(0)$ replaced by $D(n)$ for $n \neq 0$. ∎

We can write the generating function of the Virasoro operators on $\mathbf{V} = \mathbf{CM}(\mathbb{Z} + \tfrac{1}{2})$ as follows. Letting

$$\omega = D(-2)\mathbf{vac}$$

$$= \tfrac{1}{2} \sum_{1 \leq i \leq l} {}^{\circ}_{\circ} a_i(-\tfrac{1}{2}) a_i^*(-\tfrac{3}{2}) - a_i(-\tfrac{3}{2}) a_i^*(-\tfrac{1}{2}) {}^{\circ}_{\circ} \mathbf{vac}$$

$$= -\tfrac{1}{2} \sum_{1 \leq i \leq l} (a_i^*(-\tfrac{3}{2}) a_i(-\tfrac{1}{2}) + a_i(-\tfrac{3}{2}) a_i^*(-\tfrac{1}{2})) \mathbf{vac}$$

$$(3.36)$$

we find that

$$\bar{Y}(\omega, \zeta) = -\tfrac{1}{2} \sum_{1 \leq i \leq l} \bar{Y}(a_i^*(-\tfrac{3}{2}) a_i(-\tfrac{1}{2}) \mathbf{vac}, \zeta) + \bar{Y}(a_i(-\tfrac{3}{2}) a_i^*(-\tfrac{1}{2}) \mathbf{vac}, \zeta)$$

$$= -\tfrac{1}{2} \sum_{1 \leq i \leq l} {}^{\circ}_{\circ} \left(\frac{d}{d\zeta} \sum_{m \in \mathbb{Z} + \frac{1}{2}} a_i^*(m) \zeta^{-m-\frac{1}{2}} \right) \left(\sum_{n \in \mathbb{Z} + \frac{1}{2}} a_i(n) \zeta^{-n-\frac{1}{2}} \right)$$

$$+ \left(\frac{d}{d\zeta} \sum_{n \in \mathbb{Z} + \frac{1}{2}} a_i(n) \zeta^{-n-\frac{1}{2}} \right) \left(\sum_{m \in \mathbb{Z} + \frac{1}{2}} a_i^*(m) \zeta^{-m-\frac{1}{2}} \right) {}^{\circ}_{\circ}$$

$$= \tfrac{1}{2} \sum_{1 \leq i \leq l} \sum_{m,n \in \mathbb{Z} + \frac{1}{2}} \left((m + \tfrac{1}{2}) {}^{\circ}_{\circ} a_i^*(m) a_i(n) {}^{\circ}_{\circ} + (n + \tfrac{1}{2}) {}^{\circ}_{\circ} a_i(n) a_i^*(m) {}^{\circ}_{\circ} \right) \zeta^{-m-n-2}$$

$$= \tfrac{1}{2} \sum_{1 \leq i \leq l} \sum_{m,n \in \mathbb{Z} + \frac{1}{2}} (n - m) {}^{\circ}_{\circ} a_i(n) a_i^*(m) {}^{\circ}_{\circ} \zeta^{-m-n-2}$$

$$= \sum_{1 \leq i \leq l} \sum_{k \in \mathbb{Z}} \sum_{n \in \mathbb{Z} + \frac{1}{2}} (n - \tfrac{1}{2}k) {}^{\circ}_{\circ} a_i(n) a_i^*(k - n) {}^{\circ}_{\circ} \zeta^{-k-2}$$

$$= \zeta^{-2} D(\zeta)$$

$$(3.37)$$

on $\mathbf{V}$, where $\{a_1, \ldots, a_l, a_1^*, \ldots, a_l^*\}$ is any canonical basis of $\mathbf{A}$.

Computing $\bar{Y}(\omega, \zeta)$ on $\mathbf{CM}(\mathbb{Z})$ we find that the homogeneous components agree with the definitions of the Virasoro operators in (2.54)-(2.55), except that the scalar term $-l/8$ is missing from the operator which should be giving $D(0)$. The effect of leaving off that scalar term, that is, just using $\bar{Y}(\omega, \zeta) = \zeta^{-2}D(\zeta)$ to define the Virasoro operators on $\mathbf{CM}(\mathbb{Z})$, is to modify the scalar term in the bracket formula (2.62) so that $m^3 - m$ is changed to $m^3 + 2m$. Since we want to have (2.62) on $\mathbf{CM}(\mathbb{Z})$, we must add to $\bar{Y}(\omega, \zeta)$ a term which acts as $-\zeta^{-2}l/8$, so

$$\bar{Y}(\omega, \zeta) - \frac{1}{8}\zeta^{-2}l = \zeta^{-2}D(\zeta) \tag{3.38}$$

is the generating function of the Virasoro operators on $\mathbf{CM}(\mathbb{Z})$. The vertex operators $Y(v, \zeta)$ which we will define later incorporate this scalar term on $\mathbf{CM}(\mathbb{Z})$.

There are two approaches one may use to deal with generating functions of operators on $\mathbf{V}(\frac{1}{2}\mathbb{Z})$, the formal variables approach and the complex analytic approach. We deal formally with series whose variables have fractional powers in $\frac{1}{n}\mathbb{Z}$ by treating $\zeta^{1/n}$ as a formal variable. When we deal with such series analytically, we treat any expression z^r, $r \in \mathbb{Q}$, as $\exp(r \log z)$ using the principal branch of $\log z$ which is defined and analytic for

$$z \in \mathbb{C}^- = \mathbb{C} - \{x \in \mathbb{R} \mid x \le 0\}. \tag{3.39}$$

We will be dealing with series $f(z^{1/n})$, $g(z^{1/n})$ which converge to algebraic functions in some domain with $z \in \mathbb{C}^-$, such that the product $f(z^{1/n})g(z^{1/n})$ converges in that domain to a rational function of z. But then the convergence of the product is guaranteed in the domain without the condition $z \in \mathbb{C}^-$. Note that the power law

$$z^r z^s = z^{r+s}, \quad r, s \in \mathbb{Q}, \quad z \in \mathbb{C}^-, \tag{3.40}$$

is valid, but $z^r w^r$ need not equal $(zw)^r$, even for $z, w, zw \in \mathbb{C}^-$. Let

$$A(\zeta_1, \ldots, \zeta_r) = \sum_{m_1, \ldots, m_r \in \frac{1}{n}\mathbb{Z}} A_{m_1, \ldots, m_r} \zeta_1^{-m_1} \cdots \zeta_r^{-m_r} \tag{3.41}$$

be the generating function of the operators $A_{m_1, \ldots, m_r}$ on $\mathbf{V}(\frac{1}{2}\mathbb{Z})$. The identity $A(\zeta_1, \ldots, \zeta_r) = B(\zeta_1, \ldots, \zeta_r)$ means the equality of the coefficients $A_{m_1, \ldots, m_r} = B_{m_1, \ldots, m_r}$ as operators on $\mathbf{V}(\frac{1}{2}\mathbb{Z})$, for each $m_1, \ldots, m_r \in \frac{1}{n}\mathbb{Z}$. In the formal approach, such an equality might be checked by a formal manipulation of the series, and the series might involve formal expressions such as the delta function $\delta(\zeta) = \sum_{m \in \mathbb{Z}} \zeta^m$ and its derivatives. In the complex analytic approach, such an equality would be checked by showing that for any $w, w' \in \mathbf{V}(\frac{1}{2}\mathbb{Z})$, the "matrix coefficients" $(A(\zeta_1, \ldots, \zeta_r)w, w')$ and $(B(\zeta_1, \ldots, \zeta_r)w, w')$ converge to equal holomorphic functions for $(\zeta_1, \ldots, \zeta_r)$ in some open neighborhood in $\mathbb{C}^n$. This is sufficient because the Hermitian form on $\mathbf{V}(\frac{1}{2}\mathbb{Z})$ is nondegenerate. Most of our results follow because, after sometimes multiplying by an appropriate algebraic function, the matrix coefficients we need converge in some open domain in $\mathbb{C}^n$ to polynomials in the ring $\mathbb{C}[\zeta_k, \zeta_k^{-1}, (\zeta_i - \zeta_j)^{-1}; 1 \le k \le n, 1 \le i < j \le n]$.

DEFINITION. *Let $F_1(\zeta_1,\ldots,\zeta_n)$ and $F_2(\zeta_1,\ldots,\zeta_n)$ be generating functions with complex coefficients which represent rational functions of $\zeta_1,\ldots,\zeta_n$ on some domains D_1 and D_2 in $\mathbb{C}^n$. We write $F_1 \sim F_2$ if F_1 and F_2 represent the same rational function on $\mathbb{C}^n$. Note that if $F_1 \sim F_2$ and P is any linear differential operator with coefficients in $\mathbb{C}(\zeta_1,\ldots,\zeta_n)$ then $PF_1 \sim PF_2$.*

DEFINITION. *For $a_1, a_2 \in \mathbf{A}$, $0 \le n_1, n_2 \in \mathbb{Z}$ define the fermionic contraction*

$$\underbrace{a_1^{(n_1)}(\zeta_1)a_2^{(n_2)}(\zeta_2)} = a_1^{(n_1)}(\zeta_1)a_2^{(n_2)}(\zeta_2) - {}_\circ^\circ a_1^{(n_1)}(\zeta_1)a_2^{(n_2)}(\zeta_2){}_\circ^\circ$$

$$= \begin{cases} [a_1^{(n_1)}(\zeta_1), a_2^{(n_2)}(\zeta_2)^-]_+ & \text{on } \mathbf{CM}(\mathbb{Z}+\tfrac{1}{2}) \\[2ex] [a_1^{(n_1)}(\zeta_1), a_2^{(n_2)}(\zeta_2)^- + \tfrac{1}{2}a_2^{(n_2)}(\zeta_2)^0]_+ & \text{on } \mathbf{CM}(\mathbb{Z}). \end{cases} \qquad (3.42)$$

LEMMA 3.2. *For $a_1, a_2 \in \mathbf{A}$, $0 \le n_1, n_2 \in \mathbb{Z}$, as formal series in ζ_1 and ζ_2, on $\mathbf{CM}(\mathbb{Z}+\tfrac{1}{2})$ we have*

$$\underbrace{a_1^{(n_1)}(\zeta_1)a_2^{(n_2)}(\zeta_2)}$$

$$= (-1)^{n_1}\langle a_1, a_2\rangle \zeta_1^{-n_1-n_2-1}\binom{n_1+n_2}{n_1}\sum_{0 \le k \in \mathbb{Z}}\binom{n_1+n_2+k}{k}\left(\frac{\zeta_2}{\zeta_1}\right)^k$$

$$= n_1!^{-1}n_2!^{-1}(\partial/\partial\zeta_1)^{n_1}(\partial/\partial\zeta_2)^{n_2}\langle a_1, a_2\rangle\zeta_1^{-1}\sum_{0 \le m \in \mathbb{Z}}\left(\frac{\zeta_2}{\zeta_1}\right)^m$$

PROOF. This follows from (3.12), (3.13) and the anticommutation relations (2.41) in $\mathbf{Cliff}(\mathbb{Z}+\tfrac{1}{2})$. ∎

COROLLARY 3.3. *For $a_1, a_2 \in \mathbf{A}$, $0 \le n_1, n_2 \in \mathbb{Z}$, $w, w' \in \mathbf{CM}(\mathbb{Z}+\tfrac{1}{2})$, when $|\zeta_1| > |\zeta_2|$ we have the absolute convergence of*

$$(\underbrace{a_1^{(n_1)}(\zeta_1)a_2^{(n_2)}(\zeta_2)}w, w')$$

$$= \langle a_1, a_2\rangle(w, w')n_1!^{-1}n_2!^{-1}(\partial/\partial\zeta_1)^{n_1}(\partial/\partial\zeta_2)^{n_2}(\zeta_1-\zeta_2)^{-1}$$

$$= \langle a_1, a_2\rangle(w, w')\binom{-n_2-1}{n_1}(\zeta_1-\zeta_2)^{-n_1-n_2-1}$$

LEMMA 3.4. *For $a_1, a_2 \in \mathbf{A}$, $0 \le n_1, n_2 \in \mathbb{Z}$, as formal series in ζ_1 and ζ_2, on $\mathbf{CM}(\mathbb{Z})$ we have*

$$\underbrace{a_1^{(n_1)}(\zeta_1)a_2^{(n_2)}(\zeta_2)}$$

$$= \langle a_1, a_2\rangle\zeta_1^{-n_1-\frac{1}{2}}\zeta_2^{-n_2-\frac{1}{2}}\left(\frac{1}{2}\binom{-\frac{1}{2}}{n_1}\binom{-\frac{1}{2}}{n_2} + \sum_{1 \le p \in \mathbb{Z}}\binom{-p-\frac{1}{2}}{n_1}\binom{p-\frac{1}{2}}{n_2}\left(\frac{\zeta_2}{\zeta_1}\right)^p\right)$$

$$= n_1!^{-1}n_2!^{-1}(\partial/\partial\zeta_1)^{n_1}(\partial/\partial\zeta_2)^{n_2}\langle a_1, a_2\rangle\zeta_1^{-\frac{1}{2}}\zeta_2^{-\frac{1}{2}}\left(\frac{1}{2} + \sum_{1 \le p \in \mathbb{Z}}\left(\frac{\zeta_2}{\zeta_1}\right)^p\right)$$

PROOF. This follows from (3.12) - (3.14).   ∎

COROLLARY 3.5. *For $a_1, a_2 \in \mathbf{A}$, $|\zeta_1| > |\zeta_2| > 0$, $\zeta_1, \zeta_2 \in \mathbb{C}^-$, $w, w' \in$ $\mathbf{CM}(\mathbb{Z})$ and $0 \leq n_1, n_2 \in \mathbb{Z}$, we have the absolute convergence of*

$$(\underbrace{a_1^{(0)}(\zeta_1) a_2^{(0)}(\zeta_2)}w, w') = \langle a_1, a_2 \rangle (w, w') \zeta_1^{-\frac{1}{2}} \zeta_2^{-\frac{1}{2}} \tfrac{1}{2} (\zeta_1 + \zeta_2)(\zeta_1 - \zeta_2)^{-1},$$

and

$$(\underbrace{a_1^{(n_1)}(\zeta_1) a_2^{(n_2)}(\zeta_2)}w, w') = n_1!^{-1} n_2!^{-1} (\partial/\partial\zeta_1)^{n_1} (\partial/\partial\zeta_2)^{n_2} (\underbrace{a_1^{(0)}(\zeta_1) a_2^{(0)}(\zeta_2)}w, w')$$

$$= \langle a_1, a_2 \rangle (w, w') \zeta_1^{-n_1 - \frac{1}{2}} \zeta_2^{-n_2 - \frac{1}{2}} (\zeta_1 - \zeta_2)^{-n_1 - n_2 - 1} P_{n_1 n_2}(\zeta_1, \zeta_2)$$

where $P_{n_1 n_2}(\zeta_1, \zeta_2)$ is a homogeneous polynomial of degree $n_1 + n_2 + 1$.

For $v_1, \ldots, v_n \in \mathbf{V}$, $w, w' \in \mathbf{V}(\frac{1}{2}\mathbb{Z})$, using products of vertex operators we define the generating function of matrix coefficients

$$(\bar{Y}(v_1, \zeta_1) \ldots \bar{Y}(v_n, \zeta_n) w, w')$$
$$= \sum_{m_1, \ldots, m_n \in \frac{1}{2}\mathbb{Z}} (\bar{Y}_{m_1}(v_1) \ldots \bar{Y}_{m_n}(v_n) w, w') \zeta_1^{-m_1 - wt(v_1)} \ldots \zeta_n^{-m_n - wt(v_n)}. \quad (3.43)$$

Later we will find a domain in which this series converges absolutely.

We will now show how to express the vertex operators $\bar{Y}(v, \zeta)$ in terms of normally ordered exponentials of various quadratic operators on various $\mathbb{Z}_2$-graded tensor products whose factors are $\mathbf{CM}(\mathbb{Z} + \frac{1}{2})$ or $\mathbf{CM}(\mathbb{Z})$. Let $\mathbf{A}_1, \ldots, \mathbf{A}_r$ be copies of $\mathbf{A}$ with its polarization and let each of $Z_1, \ldots, Z_r$ be either $\mathbb{Z} + \frac{1}{2}$ or $\mathbb{Z}$. For $1 \leq i \leq r$ let $\mathbf{A}_i(Z_i)$ be spanned by $\{a_i(m) \mid a_i \in \mathbf{A}_i, \ m \in Z_i\}$ and let $\mathbf{Cliff}(Z_1, \ldots, Z_r)$ be the Clifford algebra generated by $\mathbf{A}_1(Z_1) \oplus \ldots \oplus \mathbf{A}_r(Z_r)$ and the form $\langle a_i(m), b_j(n) \rangle = \langle a_i, b_j \rangle \delta_{ij} \delta_{m,-n}$. Let $\mathfrak{I}(Z_1, \ldots, Z_r)$ be the left ideal in $\mathbf{Cliff}(Z_1, \ldots, Z_r)$ generated by $\mathbf{A}_1(Z_1)^+ \oplus \ldots \oplus \mathbf{A}_r(Z_r)^+$ (see (2.37)-(2.47)). Then the $\mathbb{Z}_2$-graded tensor product $\mathbf{CM}(Z_1) \otimes \ldots \otimes \mathbf{CM}(Z_r)$ is isomorphic to the Clifford module $\mathbf{CM}(Z_1, \ldots, Z_r) = \mathbf{Cliff}(Z_1, \ldots, Z_r)/\mathfrak{I}(Z_1, \ldots, Z_r)$. The point of the $\mathbb{Z}_2$-grading is that for $v_i \in \mathbf{CM}(Z_i)^{\alpha_i}$, $\alpha_i \in \{0, 1\}$, $1 \leq i \leq r$, $a_i(n) \in \mathbf{A}_i(Z_i)$, we have

$$a_i(n)(v_1 \otimes \ldots \otimes v_r) = (-1)^{\alpha_1 + \ldots + \alpha_{i-1}} v_1 \otimes \ldots \otimes a_i(n) v_i \otimes \ldots \otimes v_r. \quad (3.44)$$

Let $\{a_1, \ldots, a_l, a_1^*, \ldots, a_l^*\}$ be a canonical basis of $\mathbf{A}$ and for $1 \leq j \leq r$ denote by $\{a_{j1}, \ldots, a_{jl}, a_{j1}^*, \ldots, a_{jl}^*\}$ a copy of that canonical basis in $\mathbf{A}_j$. For

$$x = \sum_{1 \leq i \leq l} (x_i a_i + x_i^* a_i^*) \in \mathbf{A}, \quad x_i, x_i^* \in \mathbb{C}, \quad (3.45)$$

we sometimes abuse notation, denoting by x the corresponding vector

$$\sum_{1 \leq i \leq l} (x_i a_{ji} + x_i^* a_{ji}^*) \in \mathbf{A}_j. \quad (3.46)$$

Letting $1 \leq j \leq r$ and $u_j \in \mathbf{CM}(Z_j)$, the Hermitian form $(\ ,\ )$ on $\mathbf{CM}(Z_j)$ provides a projection from $\mathbf{CM}(Z_1) \otimes \ldots \otimes \mathbf{CM}(Z_r)$ onto $\mathbf{CM}(Z_1) \otimes \ldots \otimes \mathbf{CM}(Z_{j-1}) \otimes \mathbf{CM}(Z_{j+1}) \otimes \ldots \otimes \mathbf{CM}(Z_r)$ by

$$\pi_j(u_j)(v_1 \otimes \ldots \otimes v_r) = (v_j, u_j)(v_1 \otimes \ldots v_{j-1} \otimes v_{j+1} \otimes \ldots \otimes v_r). \tag{3.47}$$

We usually use the projection $\pi_j(\mathbf{vac}(Z_j))$, which we denote just by π_j.

Fix $1 \leq p \neq q \leq r$ with $Z_p = \mathbb{Z} + \frac{1}{2}$. Define the generating function of operators on $\mathbf{CM}(Z_1) \otimes \ldots \otimes \mathbf{CM}(Z_r)$,

$$\begin{aligned}
J_{pq}(\zeta) &= \sum_{\substack{1 \leq i \leq l \\ 0 \leq m \in \mathbb{Z}, n \in Z_q}} \binom{-n - \frac{1}{2}}{m} (a_{qi}(n)a^*_{pi}(m + \tfrac{1}{2}) + a^*_{qi}(n)a_{pi}(m + \tfrac{1}{2}))\zeta^{-n-m-\frac{1}{2}} \\
&= \sum_{\substack{1 \leq i \leq l \\ 0 \leq m \in \mathbb{Z}}} (a_{qi}^{(m)}(\zeta)a^*_{pi}(m + \tfrac{1}{2}) + a_{qi}^{*(m)}(\zeta)a_{pi}(m + \tfrac{1}{2}))
\end{aligned}$$

$$\tag{3.48}$$

and write

$$J_{pq}(\zeta) = J_{pq}^-(\zeta) + J_{pq}^0(\zeta) + J_{pq}^+(\zeta) \tag{3.49}$$

where the superscript corresponds to taking the summation in the first line of (3.48) over $n < 0$, $n = 0$ (if $0 \in Z_q$) or $n > 0$, respectively, and in the second line to replacing the generating functions by sums as in (3.12)-(3.14).

Define the normally ordered exponential of $J_{pq}(\zeta)$ by

$$\begin{aligned}
: \exp(J_{pq}(\zeta)): &= \exp(J_{pq}^-(\zeta)) \exp(J_{pq}^0(\zeta)) \exp(J_{pq}^+(\zeta)) \\
&= \exp(J_{pq}^-(\zeta) + \tfrac{1}{2}J_{pq}^0(\zeta)) \exp(J_{pq}^+(\zeta) + \tfrac{1}{2}J_{pq}^0(\zeta))
\end{aligned} \tag{3.50}$$

since $J_{pq}^0(\zeta)$ commutes with $J_{pq}^-(\zeta)$ and $J_{pq}^+(\zeta)$. It is easy to see that this is a generating function of well-defined operators on $\mathbf{CM}(Z_1) \otimes \ldots \otimes \mathbf{CM}(Z_r)$.

THEOREM 3.6. *For $v_1 \otimes v_2 \in \mathbf{CM}(\mathbb{Z} + \frac{1}{2}) \otimes \mathbf{CM}(Z_2)$ we have*

$$\pi_1(: \exp(J_{12}(\zeta)): v_1 \otimes v_2) = \bar{Y}(v_1, \zeta)v_2.$$

PROOF. Let $v_1 = x_1(-m_1 - \frac{1}{2}) \ldots x_k(-m_k - \frac{1}{2})\mathbf{vac} \in \mathbf{CM}(\mathbb{Z} + \frac{1}{2})^{\alpha_1}$, so $\alpha_1 \equiv k$ (mod 2), and for $1 \leq i_1, \ldots, i_n \leq k$ distinct but not necessarily in order, let $v_1(i_1, \ldots, i_n)$ be the vector obtained from v_1 by removing the $i_1^{\text{th}}, \ldots, i_n^{\text{th}}$ factors. Let $sgn(k; i_1, \ldots, i_n)$ be the sign of the permutation $1 \ldots \hat{i}_1 \ldots \hat{i}_n \ldots k i_1 \ldots i_n$ in S_k which has $i_1 \ldots i_n$ removed from their positions in the identity permutation $1 \ldots k$ and placed at the end after k. In particular, for $n = 1$, $sgn(k; i_1) = (-1)^{k-i_1}$. For $1 \leq r \leq n$ let $j_r = i_r - \mathrm{Card}\{t \mid 1 \leq t \leq n,\ i_t < i_r\}$. Then we see that

$$\begin{aligned}
sgn(k; i_1, \ldots, i_n) &= sgn(k; i_1, \ldots, i_{n-1})sgn(k; j_n) \\
&= sgn(k; i_2, \ldots, i_n)sgn(k - n + 1; j_1)
\end{aligned} \tag{3.51}$$

and for any permutation $\sigma \in S_n$,

$$sgn(k; i_{\sigma 1}, \ldots, i_{\sigma n}) = sgn(\sigma)sgn(k; i_1, \ldots, i_n). \tag{3.52}$$

Writing $J = J_{12}$, for $\# = {}^-, {}^0$ (if $Z_2 = \mathbb{Z}$), or $^+$ and $1 \leq n \in \mathbb{Z}$, from the definitions we have

$$(J^\#(\zeta))^n v_1 \otimes v_2$$

$$= \sum_{1 \leq i_1 \neq \ldots \neq i_n \leq k} sgn(k; i_1, \ldots, i_n) v_1(i_1, \ldots, i_n) \otimes x_{i_1}^{(m_{i_1})}(\zeta)^\# \ldots x_{i_n}^{(m_{i_n})}(\zeta)^\# v_2$$

$$= n! \sum_{1 \leq i_1 < \ldots < i_n \leq k} sgn(k; i_1, \ldots, i_n) v_1(i_1, \ldots, i_n) \otimes {}^\circ_\circ x_{i_1}^{(m_{i_1})}(\zeta)^\# \ldots x_{i_n}^{(m_{i_n})}(\zeta)^\# {}^\circ_\circ v_2.$$

$$(3.53)$$

Then, since $J^-(\zeta)^a J^0(\zeta)^b J^+(\zeta)^c v_1 \otimes v_2 = 0$ if $a + b + c > k$, and $\pi_1(J^-(\zeta)^a J^0(\zeta)^b J^+(\zeta)^c v_1 \otimes v_2) = 0$ if $a + b + c < k$, we have

$$\pi_1(\exp(J^-(\zeta)) \exp(J^0(\zeta)) \exp(J^+(\zeta)) v_1 \otimes v_2)$$

$$= \pi_1 \left(\sum_{\substack{a+b+c=k \\ 0 \leq a,b,c \in \mathbb{Z}}} \frac{J^-(\zeta)^a \, J^0(\zeta)^b \, J^+(\zeta)^c}{a!b!c!} v_1 \otimes v_2 \right). \tag{3.54}$$

This matches the definition of $\bar{Y}(v_1, \zeta)v_2$ as in (3.24) if $Z_2 = \mathbb{Z}$ by using (3.53) with $\# = {}^-, {}^0$, and $^+$, and matches the definition of $\bar{Y}(v_1, \zeta)v_2$ as in (3.23) if $Z_2 = \mathbb{Z} + \frac{1}{2}$ by using (3.53) with $\# = {}^-$ and $^+$. ∎

COROLLARY 3.7. *For $v_i \in \mathbf{CM}(\mathbb{Z} + \frac{1}{2})$, $1 \leq i \leq r$, $w \in \mathbf{CM}(Z)$ we have*

$$\pi_1 \ldots \pi_r(: \exp(J_{1(r+1)}(\zeta_1)): \ldots : \exp(J_{r(r+1)}(\zeta_r)): v_1 \otimes \ldots \otimes v_r \otimes w)$$

$$= \bar{Y}(v_1, \zeta_1) \ldots \bar{Y}(v_r, \zeta_r)w.$$

COROLLARY 3.8. *For $v_i \in \mathbf{CM}(Z_i)^{\alpha_i}$, $\alpha_i \in \{0,1\}$, $1 \leq i \leq r$, $1 \leq p \neq q \leq r$, $Z_p = \mathbb{Z} + \frac{1}{2}$, we have*

$$\pi_p(: \exp(J_{pq}(\zeta)): v_1 \otimes \ldots \otimes v_r)$$

$$= \begin{cases} (-1)^{\alpha_p(\alpha_{p+1}+\ldots+a_{q-1})} v_1 \otimes \ldots \otimes \hat{v}_p \otimes \ldots \otimes \bar{Y}(v_p, \zeta)v_q \otimes \ldots \otimes v_r & \text{if } p < q \\\\ (-1)^{\alpha_p(\alpha_q+\ldots+\alpha_{p-1})} v_1 \otimes \ldots \otimes \bar{Y}(v_p, \zeta)v_q \otimes \ldots \otimes \hat{v}_p \otimes \ldots \otimes v_r & \text{if } q < p \end{cases}$$

where ^ indicates removal.

PROOF. This follows from the proof of Theorem 3.6, using (3.44) to keep track of the signs. ∎

DEFINITION. *For $0 \leq m, n \in \mathbb{Z}$, let*

$$C_{mn} = \frac{1}{2} \frac{m-n}{m+n+1} \binom{-\frac{1}{2}}{m} \binom{-\frac{1}{2}}{n}, \tag{3.55}$$

and note that

$$(m+1)C_{(m+1)n} + (n+1)C_{m(n+1)} = -(m+n+1)C_{mn}. \tag{3.56}$$

LEMMA 3.9. *For $0 \leq k, r, t \in \mathbb{Z}$ the coefficients defined in (3.55) satisfy*

$$\sum_{0 \leq m \leq k} \binom{m+r}{r}\binom{k-m+t}{t} C_{(m+r)(k-m+t)} = \binom{-r-t-1}{k} C_{rt}.$$

PROOF. The proof is by induction on k, using (3.56). ∎

We now define a generating function $\Delta(\zeta)$ of quadratic operators on $\mathbf{CM}(\mathbb{Z}+\frac{1}{2})$ which allows us to define the operators $Y(v,\zeta)$ we want.

DEFINITION. *For $\{a_1,\ldots,a_l,a_1^*,\ldots,a_l^*\}$ a canonical basis of $\mathbf{A}$, define*

$$\Delta(\zeta) = \sum_{1 \leq i \leq l} \sum_{0 \leq m,n \in \mathbb{Z}} C_{mn} a_i(m+\tfrac{1}{2}) a_i^*(n+\tfrac{1}{2}) \zeta^{-m-n-1} \qquad (3.57)$$

on $\mathbf{CM}(\mathbb{Z}+\frac{1}{2})$. Since $\Delta(\zeta)$ is quadratic, it preserves $\mathbf{V}_0$ and $\mathbf{V}_1$.

DEFINITION. *For $v \in \mathbf{CM}(\mathbb{Z}+\frac{1}{2})$, let*

$$Y(v,\zeta) = \begin{cases} \bar{Y}(v,\zeta) & \text{on } \mathbf{CM}(\mathbb{Z}+\tfrac{1}{2}) \\[2ex] \bar{Y}(\exp(\Delta(\zeta))v,\zeta) & \text{on } \mathbf{CM}(\mathbb{Z}). \end{cases} \qquad (3.58)$$

For $v \in (\mathbf{V})_k$, define the homogeneous components $Y_m(v)$ of $Y(v,\zeta)$ and the notation $\{v\}_n$ by

$$Y(v,\zeta) = \sum_{n \in \frac{1}{2}\mathbb{Z}} Y_{n+1-\mathrm{wt}(v)}(v)\zeta^{-n-1} = \sum_{n \in \frac{1}{2}\mathbb{Z}} \{v\}_n \zeta^{-n-1}. \qquad (3.59)$$

Since $\Delta(\zeta)$ preserves $\mathbf{V}_0$ and $\mathbf{V}_1$, the discussion after (3.25) of what coefficients of (3.25) are zero remains valid for (3.59). Let us use the notation

$$\bar{Y}'(v,\zeta) = (d/d\zeta)\bar{Y}(v,\zeta). \qquad (3.60)$$

LEMMA 3.10. *For $a \in \mathbf{A}$, $0 \leq n \in \mathbb{Z}$, on $\mathbf{CM}(\mathbb{Z}+\frac{1}{2})$, we have*

(a) $[\Delta(\zeta), a(-n-\tfrac{1}{2})] = \displaystyle\sum_{0 \leq m \in \mathbb{Z}} C_{mn} a(m+\tfrac{1}{2})\zeta^{-m-n-1}$,

(b) *If $v = b_1(-n_1 - \tfrac{1}{2})\ldots b_k(-n_k - \tfrac{1}{2})\mathbf{vac} \in \mathbf{CM}(\mathbb{Z}+\tfrac{1}{2})$, and v_{rs}, $1 \leq r < s \leq k$, is the vector obtained from v by removing $b_r(-n_r - \tfrac{1}{2})$ and $b_s(-n_s - \tfrac{1}{2})$, then*

$$\Delta(\zeta)v = -\sum_{1 \leq r < s \leq k} (-1)^{r+s+1} C_{n_r n_s} \langle b_r, b_s \rangle \zeta^{-n_r - n_s - 1} v_{rs},$$

(c) $[D(-1), \Delta(\zeta)] = \Delta'(\zeta)$ *on $\mathbf{CM}(\mathbb{Z}+\tfrac{1}{2})$, and $\Delta'(\zeta) = (d/d\zeta)\Delta(\zeta)$ commutes with $\Delta(\zeta)$,*

(d) $[D(-1), \exp(\Delta(\zeta))] = \Delta'(\zeta)\exp(\Delta(\zeta)) = (d/d\zeta)\exp(\Delta(\zeta))$,

(e) $[D(0), \Delta(\zeta)] = -\zeta\Delta'(\zeta)$,

(f) $[D(0), \exp(\Delta(\zeta))] = -\zeta\Delta'(\zeta)\exp(\Delta(\zeta))$.

PROOF. Part (a) follows from the anticommutation relations in $\mathbf{Cliff}(\mathbb{Z} + \frac{1}{2})$, and (b) follows from (a) by induction. Parts (c) and (e) follow from the definitions using (3.56), and (d) and (f) follow from (c) and (e). ∎

Note that

$$Y(D(-2)\mathbf{vac}, \zeta) = \zeta^{-2}D(\zeta) \tag{3.61}$$

includes the correction term $-l/8$ to the operator $D(0)$ on $\mathbf{CM}(\mathbb{Z})$, since from (3.36) we get

$$\Delta(\zeta)D(-2)\mathbf{vac} = -\tfrac{1}{8}l\ \mathbf{vac} \tag{3.62}$$

and higher powers of $\Delta(\zeta)$ kill $D(-2)\mathbf{vac}$.

For $1 \leq j \leq r$, if $Z_j = \mathbb{Z} + \frac{1}{2}$ let

$$\Delta_j(\zeta) = \sum_{\substack{1 \leq i \leq l \\ 0 \leq m,n \in \mathbb{Z}}} C_{mn} a_{ji}(m + \tfrac{1}{2})a_{ji}^*(n + \tfrac{1}{2})\zeta^{-m-n-1} \tag{3.63}$$

on $\mathbf{CM}(Z_1) \otimes \ldots \otimes \mathbf{CM}(Z_r)$. It is clear that

$$[\Delta_j(\zeta), J_{pq}^{\#}(z)] = 0 \quad \text{for} \quad q \neq j \quad \text{and} \quad {}^{\#} = {}^-, {}^0 \text{ or } {}^+ \tag{3.64}$$

and

$$[\Delta_q(\zeta), J_{pq}^+(z)] = 0. \tag{3.65}$$

PROPOSITION 3.11. *For* $1 \leq p \neq q \leq r$, $Z_p = Z_q = \mathbb{Z} + \frac{1}{2}$, *on* $\mathbf{CM}(Z_1) \otimes \ldots \otimes \mathbf{CM}(Z_r)$ *we have*

(a) $[\Delta_q(\zeta), J_{pq}^-(z)]$

$$= \sum_{\substack{1 \leq i \leq l \\ 0 \leq m,t \in \mathbb{Z}}} (a_{qi}(m + \tfrac{1}{2})a_{pi}^*(t + \tfrac{1}{2}) + a_{qi}^*(m + \tfrac{1}{2})a_{pi}(t + \tfrac{1}{2}))\zeta^{-m-t-1}$$

$$\cdot \sum_{0 \leq n \in \mathbb{Z}} \binom{n}{t} C_{mn} \left(\frac{z}{\zeta}\right)^{n-t},$$

(b) $[[\Delta_q(\zeta), J_{pq}^-(z)], J_{pq}^-(z)]$

$$= \sum_{\substack{1 \leq i \leq l \\ 0 \leq r,t \in \mathbb{Z}}} a_{pi}(r + \tfrac{1}{2})a_{pi}^*(t + \tfrac{1}{2})\zeta^{-r-t-1}2C_{rt} \sum_{0 \leq k \in \mathbb{Z}} \binom{-r-t-1}{k}\left(\frac{z}{\zeta}\right)^k$$

$$= 2\Delta_p(\zeta + z) \ \text{expanded in nonnegative powers of } z.$$

PROOF. These are straightforward calculations using the definitions and (3.44). Lemma 3.9 is required for part (b). ∎

COROLLARY 3.12. *For* $1 \leq p \neq q \leq r$, $Z_p = Z_q = \mathbb{Z} + \frac{1}{2}$, *on* $\mathbf{CM}(Z_1) \otimes \ldots \otimes \mathbf{CM}(Z_r)$ *we have*

$$\exp(\Delta_q(\zeta))\colon \exp(J_{pq}(z))\colon$$

$$= \colon \exp(J_{pq}(z))\colon \exp(\Delta_q(\zeta) + [\Delta_q(\zeta), J_{pq}^-(z)] + \Delta_p(\zeta + z))$$

with $\Delta_p(\zeta + z)$ expanded in nonnegative powers of z.

PROOF. For operators A, B on a vector space such that B is locally nilpotent and $(ad_B)^n A = 0$ for some n, one has the well-known formula

$$\exp(B)A\exp(-B) = \exp(ad_B)A \tag{3.66}$$

where $ad_B(X) = [B, X]$. Letting $A = \Delta_q(\zeta)$ and $B = -J_{pq}^-(z)$ we get

$$\exp(-J_{pq}^-(z))\Delta_q(\zeta)\exp(J_{pq}^-(z)) =$$
$$\Delta_q(\zeta) + [\Delta_q(\zeta), J_{pq}^-(z)] + \tfrac{1}{2}[[\Delta_q(\zeta), J_{pq}^-(z)], J_{pq}^-(z)] \tag{3.67}$$

and exponentiating both sides gives

$$\exp(-J_{pq}^-(z))\exp(\Delta_q(\zeta))\exp(J_{pq}^-(z)) =$$
$$\exp(\Delta_q(\zeta) + [\Delta_q(\zeta), J_{pq}^-(z)] + \tfrac{1}{2}[[\Delta_q(\zeta), J_{pq}^-(z)], J_{pq}^-(z)]). \tag{3.68}$$

From the last proposition, $[\Delta_q(\zeta), J_{pq}^-(z)]$ commutes with $\Delta_q(\zeta)$, $[[\Delta_q(\zeta), J_{pq}^-(z)], J_{pq}^-(z)] = \Delta_p(\zeta+z)$ commutes with both $\Delta_q(\zeta)$ and $J_{pq}^-(z)$, and $J_{pq}^+(z)$ commutes with $\Delta_q(\zeta)$, $[\Delta_q(\zeta), J_{pq}^-(z)]$, and $\Delta_p(\zeta + z)$. Under the assumptions we have

$$: \exp(J_{pq}(z)) := \exp(J_{pq}^-(z))\exp(J_{pq}^+(z)) \tag{3.69}$$

so the result follows. ∎

For $1 \leq p \leq r$, define the Virasoro operators $D_p(m)$, $m \in \mathbb{Z}$, on the pth tensor factor of $\mathbf{CM}(Z_1) \otimes \ldots \otimes \mathbf{CM}(Z_r)$ by the usual formulas (2.54)-(2.56) using the canonical basis in $\mathbf{A}_p$. For $1 \leq p \neq q \leq r$, let $D_{pq}(m) = D_p(m) + D_q(m)$. From Lemma 3.10 (c) - (f) we have

$$[D_p(-1), \Delta_p(z)] = \Delta_p'(z), \tag{3.70}$$

$\Delta_p'(z)$ commutes with $\Delta_p(z)$, and

$$[D_p(-1), \exp(\Delta_p(z))] = \Delta_p'(z)\exp(\Delta_p(z)) = (d/dz)\exp(\Delta_p(z)), \tag{3.71}$$

where $\Delta_p'(z) = (d/dz)\Delta_p(z)$.

PROPOSITION 3.13. *For $1 \leq p \neq q \leq r$, on $\mathbf{CM}(Z_1) \otimes \ldots \otimes \mathbf{CM}(Z_r)$ we have*

(a) $[D_p(-1), J_{pq}^\#(z)] = (d/dz)J_{pq}^\#(z)$ *for* $\# = {}^-, {}^0$ *(if $Z_q = \mathbb{Z}$), or* $^+$,

(b) $[D_q(-1), J_{pq}^\#(z)] = -(d/dz)J_{pq}^\#(z)$ *for* $\# = {}^-$ *or* $^+$ *if $Z_q = \mathbb{Z} + \tfrac{1}{2}$,*

(c) $[D_q(-1), J_{pq}^-(z)] = -(d/dz)J_{pq}^-(z) +$

$$\tfrac{1}{2} \sum_{\substack{1 \leq i \leq l \\ 0 \leq m \in \mathbb{Z}}} \binom{-\frac{1}{2}}{m} (a_{qi}(-1)a_{pi}^*(m+\tfrac{1}{2}) + a_{qi}^*(-1)a_{pi}(m+\tfrac{1}{2}))z^{-m-\frac{1}{2}} \text{ if } Z_q = \mathbb{Z},$$

(d) $[D_q(-1), J_{pq}^0(z)]$

$$= -\tfrac{1}{2} \sum_{\substack{1 \leq i \leq l \\ 0 \leq m \in \mathbb{Z}}} \binom{-\frac{1}{2}}{m} (a_{qi}(-1)a_{pi}^*(m + \tfrac{1}{2}) + a_{qi}^*(-1)a_{pi}(m + \tfrac{1}{2}))z^{-m-\frac{1}{2}}$$

if $Z_q = \mathbb{Z}$,

(e) $[D_q(-1), J_{pq}^+(z)] = -(d/dz)J_{pq}^+(z) - (d/dz)J_{pq}^0(z)$ *if $Z_q = \mathbb{Z}$.*

PROOF. Part (a) is immediate from the definition of $J_{pq}(z)$ using $[D(-1), a(m + \frac{1}{2})] = ma(m - \frac{1}{2})$. This also gives

$$[D_q(-1), J_{pq}^{\#}(z)] =$$

$$\sum_{\substack{1 \leq i \leq l \\ 0 \leq m \in \mathbb{Z} \\ n \in Z_q^{\#}}} \binom{-n - \frac{1}{2}}{m}(n - \tfrac{1}{2})(a_{qi}(n - 1)a_{pi}^{*}(m + \tfrac{1}{2}) + a_{qi}^{*}(n - 1)a_{pi}(m + \tfrac{1}{2}))z^{-n-m-\frac{1}{2}}.$$

$$(3.72)$$

Using

$$\binom{-n - \frac{1}{2}}{m}(n - \tfrac{1}{2}) = -\binom{-n + \frac{1}{2}}{m}(-n - m + \tfrac{1}{2})$$

and replacing n by $n + 1$, we get (b)-(e). ∎

COROLLARY 3.14. *For $1 \leq p \neq q \leq r$ and $Z_p = \mathbb{Z} + \frac{1}{2}$ on $\mathbf{CM}(Z_1) \otimes \ldots \otimes \mathbf{CM}(Z_r)$ we have*

(a) $[D_{pq}(-1), J_{pq}^{\#}(z)] = 0$ *for $\# = ^- $ or $^+$ if $Z_q = \mathbb{Z} + \frac{1}{2}$,*

(b) $[D_{pq}(-1), J_{pq}^{+}(z)] = -(d/dz)J_{pq}^{0}(z)$ *if $Z_q = \mathbb{Z}$, which commutes with $J_{pq}^{+}(z)$ and $J_{pq}^{-}(z)$,*

(c) $[D_{pq}(-1), J_{pq}^{-}(z)] =$

$$\frac{1}{2}\sum_{\substack{1 \leq i \leq l \\ 0 \leq m \in \mathbb{Z}}} \binom{-\frac{1}{2}}{m}(a_{qi}(-1)a_{pi}^{*}(m + \tfrac{1}{2}) + a_{qi}^{*}(-1)a_{pi}(m + \tfrac{1}{2}))z^{-m-\frac{1}{2}}$$

if $Z_q = \mathbb{Z}$, which commutes with $J_{pq}^{-}(z)$ and $J_{pq}^{0}(z)$,

(d) $[D_{pq}(-1), J_{pq}^{0}(z)] = (d/dz)J_{pq}^{0}(z) -$

$$\frac{1}{2}\sum_{\substack{1 \leq i \leq l \\ 0 \leq m \in \mathbb{Z}}} \binom{-\frac{1}{2}}{m}(a_{qi}(-1)a_{pi}^{*}(m + \tfrac{1}{2}) + a_{qi}^{*}(-1)a_{pi}(m + \tfrac{1}{2}))z^{-m-\frac{1}{2}}$$

if $Z_q = \mathbb{Z}$, which commutes with $J_{pq}^{-}(z)$,

(e) $[[D_{pq}(-1), J_{pq}^{0}(z)], J_{pq}^{0}(z)] = -2\Delta_p'(z)$ *if $Z_q = \mathbb{Z}$, which commutes with $J_{pq}^{\#}(z)$ for $\# = ^-, ^0$ or $^+$, and with $\Delta_p(z)$,*

(f) $[D_{pq}(-1), \exp(J_{pq}^{0}(z))] = \exp(J_{pq}^{0}(z))([D_{pq}(-1), J_{pq}^{0}(z)] - \Delta_p'(z))$ *if $Z_q = \mathbb{Z}$.*

PROOF. Part (a) follows from parts (a) and (b) of the last proposition. Parts (b)-(d) follow from parts (a) and (c)-(e) of the last proposition. Part (e) follows from part (d) and the definitions. Part (f) follows from part (e) and the formula (3.66) with $A = D_{pq}(-1)$ and $B = -J_{pq}^{0}(z)$. ∎

THEOREM 3.15. *For $1 \leq p \neq q \leq r$ and $Z_p = \mathbb{Z} + \frac{1}{2}$ on $\mathbf{CM}(Z_1) \otimes \ldots \otimes \mathbf{CM}(Z_r)$ we have*

$$[D_{pq}(-1), \, :\exp(J_{pq}(z)):] = \begin{cases} 0 & \text{if } Z_q = \mathbb{Z} + \frac{1}{2} \\[2mm] -\Delta_p'(z) :\exp(J_{pq}(z)): & \text{if } Z_q = \mathbb{Z} \end{cases}$$

$$(3.73)$$

so

$$[D_{pq}(-1), : \exp(J_{pq}(z)) : \exp(\Delta_p(z))] = 0 \quad \text{if } Z_q = \mathbb{Z}. \tag{3.74}$$

PROOF. The result follows from the last corollary and (3.71). ∎

COROLLARY 3.16. *For $1 \leq p \neq q \leq r$ and $Z_p = \mathbb{Z} + \frac{1}{2}$ on $\mathbf{CM}(Z_1) \otimes \ldots \otimes \mathbf{CM}(Z_r)$ we have*

$$\exp(\zeta D_{pq}(-1)) : \exp(J_{pq}(z)) : \exp(-\zeta D_{pq}(-1)) =: \exp(J_{pq}(z)): \tag{3.75}$$

if $Z_q = \mathbb{Z} + \frac{1}{2}$, which gives

$$\exp(\zeta D(-1))Y(v, z) \exp(-\zeta D(-1))w = Y(\exp(\zeta D(-1))v, z)w \tag{3.76}$$

for $v, w \in \mathbf{CM}(\mathbb{Z} + \frac{1}{2})$, and

$$\begin{aligned}
\exp(\zeta D_{pq}(-1)) &: \exp(J_{pq}(z)) : \exp(\Delta_p(z)) \exp(-\zeta D_{pq}(-1)) = \\
&: \exp(J_{pq}(z)) : \exp(\Delta_p(z))
\end{aligned} \tag{3.77}$$

if $Z_q = \mathbb{Z}$, which gives

$$\exp(\zeta D(-1))Y(v, z) \exp(-\zeta D(-1))w = Y(\exp(\zeta D(-1))v, z)w \tag{3.78}$$

for $v \in \mathbf{CM}(\mathbb{Z} + \frac{1}{2})$ and $w \in \mathbf{CM}(\mathbb{Z})$.

PROPOSITION 3.17. *For $v \in \mathbf{V}$, on $\mathbf{V}(\frac{1}{2}\mathbb{Z})$ we have*

(a) $[D(-1), Y(v, z)] = Y(D(-1)v, z) = -(d/dz)Y(v, z)$,
(b) $[D(0), Y(v, z)] = -z(d/dz)Y(v, z) + Y(D(0)v, z)$.

PROOF. (a) The first equality comes from (3.76) and (3.78) by taking the derivative with respect to ζ and then setting $\zeta = 0$. Proposition 3.1 (b) gives the second equality on $\mathbf{CM}(\mathbb{Z} + \frac{1}{2})$, and with (3.71) on $\mathbf{CM}(\mathbb{Z})$ gives

$$\begin{aligned}
-(d/dz)Y(v, z) &= -(d/dz)\bar{Y}(\exp(\Delta(z))v, z) \\
&= -\bar{Y}'(\exp(\Delta(z))v, z) - \bar{Y}((d/dz)\exp(\Delta(z))v, z) \\
&= -\bar{Y}'(\exp(\Delta(z))v, z) - \bar{Y}(\Delta'(z)\exp(\Delta(z))v, z) \\
&= \bar{Y}(D(-1)\exp(\Delta(z))v, z) - \bar{Y}([D(-1), \exp(\Delta(z))]v, z) \\
&= \bar{Y}(\exp(\Delta(z))D(-1)v, z) \\
&= Y(D(-1)v, z).
\end{aligned}$$

(b) From Proposition 3.1 (c) and Lemma 3.10 (f) we have

$$\begin{aligned}
&- z(d/dz)Y(v, z) + Y(D(0)v, z) \\
&= - z(d/dz)\bar{Y}(\exp(\Delta(z))v, z) + \bar{Y}(\exp(\Delta(z))D(0)v, z) \\
&= - z\bar{Y}'(\exp(\Delta(z))v, z) - z\bar{Y}(\Delta'(z)\exp(\Delta(z))v, z) \\
&\quad + \bar{Y}(D(0)\exp(\Delta(z))v, z) - \bar{Y}([D(0), \exp(\Delta(z))]v, z) \\
&= [D(0), \bar{Y}(\exp(\Delta(z))v, z)] \\
&= [D(0), Y(v, z)].
\end{aligned}$$

∎

COROLLARY 3.18. *For $v \in \mathbf{V}$, on $\mathbf{V}(\frac{1}{2}\mathbb{Z})$ we have*

$$\exp(\zeta D(0))Y(v, z)\exp(-\zeta D(0)) = \exp(-\zeta z(d/dz))Y(\exp(\zeta D(0))v, z)$$
$$= Y(\exp(\zeta D(0))v, ze^{-\zeta})$$

where the last series is expanded in nonnegative powers of ζ.

PROOF. Using (3.66), the first expression equals

$$\exp(ad_{\zeta D(0)})Y(v, z) = \sum_{0 \leq n \in \mathbb{Z}} \frac{\zeta^n}{n!}(ad_{D(0)})^n Y(v, z)$$

$$= \sum_{0 \leq n \in \mathbb{Z}} \frac{\zeta^n}{n!} \sum_{0 \leq k \leq n} \binom{n}{k} \left(-z\frac{d}{dz}\right)^k Y(D(0)^{n-k}v, z)$$

$$= \sum_{0 \leq k,l \in \mathbb{Z}} \frac{\zeta^k \zeta^l}{k!l!} \left(-z\frac{d}{dz}\right)^k Y(D(0)^l v, z)$$

$$= \exp(-\zeta z(d/dz))Y(\exp(\zeta D(0))v, z)$$

$$= Y(\exp(\zeta D(0))v, ze^{-\zeta}). \qquad \blacksquare$$

PROPOSITION 3.19. *For $1 \leq p \neq q \leq r$, $Z_p = \mathbb{Z} + \frac{1}{2}$ and $\# = {}^-, {}^0$ or ${}^+$, on $\mathbf{CM}(Z_1) \otimes \ldots \otimes \mathbf{CM}(Z_r)$ we have*

$$\exp(\zeta D_p(-1))J_{pq}^{\#}(z)\exp(-\zeta D_p(-1)) = J_{pq}^{\#}(z + \zeta) \tag{3.79}$$

and

$$\exp(\zeta D_p(-1))\Delta_p(z)\exp(-\zeta D_p(-1)) = \Delta_p(z + \zeta) \tag{3.80}$$

where $J_{pq}^{\#}(z + \zeta)$ and $\Delta_p(z + \zeta)$ are expanded in nonnegative powers of ζ.

PROOF. By (3.66) the left side of the first equation equals

$$\exp(ad_{\zeta D_p(-1)})J_{pq}^{\#}(z) = \sum_{n \geq 0} n!^{-1}(ad_{\zeta D_p(-1)})^n J_{pq}^{\#}(z)$$

$$= \sum_{n \geq 0} \frac{\zeta^n}{n!} \left(\frac{d}{dz}\right)^n J_{pq}^{\#}(z) = J_{pq}^{\#}(z + \zeta)$$

as a formal Taylor series in nonnegative powers of ζ. The second equation follows similarly using (3.70). $\qquad \blacksquare$

COROLLARY 3.20. *For $1 \leq p \neq q \leq r$ and $Z_p = \mathbb{Z} + \frac{1}{2}$, on $\mathbf{CM}(Z_1) \otimes \ldots \otimes \mathbf{CM}(Z_r)$ we have*

$$\exp(\zeta D_p(-1)){:}\exp(J_{pq}(z)){:}\exp(-\zeta D_p(-1)) =: \exp(J_{pq}(z + \zeta)){:} \tag{3.81}$$

and

$$\exp(\zeta D_p(-1))\exp(\Delta_p(z))\exp(-\zeta D_p(-1)) =: \exp(\Delta_p(z + \zeta)){:} \tag{3.82}$$

so that

$$Y(\exp(-\zeta D(-1))v, z)w = Y(v, z + \zeta)w \tag{3.83}$$

for $v \in \mathbf{CM}(\mathbb{Z} + \frac{1}{2})$, $w \in \mathbf{V}(\frac{1}{2}\mathbb{Z})$, with $Y(v, z + \zeta)w$ expanded as a series in nonnegative powers of ζ.

PROPOSITION 3.21. *For $1 \leq p_1, p_2, q \leq r$ distinct and $Z_{p_1} = Z_{p_2} = \mathbb{Z} + \frac{1}{2}$, on $\mathbf{CM}(Z_1) \otimes \ldots \otimes \mathbf{CM}(Z_r)$ we have*

$$[J_{p_1 q}^{\#1}(\zeta_1), J_{p_2 q}^{\#2}(\zeta_2)] = \sum_{\substack{1 \leq i \leq l \\ 0 \leq m, n \in \mathbb{Z}}} (a_{p_2 i}(n + \tfrac{1}{2}) a_{p_1 i}^{*}(m + \tfrac{1}{2}) + a_{p_2 i}^{*}(n + \tfrac{1}{2}) a_{p_1 i}(m + \tfrac{1}{2}))$$

$$\cdot [a_{qi}^{(m)}(\zeta_1)^{\#1}, a_{qi}^{*(n)}(\zeta_2)^{\#2}]_{+}$$

and the anticommutator $[a_{qi}^{(m)}(\zeta_1)^{\#1}, a_{qi}^{*(n)}(\zeta_2)^{\#2}]_{+}$ is a formal series independent of i and q, depending only on m, n and whether $Z_q = \mathbb{Z}$ or $\mathbb{Z} + \frac{1}{2}$.

COROLLARY 3.22. *For $1 \leq p_1, p_2, q \leq r$ distinct, $Z_{p_1} = Z_{p_2} = \mathbb{Z} + \frac{1}{2}$ and $Z_q = \mathbb{Z}$, on $\mathbf{CM}(Z_1) \otimes \ldots \otimes \mathbf{CM}(Z_r)$ we have*

$$[J_{p_1 q}^{+}(\zeta_1), J_{p_2 q}^{-}(\zeta_2)] + \tfrac{1}{2}[J_{p_1 q}^{0}(\zeta_1), J_{p_2 q}^{0}(\zeta_2)] =$$

$$\sum_{\substack{1 \leq i \leq l \\ 0 \leq m, n \in \mathbb{Z}}} (a_{p_2 i}(n + \tfrac{1}{2}) a_{p_1 i}^{*}(m + \tfrac{1}{2}) + a_{p_2 i}^{*}(n + \tfrac{1}{2}) a_{p_1 i}(m + \tfrac{1}{2})) (\underbrace{a_{qi}^{(m)}(\zeta_1) a_{qi}^{*(n)}(\zeta_2)})$$

and the contraction is a formal series independent of i and q, depending only on m, n, given in Lemma 3.4. If $Z_q = \mathbb{Z} + \frac{1}{2}$ then we have the same formula with only the first bracket on the left side, and the contraction given in Lemma 3.2.

DEFINITION. *For $1 \leq p_1, p_2, q \leq r$ distinct, $Z_{p_1} = Z_{p_2} = \mathbb{Z} + \frac{1}{2}$, let $K_{p_1 p_2}(\zeta_1, \zeta_2; Z_q)$ denote the generating function of operators on $\mathbf{CM}(Z_1) \otimes \ldots \otimes \mathbf{CM}(Z_r)$ given in Corollary 3.22.*

PROPOSITION 3.23. *For $1 \leq p_1, p_2, q \leq r$ distinct, $Z_{p_1} = Z_{p_2} = \mathbb{Z} + \frac{1}{2}$ and $Z_q = \mathbb{Z}$ or $Z_q = \mathbb{Z} + \frac{1}{2}$, on $\mathbf{CM}(Z_1) \otimes \ldots \otimes \mathbf{CM}(Z_r)$ we have*

$$[J_{p_1 q}^{\#}(\zeta_1), J_{p_2 p_1}^{-}(\zeta_2)] =$$

$$\sum_{\substack{1 \leq i \leq l \\ 0 \leq m \in \mathbb{Z}, n \in Z_q^{\#}}} (a_{qi}(n) a_{p_2 i}^{*}(m + \tfrac{1}{2}) + a_{qi}^{*}(n) a_{p_2 i}(m + \tfrac{1}{2})) \binom{-n - \tfrac{1}{2}}{m} \zeta_1^{-m-n-\tfrac{1}{2}}$$

$$\cdot \sum_{0 \leq t \in \mathbb{Z}} \binom{-m - n - \tfrac{1}{2}}{t} \left(\frac{\zeta_2}{\zeta_1}\right)^{t}$$

$$= J_{p_2 q}^{\#}(\zeta_1 + \zeta_2)$$

$$(3.84)$$

with all powers of $\zeta_1 + \zeta_2$ formally expanded in nonnegative powers of ζ_2. This operator commutes with $J_{p_2 p_1}^{\pm}(\zeta_2)$, so on $\mathbf{CM}(Z_1) \otimes \ldots \otimes \mathbf{CM}(Z_r)$ we have

$$\exp(J_{p_1 q}^{\#}(\zeta_1)) \exp(J_{p_2 p_1}^{-}(\zeta_2)) =$$
$$\exp(J_{p_2 p_1}^{-}(\zeta_2)) \exp(J_{p_1 q}^{\#}(\zeta_1) + J_{p_2 q}^{\#}(\zeta_1 + \zeta_2)).$$

$$(3.85)$$

PROOF. The definitions give

$$[J^{\#}_{p_1 q}(\zeta_1), J^{-}_{p_2 p_1}(\zeta_2)]$$

$$= \sum_{\substack{1\le i\le l \\ 0\le m,k\in\mathbb{Z}}} (a^{(k)\#}_{qi}(\zeta_1)a^*_{p_2 i}(m+\tfrac{1}{2}) + a^{*(k)\#}_{qi}(\zeta_1)a_{p_2 i}(m+\tfrac{1}{2})) \binom{k}{m}\zeta_2^{k-m}$$

$$= \sum_{\substack{1\le i\le l \\ 0\le m\in\mathbb{Z},n\in Z^{\#}_q}} (a_{qi}(n)a^*_{p_2 i}(m+\tfrac{1}{2}) + a^*_{qi}(n)a_{p_2 i}(m+\tfrac{1}{2})) \qquad (3.86)$$

$$\cdot \sum_{m\le k\in\mathbb{Z}} \binom{k}{m}\binom{-n-\tfrac{1}{2}}{k}\zeta_2^{k-m}\zeta_1^{-n-k-\tfrac{1}{2}}.$$

Then using

$$\sum_{m\le k\in\mathbb{Z}} \binom{k}{m}\binom{-n-\tfrac{1}{2}}{k}\zeta_2^{k-m}\zeta_1^{-n-k-\tfrac{1}{2}}$$

$$= \zeta_1^{-m-n-\tfrac{1}{2}} \sum_{0\le t\in\mathbb{Z}} \binom{m+t}{m}\binom{-n-\tfrac{1}{2}}{m+t}\left(\frac{\zeta_2}{\zeta_1}\right)^t$$

$$= \binom{-n-\tfrac{1}{2}}{m}\zeta_1^{-m-n-\tfrac{1}{2}} \sum_{0\le t\in\mathbb{Z}} \binom{-m-n-\tfrac{1}{2}}{t}\left(\frac{\zeta_2}{\zeta_1}\right)^t \qquad (3.87)$$

$$= \binom{-n-\tfrac{1}{2}}{m}\zeta_1^{-m-n-\tfrac{1}{2}}\left(1+\frac{\zeta_2}{\zeta_1}\right)^{-m-n-\tfrac{1}{2}}$$

we get (3.84). The rest is clear. ∎

DEFINITION. *For* $1 \le p_1,\ldots,p_t,q \le r$ *distinct,* $Z_{p_i} = \mathbb{Z} + \tfrac{1}{2}$, $1 \le i \le t$, *let* $J_i = J_{p_i q}(\zeta_i)$, $J^{\#}_i = J^{\#}_{p_i q}(\zeta_i)$ *for* $\# = {}^-, {}^0$ *or* ${}^+$, *and* $K_{ij} = K_{p_i p_j}(\zeta_i, \zeta_j; Z_q)$. *Define*

$$\begin{aligned} &: \exp(J_1)\ldots\exp(J_t): = \\ &\exp(J^-_1 + \ldots + J^-_t)\exp(J^0_1 + \ldots + J^0_t)\exp(J^+_1 + \ldots + J^+_t). \end{aligned} \qquad (3.88)$$

Note that Proposition 3.21 gives

$$[J^-_i, J^-_j] = 0 = [J^+_i, J^+_j] \quad \text{for} \ \ 1 \le i,j \le t \qquad (3.89)$$

and

$$[[J^{\#1}_i, J^{\#2}_j], J^{\#3}_k] = 0 \quad \text{for} \ \ 1 \le i,j,k \le t. \qquad (3.90)$$

Also note that for any permutation $\sigma \in S_t$, we have

$$: \exp(J_{\sigma 1})\ldots\exp(J_{\sigma t}): = : \exp(J_1)\ldots\exp(J_t): . \qquad (3.91)$$

LEMMA 3.24. *With notation as above, on $\mathbf{CM}(Z_1) \otimes \ldots \otimes \mathbf{CM}(Z_r)$ we have*

$$\exp(J_1^0 + \ldots + J_t^0) \exp\left(\tfrac{1}{2} \sum_{1 \leq i < j \leq t} [J_i^0, J_j^0]\right) = \exp(J_1^0) \ldots \exp(J_t^0).$$

COROLLARY 3.25. *With notation as above, on $\mathbf{CM}(Z_1) \otimes \ldots \otimes \mathbf{CM}(Z_r)$ we have*

$$: \exp(J_1) : \ldots : \exp(J_t) : \; =: \exp(J_1) \ldots \exp(J_t) : \exp\left(\sum_{1 \leq i < j \leq t} K_{ij}\right).$$

THEOREM 3.26. *With notation as above, let $v_i \in \mathbf{CM}(Z_i)^{\alpha_i}$, $1 \leq i \leq r$, $w' \in \mathbf{CM}(Z_q)$, let $\mathbf{s}(v_{p_1}, \ldots, v_{p_t}; v_q) = (s_1, \ldots, s_t)$ be defined by*

$$s_i = \begin{cases} \tfrac{1}{2} & \text{if } \alpha_{p_i} = 1 \text{ and } Z_q = \mathbb{Z}, \\[2ex] 0 & \text{if } \alpha_{p_i} = 0 \text{ or } Z_q = \mathbb{Z} + \tfrac{1}{2}, \end{cases}$$

and let $\zeta^{\mathbf{s}} = \prod_{1 \leq i \leq t} \zeta_i^{s_i}$. Then we have the matrix coefficient

$$\zeta^{\mathbf{s}} \pi_1 \ldots \pi_q(w') \ldots \pi_r(: \exp(J_1) \ldots \exp(J_t): v_1 \otimes \ldots \otimes v_r) \in \mathbb{C}[\zeta_i, \zeta_i^{-1}, \; 1 \leq i \leq t].$$

PROOF. From (3.48) the powers of ζ_i in J_i are in $\mathbb{Z}$ if $Z_q = \mathbb{Z} + \tfrac{1}{2}$, but they are in $\mathbb{Z} + \tfrac{1}{2}$ if $Z_q = \mathbb{Z}$. We may assume v_{p_i} is a vector with k_i creation operators applied to $\mathbf{vac}(Z_{p_i})$, so $k_i \equiv \alpha_{p_i} \pmod{2}$. Let $J^\# = J_1^\# + \ldots + J_t^\#$, $\# = {}^-, {}^0$ or ${}^+$, and let $k = k_1 + \ldots + k_t$. Then, as in (3.54), we have

$$\pi_1 \ldots \pi_q(w') \ldots \pi_r(: \exp(J_1) \ldots \exp(J_t): v_1 \otimes \ldots \otimes v_r) =$$

$$\pi_1 \ldots \pi_q(w') \ldots \pi_r \left(\sum_{\substack{a+b+c=k \\ 0 \leq a,b,c \in \mathbb{Z}}} \frac{(J^-)^a (J^0)^b (J^+)^c}{a! b! c!} v_1 \otimes \ldots \otimes v_r \right) \tag{3.92}$$

where, expanding each term of the sum, the only contributing terms have total degree k_i in J_i^-, J_i^0 and J_i^+, $1 \leq i \leq t$. It is clear from (3.53) that $(J^0)^b (J^+)^c v_1 \otimes \ldots \otimes v_r$ is a finite sum. Although $(J^-)^a$ applied to that finite sum gives an infinite sum of terms, all but finitely many are killed by $\pi_q(w')$. If $Z_q = \mathbb{Z}$, the total power of ζ_i in (3.92) is in $\mathbb{Z}$ if k_i is even ($\alpha_{p_i} = 0$), but it is in $\mathbb{Z} + \tfrac{1}{2}$ if k_i is odd ($\alpha_{p_i} = 1$). ∎

LEMMA 3.27. *With notation as in Theorem 3.26,*

$$\zeta^{\mathbf{s}} \pi_1 \ldots \pi_q(w') \ldots \pi_r(: \exp(J_1) : \ldots : \exp(J_t): v_1 \otimes \ldots \otimes v_r)$$

converges absolutely when $|\zeta_1| > \ldots > |\zeta_t| > 0$ to a rational function in

$$\mathbb{C}[\zeta_i, \zeta_i^{-1}, (\zeta_j - \zeta_k)^{-1}; \; 1 \leq i \leq t, \; 1 \leq j < k \leq t].$$

PROOF. The operators K_{ij}, $1 \le i < j \le t$, are given in Corollary 3.22. It is clear that they commute with each other, and for any vector in $\mathbf{CM}(Z_1) \otimes \ldots \otimes \mathbf{CM}(Z_r)$, a sufficiently large power of K_{ij} kills the vector. If $Z_q = \mathbb{Z} + \frac{1}{2}$, then Corollary 3.25, Theorem 3.26, and Corollary 3.3 combine to give the result. If $Z_q = \mathbb{Z}$ then we can similarly get the result using Corollary 3.5 in place of Corollary 3.3. The factor $\zeta^{\mathbf{s}}$ occurs when $Z_q = \mathbb{Z}$ because J_i, $1 \le i \le t$, and K_{ij}, $1 \le i < j \le t$, are series whose variables occur with powers in $\mathbb{Z} + \frac{1}{2}$. If v_{p_i} has k_i creation operators applied to $\mathbf{vac}(Z_{p_i})$, then the only terms contributing to the projection π_i have a total of k_i operators J_i, K_{ij} and K_{li} applied, so the power of ζ_i is in $\mathbb{Z}$ if k_i is even ($\alpha_{p_i} = 0$), but is in $\mathbb{Z} + \frac{1}{2}$ if k_i is odd ($\alpha_p = 1$). $\blacksquare$

THEOREM 3.28. *(Rationality) Let* $v_1, \ldots, v_n \in \mathbf{V}_0 \cup \mathbf{V}_1$, $w, w' \in \mathbf{CM}(Z)$, $\mathbf{s} = \mathbf{s}(v_1, \ldots, v_n; w) = (s_1, \ldots, s_n)$. *Then the series* $\zeta^{\mathbf{s}}(Y(v_1, \zeta_1) \ldots Y(v_n, \zeta_n)w, w')$ *converges absolutely when* $|\zeta_1| > \ldots > |\zeta_n| > 0$ *to a rational function in* $\mathbb{C}[\zeta_i, \zeta_i^{-1}, (\zeta_j - \zeta_k)^{-1}; 1 \le i \le n, 1 \le j < k \le n]$.

PROOF. For $Z = \mathbb{Z} + \frac{1}{2}$ this follows from Corollary 3.7, (3.58) and Lemma 3.27. For $Z = \mathbb{Z}$, one needs in addition the fact that for $1 \le i \le n$, $\exp(\Delta(\zeta_i))v_i$ is a finite sum of vectors (see Lemma 3.10 (b)) from $\mathbf{V}_0$ if $v_i \in \mathbf{V}_0$, or from $\mathbf{V}_1$ if $v_i \in \mathbf{V}_1$. So each $Y(v_i, \zeta_i)$ is a finite sum of $\bar{Y}$ terms. $\blacksquare$

Define the right action of $\sigma \in S_n$ on $\{0,1\}^n$ by

$$\mathbf{m}\sigma = (m_{\sigma 1}, \ldots, m_{\sigma n}) \tag{3.93}$$

where $\mathbf{m} = (m_1, \ldots, m_n) \in \{0,1\}^n$. We call a transposition $\tau \in S_n$ an *adjacent transposition* if it is of the form $\tau = (i, i+1)$ for some $1 \le i \le n-1$. For $\tau = (i, i+1)$ and $\mathbf{m} = (m_1, \ldots, m_n)$ we define the "fermionic sign"

$$fsgn(\mathbf{m}, \tau) = (-1)^{m_i m_{i+1}}. \tag{3.94}$$

With $\sigma \in S_n$ written as a product of adjacent transpositions $\sigma = \tau_1 \ldots \tau_k$, let

$$fsgn(\mathbf{m}, \sigma) = fsgn(\mathbf{m}, \tau_1)fsgn(\mathbf{m}\tau_1, \tau_2) \ldots fsgn(\mathbf{m}\tau_1 \ldots \tau_{k-1}, \tau_k). \tag{3.95}$$

We leave it to the reader to check that this does not depend on the particular representation of σ as a product of adjacent transpositions. (See [**CM**], p. 63.) It follows from (3.95) that

$$fsgn(\mathbf{m}, \sigma_1\sigma_2) = fsgn(\mathbf{m}, \sigma_1)fsgn(\mathbf{m}\sigma_1, \sigma_2) \tag{3.96}$$

for any $\sigma_1, \sigma_2 \in S_n$. If $v \in \mathbf{V}_0$ or $v \in \mathbf{V}_1$, define

$$|v| = \begin{cases} 0 & \text{if } v \in \mathbf{V}_0 \\ 1 & \text{if } v \in \mathbf{V}_1. \end{cases} \tag{3.97}$$

For $1 \le i \le n$, if $v_i \in \mathbf{V}_0$ or $v_i \in \mathbf{V}_1$, define

$$\mathbf{m} = \mathbf{m}(v_1, \ldots, v_n) = (m_1, \ldots, m_n) = (|v_1|, \ldots, |v_n|) \in \{0,1\}^n. \tag{3.98}$$

THEOREM 3.29. *(Permutability) Let $v_i \in \mathbf{CM}(\mathbb{Z} + \frac{1}{2})^{\alpha_i}$, $1 \le i \le r$, $w, w' \in \mathbf{CM}(Z)$, let $\mathbf{s} = \mathbf{s}(v_1, \dots, v_r; w) = (s_1, \dots, s_r)$ and $\zeta^{\mathbf{s}} = \zeta_1^{s_1} \dots \zeta_r^{s_r}$ be defined as in Theorem 3.26, and note that $\mathbf{m} = \mathbf{m}(v_1, \dots, v_r) = (\alpha_1, \dots, \alpha_r)$ as defined in (3.97)-(3.98). Then for any $\sigma \in S_r$ we have*

$$\zeta^{\mathbf{s}}(Y(v_1, \zeta_1) \dots Y(v_r, \zeta_r)w, w') \sim fsgn(\mathbf{m}, \sigma) \zeta^{\mathbf{s}}(Y(v_{\sigma 1}, \zeta_{\sigma 1}) \dots Y(v_{\sigma r}, \zeta_{\sigma r})w, w').$$

PROOF. Let $J_i = J_{i(r+1)}(\zeta_i)$ and $\Delta_i = \Delta_i(\zeta_i)$ for $1 \le i \le r$, and $K_{ij} = K_{ij}(\zeta_i, \zeta_j; Z)$ for $1 \le i \ne j \le r$. It is clear from (3.96) that it suffices to establish the result for an adjacent transposition $\sigma = (t, t+1)$, $1 \le t < r$, in which case $fsgn(\mathbf{m}, \sigma) = (-1)^{\alpha_t \alpha_{t+1}}$ from (3.94). From Corollary 3.7 we see that when $Z = \mathbb{Z} + \frac{1}{2}$ we have

$$\zeta^{\mathbf{s}} \pi_1 \dots \pi_r \pi_{r+1}(w')(:\exp(J_1):\dots:\exp(J_r): v_1 \otimes \dots \otimes v_r \otimes w)$$
$$= \zeta^{\mathbf{s}}(Y(v_1, \zeta_1) \dots Y(v_r, \zeta_r)w, w') \tag{3.99}$$

but

$$\zeta^{\mathbf{s}} \pi_1 \dots \pi_r \pi_{r+1}(w')(:\exp(J_{\sigma 1}):\dots:\exp(J_{\sigma r}): v_1 \otimes \dots \otimes v_r \otimes w)$$
$$= (-1)^{\alpha_t \alpha_{t+1}} \zeta^{\mathbf{s}}(Y(v_{\sigma 1}, \zeta_{\sigma 1}) \dots Y(v_{\sigma r}, \zeta_{\sigma r})w, w'). \tag{3.100}$$

When $Z = \mathbb{Z}$, Corollary 3.7 gives the same result with $:\exp(J_i):$ replaced by $:\exp(J_i):\exp(\Delta_i)$ for $1 \le i \le r$. From Corollary 3.25, on $\mathbf{CM} = \mathbf{CM}(\mathbb{Z} + \frac{1}{2}) \otimes \dots \otimes \mathbf{CM}(\mathbb{Z} + \frac{1}{2}) \otimes \mathbf{CM}(Z)$ we have

$$:\exp(J_1):\dots:\exp(J_r):$$
$$=:\exp(J_1) \dots \exp(J_r): \exp\left(\sideset{}{'}\sum_{1 \le i < j \le r} K_{ij} \right) \exp(K_{t(t+1)}) \tag{3.101}$$

and, using (3.87),

$$:\exp(J_{\sigma 1}):\dots:\exp(J_{\sigma r}):$$
$$=:\exp(J_1) \dots \exp(J_r): \exp\left(\sideset{}{'}\sum_{1 \le i < j \le r} K_{ij} \right) \exp(K_{(t+1)t}) \tag{3.102}$$

where $\sum'$ indicates that the summation excludes $(i, j) = (t, t+1)$. Note that the operators Δ_i, $1 \le i \le r$, commute with each other, and with the J and K operators, so they are unaffected by σ. From (3.101) and (3.102) the only issue is the relationship between $\exp(K_{t(t+1)})$ and $\exp(K_{(t+1)t})$. When they are applied to any vector in $\mathbf{CM}$, they each give a finite sum of vectors with coefficients which are series in ζ_t and ζ_{t+1}. From Corollary 3.22 we see that the operator part of $K_{t(t+1)}$ is the negative of the operator part of $K_{(t+1)t}$. Thus, it suffices to show that the function to which the contraction part of $K_{t(t+1)}$ converges (in the appropriate domain) is the negative of the function to which the contraction part of $K_{(t+1)t}$ converges (in the appropriate domain). When $Z = \mathbb{Z} + \frac{1}{2}$ this is evident from Corollary 3.3, and when $Z = \mathbb{Z}$ it follows from Corollary 3.5. ∎

For $v_1, \ldots, v_n, w, w' \in \mathbf{V}$, using compositions of vertex operators we define the generating function of matrix coefficients

$$(Y(\ldots Y(Y(v_1, \zeta_1)v_2, \zeta_2)\ldots v_n, \zeta_n)w, w')$$

$$= \sum_{m_1, \ldots, m_n \in \frac{1}{2}\mathbb{Z}} (Y_{m_n}(\ldots Y_{m_2}(Y_{m_1}(v_1)v_2)v_3 \ldots v_n)w, w') \tag{3.103}$$

$$\cdot \zeta_1^{-m_1 - wt(v_1)} \zeta_2^{-m_2 + m_1 - wt(v_2)} \ldots \zeta_n^{-m_n + m_{n-1} - wt(v_n)}.$$

THEOREM 3.30. *(Associativity) Let $v_i \in \mathbf{CM}(\mathbb{Z} + \frac{1}{2})^{\alpha_i}$, $i = 1, 2$, $w, w' \in \mathbf{CM}(Z)$ and let $(s_1, s_2) = \mathbf{s}(v_1, v_2; w)$ be as defined in Theorem 3.26. Then we have*

$$\zeta_1^{s_1} \zeta_2^{s_2} (Y(v_1, \zeta_1)Y(v_2, \zeta_2)w, w') \sim$$

$$\zeta_2^{s_1 + s_2} (1 + (\zeta_1 - \zeta_2)/\zeta_2)^{s_1} (Y(Y(v_1, \zeta_1 - \zeta_2)v_2, \zeta_2)w, w')$$

where the series on the right side converges absolutely when $0 < |\zeta_1 - \zeta_2| < |\zeta_2|$.

PROOF. There are two cases: (a) $Z = \mathbb{Z} + \frac{1}{2}$, in which case $s_1 = s_2 = 0$; (b) $Z = \mathbb{Z}$, in which case $s_1 = \frac{1}{2}\alpha_1$ and $s_2 = \frac{1}{2}\alpha_2$.

(a) From Theorem 3.28 we know that the series in ζ_1 and ζ_2,

$$(Y(v_1, \zeta_1)Y(v_2, \zeta_2)w, w') = \pi_1\pi_2\pi_3(w')(:\exp(J_{13}(\zeta_1)): :\exp(J_{23}(\zeta_2)): v_1 \otimes v_2 \otimes w)$$

converges absolutely when $|\zeta_1| > |\zeta_2| > 0$ to a rational function in $\mathbb{C}[\zeta_1, \zeta_1^{-1}, \zeta_2, \zeta_2^{-1}, (\zeta_1 - \zeta_2)^{-1}]$. We will show that the series in $\zeta_1 - \zeta_2$ and ζ_2,

$$(Y(Y(v_1, \zeta_1 - \zeta_2)v_2, \zeta_2)w, w')$$

$$= \pi_1\pi_2\pi_3(w')(:\exp(J_{23}(\zeta_2)): :\exp(J_{12}(\zeta_1 - \zeta_2)): v_1 \otimes v_2 \otimes w)$$

converges absolutely when $0 < |\zeta_1 - \zeta_2| < |\zeta_2|$ to the same rational function.

We begin in the middle by applying the "normally ordered" expression

$$\exp(J_{12}^-(\zeta)) \exp(J_{13}^-(\zeta_2 + \zeta) + J_{23}^-(\zeta_2)) \exp(J_{13}^+(\zeta_2 + \zeta) + J_{23}^+(\zeta_2) + J_{12}^+(\zeta)) \tag{3.104}$$

to $v_1 \otimes v_2 \otimes w$ and then applying the projections $\pi_1(u_1)\pi_2(u_2)\pi_3(u_3)$ for arbitrary vectors $u_1, u_2, u_3 \in \mathbf{CM}(\mathbb{Z} + \frac{1}{2})$. The result is a rational function $f(\zeta, \zeta_2)$ in $\mathbb{C}[\zeta, \zeta^{-1}, \zeta_2, \zeta_2^{-1}, (\zeta_2 + \zeta)^{-1}]$. Note that $J_{13}^+(\zeta_2 + \zeta)$, $J_{23}^+(\zeta_2)$ and $J_{12}^+(\zeta)$ commute, and $J_{13}^-(\zeta_2 + \zeta)$ commutes with $J_{12}^-(\zeta)$ and $J_{23}^-(\zeta_2)$, but $J_{12}^-(\zeta)$ does not commute with $J_{23}^-(\zeta_2)$. If in (3.104) we expand $J_{13}^-(\zeta_2 + \zeta)$ and $J_{13}^+(\zeta_2 + \zeta)$ in nonnegative powers of ζ, then the resulting series in ζ and ζ_2 converges absolutely to $f(\zeta, \zeta_2)$ when $0 < |\zeta| < |\zeta_2|$. From Propostion 3.23 we have

$$\exp(J_{12}^-(\zeta)) \exp(J_{13}^-(\zeta_2 + \zeta) + J_{23}^-(\zeta_2)) = \exp(J_{23}^-(\zeta_2)) \exp(J_{12}^-(\zeta)) \tag{3.105}$$

and then

$$\exp(J_{12}^-(\zeta)) \exp(J_{13}^+(\zeta_2 + \zeta) + J_{23}^+(\zeta_2)) = \exp(J_{23}^+(\zeta_2)) \exp(J_{12}^-(\zeta)) \tag{3.106}$$

so that

$$\pi_1(u_1)\pi_2(u_2)\pi_3(u_3)(:\exp(J_{23}(\zeta_2)): :\exp(J_{12}(\zeta)): v_1 \otimes v_2 \otimes w) \tag{3.107}$$

converges absolutely to $f(\zeta, \zeta_2)$ when $0 < |\zeta| < |\zeta_2|$.

Returning to expression (3.104), substituting $\zeta_1 = \zeta_2 + \zeta$ gives the rational function $f(\zeta_1 - \zeta_2, \zeta_2)$. From Corollary 3.3, the contraction part of each term in $K_{12}(\zeta_1, \zeta_2; \mathbb{Z} + \frac{1}{2})$ given in Corollary 3.22 converges absolutely when $|\zeta_1| > |\zeta_2|$ to $\binom{-n-1}{m}(\zeta_1 - \zeta_2)^{-m-n-1}$, so we get $K_{12}(\zeta_1, \zeta_2; \mathbb{Z} + \frac{1}{2}) = J_{12}^+(\zeta_1 - \zeta_2)$ when expanded in nonnegative powers of ζ_2. Then Corollary 3.25 says that

$$\pi_1(u_1)\pi_2(u_2)\pi_3(u_3)(\exp(J_{12}^-(\zeta_1 - \zeta_2)){:}\exp(J_{13}(\zeta_1)){:}{:}\exp(J_{23}(\zeta_2)){:} v_1 \otimes v_2 \otimes w)$$
$$(3.108)$$

converges absolutely to $f(\zeta_1 - \zeta_2, \zeta_2)$ when $|\zeta_1| > |\zeta_2| > 0$. For any $v_1, v_2, w, w' \in \mathbf{CM}(\mathbb{Z} + \frac{1}{2})$ it is easy to see that with $u_1 = u_2 = \mathbf{vac}(\mathbb{Z} + \frac{1}{2})$, we have

$$\pi_1 \pi_2 \pi_3(w')(J_{12}^-(\zeta_1 - \zeta_2)v_1 \otimes v_2 \otimes w) = 0. \qquad (3.109)$$

Therefore,

$$\pi_1 \pi_2 \pi_3(w')({:}\exp(J_{13}(\zeta_1)){:}{:}\exp(J_{23}(\zeta_2)){:} v_1 \otimes v_2 \otimes w) \qquad (3.110)$$

converges absolutely to $f(\zeta_1 - \zeta_2, \zeta_2)$ when $|\zeta_1| > |\zeta_2| > 0$, and from (3.107),

$$\pi_1 \pi_2 \pi_3(w')({:}\exp(J_{23}(\zeta_2)){:}{:}\exp(J_{12}(\zeta_1 - \zeta_2)){:} v_1 \otimes v_2 \otimes w) \qquad (3.111)$$

converges absolutely to $f(\zeta_1 - \zeta_2, \zeta_2)$ when $0 < |\zeta_1 - \zeta_2| < |\zeta_2|$.

(b) From Theorem 3.28 we know that the series in ζ_1 and ζ_2,

$$\zeta_1^{s_1} \zeta_2^{s_2} (Y(v_1, \zeta_1)Y(v_2, \zeta_2)w, w')$$
$$= \zeta_1^{s_1} \zeta_2^{s_2} \pi_1 \pi_2 \pi_3(w')({:}\exp(J_{12}(\zeta_1)){:}\exp(\Delta_1(\zeta_1)) \qquad (3.112)$$
$$\cdot {:}\exp(J_{23}(\zeta_2)){:}\exp(\Delta_2(\zeta_2)v_1 \otimes v_2 \otimes w)$$

converges absolutely when $|\zeta_1| > |\zeta_2| > 0$ to a rational function $f(\zeta_1, \zeta_2)$ in $\mathbb{C}[\zeta_1, \zeta_1^{-1}, \zeta_2, \zeta_2^{-1}, (\zeta_1 - \zeta_2)^{-1}]$. We will show that the series in $\zeta = \zeta_1 - \zeta_2$ and ζ_2,

$$\zeta_2^{s_1 + s_2} (1 + \zeta/\zeta_2)^{s_1} (Y(Y(v_1, \zeta)v_2, \zeta_2)w, w')$$
$$= \zeta_2^{s_1 + s_2} (1 + \zeta/\zeta_2)^{s_1} \pi_1 \pi_2 \pi_3(w')({:}\exp(J_{23}(\zeta_2)){:}\exp(\Delta_2(\zeta_2)) \qquad (3.113)$$
$$\cdot {:}\exp(J_{12}(\zeta)){:} v_1 \otimes v_2 \otimes w)$$

converges absolutely when $0 < |\zeta| < |\zeta_2|$ to $f(\zeta_2 + \zeta, \zeta_2)$.

From Corollary 3.5, the contraction part of each term in $K_{12}(\zeta_1, \zeta_2; \mathbb{Z})$ converges absolutely when $|\zeta_1| > |\zeta_2| > 0$, $\zeta_1, \zeta_2 \in \mathbb{C}^-$ to

$$\zeta_1^{-m-\frac{1}{2}} \zeta_2^{-n-\frac{1}{2}} (\zeta_1 - \zeta_2)^{-m-n-1} P_{mn}(\zeta_1, \zeta_2) \qquad (3.114)$$

for some homogeneous polynomial $P_{mn}(\zeta_1, \zeta_2)$ of degree $m + n + 1$. Let $\bar{K}_{12}(\zeta_1, \zeta_2; \mathbb{Z})$ be the expression for $K_{12}(\zeta_1, \zeta_2; \mathbb{Z})$ with each contraction part replaced by the corresponding algebraic function (3.114). Then

$$\exp(\bar{K}_{12}(\zeta_1, \zeta_2; \mathbb{Z}))\exp(\Delta_2(\zeta_2))\exp(\Delta_1(\zeta_1))v_1 \otimes v_2 \otimes w \qquad (3.115)$$

is a finite sum of vectors with algebraic function coefficients analytic for $\zeta_1, \zeta_2 \in \mathbb{C}^-$, $\zeta_1 \neq \zeta_2$, so for any $u_1, u_2 \in \mathbf{CM}(\mathbb{Z} + \frac{1}{2})$, $u_3 \in \mathbf{CM}(\mathbb{Z})$,

$$\pi_1(u_1)\pi_2(u_2)\pi_3(u_3)(\exp(J_{12}^-(\zeta_1 - \zeta_2)) : \exp(J_{13}(\zeta_1)) \exp(J_{23}(\zeta_2)) :$$
$$\cdot \exp(\bar{K}_{12}(\zeta_1, \zeta_2; \mathbb{Z})) \exp(\Delta_2(\zeta_2)) \exp(\Delta_1(\zeta_1)) v_1 \otimes v_2 \otimes w) \tag{3.116}$$

is a polynomial in $\mathbb{C}[\zeta_1^{1/2}, \zeta_1^{-1/2}, \zeta_2^{1/2}, \zeta_2^{-1/2}, (\zeta_1 - \zeta_2)^{-1}]$ which is analytic for $\zeta_1, \zeta_2 \in \mathbb{C}^-$, $\zeta_1 \neq \zeta_2$. By Corollary 3.25, when $|\zeta_1| > |\zeta_2| > 0$, $\zeta_1, \zeta_2 \in \mathbb{C}^-$,

$$\pi_1(u_1)\pi_2(u_2)\pi_3(u_3)(\exp(J_{12}^-(\zeta_1 - \zeta_2)) : \exp(J_{13}(\zeta_1)) : : \exp(J_{23}(\zeta_2)) :$$
$$\cdot \exp(\Delta_2(\zeta_2)) \exp(\Delta_1(\zeta_1)) v_1 \otimes v_2 \otimes w) \tag{3.117}$$

converges absolutely to the algebraic function (3.116).

Letting $\zeta_1 = \zeta_2 + \zeta$ in (3.116), we get an algebraic function

$$\pi_1(u_1)\pi_2(u_2)\pi_3(u_3) \exp(J_{12}^-(\zeta)) \exp(J_{13}^-(\zeta_2 + \zeta) + J_{23}^-(\zeta_2))$$
$$\cdot \exp(J_{13}^0(\zeta_2 + \zeta) + J_{23}^0(\zeta_2)) \exp(J_{13}^+(\zeta_2 + \zeta) + J_{23}^+(\zeta_2)) \tag{3.118}$$
$$\cdot \exp(\bar{K}_{12}(\zeta_2 + \zeta, \zeta_2; \mathbb{Z})) \exp(\Delta_2(\zeta_2)) \exp(\Delta_1(\zeta_2 + \zeta)) v_1 \otimes v_2 \otimes w)$$

analytic for $\zeta_2 + \zeta$, $\zeta_2 \in \mathbb{C}^-$, $\zeta \neq 0$. If we further restrict to the domain

$$\zeta_2 \in \mathbb{C}^-, \quad 0 < |\zeta| < |\zeta_2| \qquad \text{if } Re(\zeta_2) \geq 0,$$
$$0 < |\zeta| < |Im(\zeta_2)| \quad \text{if } Re(\zeta_2) < 0, \tag{3.119}$$

then we have

$$(\zeta_2 + \zeta)^{1/2} = \zeta_2^{1/2}(1 + \zeta/\zeta_2)^{1/2} \tag{3.120}$$

and in (3.118) we can expand all powers of $\zeta_2 + \zeta$ in nonnegative powers of ζ. Then Proposition 3.23 gives

$$\exp(J_{12}^-(\zeta)) \exp(J_{13}^-(\zeta_2 + \zeta) + J_{23}^-(\zeta_2)) = \exp(J_{23}^-(\zeta_2)) \exp(J_{12}^-(\zeta)), \tag{3.121}$$

$$\exp(J_{12}^-(\zeta)) \exp(J_{13}^0(\zeta_2 + \zeta) + J_{23}^0(\zeta_2)) = \exp(J_{23}^0(\zeta_2)) \exp(J_{12}^-(\zeta)), \tag{3.122}$$

and

$$\exp(J_{12}^-(\zeta)) \exp(J_{13}^+(\zeta_2 + \zeta) + J_{23}^+(\zeta_2)) = \exp(J_{23}^+(\zeta_2)) \exp(J_{12}^-(\zeta)). \tag{3.123}$$

So

$$\pi_1(u_1)\pi(u_2)\pi_3(u_3)(: \exp(J_{23}(\zeta_2)) : \exp(J_{12}^-(\zeta)) \exp(\bar{K}_{12}(\zeta_2 + \zeta, \zeta_2; \mathbb{Z}))$$
$$\cdot \exp(\Delta_2(\zeta_2)) \exp(\Delta_1(\zeta_2 + \zeta)) v_1 \otimes v_2 \otimes w) \tag{3.124}$$

converges absolutely in (3.119) to the algebraic function in (3.118). From Lemma 3.31 (proven below) and Proposition 3.11 (a), we have

$$\bar{K}_{12}(\zeta_2 + \zeta, \zeta_2; \mathbb{Z}) = [\Delta_2(\zeta_2), J_{12}^-(\zeta)] + J_{12}^+(\zeta) \tag{3.125}$$

in the domain (3.119). From (3.125) and Corollary 3.12 we get that (3.124) equals

$$\pi_1(u_1)\pi_2(u_2)\pi_3(u_3)(:\exp(J_{23}(\zeta_2)): :\exp(J_{12}(\zeta)):$$
$$\cdot \exp(\Delta_2(\zeta_2) + [\Delta_2(\zeta_2), J_{12}^-(\zeta)] + \Delta_1(\zeta_2 + \zeta))v_1 \otimes v_2 \otimes w)$$
$$= \pi_1(u_1)\pi_2(u_2)\pi_3(u_3)(:\exp(J_{23}(\zeta_2)):\exp(\Delta_2(\zeta_2)):\exp(J_{12}(\zeta)): v_1 \otimes v_2 \otimes w)$$

$$(3.126)$$

in the domain (3.119). Note that this converges absolutely to a polynomial in $\mathbb{C}[(1 + \zeta/\zeta_2)^{1/2}, (1 + \zeta/\zeta_2)^{-1/2}, \zeta_2^{1/2}, \zeta_2^{-1/2}, \zeta^{-1}]$, and therefore has the larger domain $0 < |\zeta| < |\zeta_2|$, $\zeta_2 \in \mathbb{C}^-$.

Letting $u_1 = u_2 = \mathbf{vac}(\mathbb{Z} + \frac{1}{2})$ and $u_3 = w' \in \mathbf{CM}(\mathbb{Z})$ in (3.117), only the constant term of the factor $\exp(J_{12}^-(\zeta_1 - \zeta_2))$ contributes, so after multiplying by $\zeta_1^{s_1}\zeta_2^{s_2}$, we get the rational function $f(\zeta_1, \zeta_2)$. Therefore, with the same values for u_1, u_2 and u_3 in (3.124), and with $\zeta_1 = \zeta_2 + \zeta$, multiplying by the algebraic function $(\zeta_2 + \zeta)^{s_1}\zeta_2^{s_2}$, yields the rational function $f(\zeta_2 + \zeta, \zeta_2)$ in the domain (3.119). Using (3.120), which is valid in that domain, and expanding $(1 + \zeta/\zeta_2)^{s_1}$ in nonnegative powers of ζ, we get that (3.113) converges absolutely to $f(\zeta_2 + \zeta, \zeta_2)$. Since this function is in $\mathbb{C}[\zeta_1, \zeta_1^{-1}, \zeta_2, \zeta_2^{-1}, (\zeta_1 - \zeta_2)^{-1}]$, that series actually converges absolutely in the larger domain $0 < |\zeta| < |\zeta_2|$. ∎

LEMMA 3.31. *For $0 \leq m, n \in \mathbb{Z}$, the series*

$$\binom{-n-1}{m}\zeta^{-m-n-1} + \zeta_2^{-m-n-1}\sum_{0 \leq k \in \mathbb{Z}}\binom{m+k}{m}C_{n(m+k)}\left(\frac{\zeta}{\zeta_2}\right)^k \qquad (3.127)$$

converges absolutely to the algebraic function

$$f_{mn}(\zeta, \zeta_2) = (\zeta_2 + \zeta)^{-m-\frac{1}{2}}\zeta_2^{-n-\frac{1}{2}}\zeta^{-m-n-1}P_{mn}(\zeta_2 + \zeta, \zeta_2) \qquad (3.128)$$

defined in Corollary 3.5, in the domain

$$\zeta_2 \in \mathbb{C}^-, \quad 0 < |\zeta| < |\zeta_2| \qquad \qquad if\ Re(\zeta_2) \geq 0,$$
$$0 < |\zeta| < |Im(\zeta_2)| \quad if\ Re(\zeta_2) < 0. \qquad (3.129)$$

PROOF. From Corollary 3.5, the contraction $\underline{a_i^{(m)}(\zeta_2 + \zeta)a_i^{*(n)}(\zeta_2)}$ converges absolutely when $|\zeta_2 + \zeta| > |\zeta_2|$ to the algebraic function $f_{mn}(\zeta, \zeta_2)$. Therefore, $\underline{a_i^{(m+1)}(\zeta_2 + \zeta)a_i^{*(n)}(\zeta_2)}$ converges absolutely in the same domain to $(m+1)^{-1}(\partial/\partial\zeta)f_{mn}(\zeta, \zeta_2)$ and $\underline{a_i^{(m)}(\zeta_2 + \zeta)a_i^{*(n+1)}(\zeta_2)}$ converges absolutely in that domain to $(n+1)^{-1}(\partial/\partial\zeta_2 - \partial/\partial\zeta)f_{mn}(\zeta, \zeta_2)$. So

$$(m+1)^{-1}(\partial/\partial\zeta)f_{mn}(\zeta, \zeta_2) = f_{(m+1)n}(\zeta, \zeta_2) \qquad (3.130)$$

and

$$(n+1)^{-1}(\partial/\partial\zeta_2 - \partial/\partial\zeta)f_{mn}(\zeta, \zeta_2) = f_{m(n+1)}(\zeta, \zeta_2). \qquad (3.131)$$

It is easy to see that applying $(m+1)^{-1}(\partial/\partial\zeta)$ to the series (3.127) yields the same series with m replaced by $m+1$. So it suffices to check the assertion for $m = 0$. Using (3.56) one finds that applying $(n+1)^{-1}(\partial/\partial\zeta_2 - \partial/\partial\zeta)$ to (3.127) with $m = 0$ yields that series with $m = 0$ and n replaced by $n+1$. So it suffices to check the assertion for $m = n = 0$.

We have

$$f_{00}(\zeta, \zeta_2) = \frac{\zeta_2 + \frac{1}{2}\zeta}{(\zeta_2 + \zeta)^{1/2}\zeta_2^{1/2}\zeta} = \frac{\zeta_2 + \frac{1}{2}\zeta}{(1 + \zeta/\zeta_2)^{1/2}\zeta_2\zeta} \tag{3.132}$$

in the domain (3.129). For $|w| < 1$,

$$(1 + w)^{-1/2} = \sum_{0 \leq k \in \mathbb{Z}} \binom{-\frac{1}{2}}{k} w^k \tag{3.133}$$

converges absolutely. Integrating from 0 to z, $|z| < 1$, and dividing by $2z$, we get

$$\frac{(1+z)^{1/2} - 1}{z} = \sum_{0 \leq k \in \mathbb{Z}} \binom{-\frac{1}{2}}{k} \frac{z^k}{2(k+1)}. \tag{3.134}$$

Differentiating and then multiplying by $-z$ yields

$$\frac{z+2}{2z(1+z)^{1/2}} - \frac{1}{z} = -\sum_{0 \leq k \in \mathbb{Z}} \binom{-\frac{1}{2}}{k} \frac{kz^k}{2(k+1)} \tag{3.135}$$

so that

$$\frac{z+2}{2z(1+z)^{1/2}} = \frac{1}{z} + \sum_{0 \leq k \in \mathbb{Z}} C_{0k}z^k. \tag{3.136}$$

Finally, substituting $z = \zeta/\zeta_2$ gives the result. ∎

Although we do not need the results, one can investigate matrix coefficients involving complicated combinations of products and compositions of vertex operators. Results such as the following can be obtained by techniques similar to those used in the last theorem.

THEOREM 3.32. *For $v_1, \ldots, v_n, w, w' \in \mathbf{V}$, we have*

$$(Y(v_1, \zeta_1 + \ldots + \zeta_n)Y(v_2, \zeta_2 + \ldots + \zeta_n)\ldots Y(v_n, \zeta_n)w, w') \sim$$
$$(Y(\ldots Y(Y(v_1, \zeta_1)v_2, \zeta_2)\ldots v_n, \zeta_n)w, w')$$

where the series on the right side converges absolutely when

$$0 < \sum_{p \leq i \leq m} |\zeta_i| < |\zeta_{m+1}| \quad for\ 1 \leq p \leq m \leq n-1.$$

THEOREM 3.33. *(The Jacobi Identity on $\mathbf{CM}(Z)$) Let $v_i \in \mathbf{CM}(\mathbb{Z}+\frac{1}{2})^{\alpha_i} = \mathbf{V}_{\alpha_i}$ for $i = 1, 2$, $w, w' \in \mathbf{CM}(Z)$, and let $f(\zeta_1, \zeta_2) \in \mathbb{C}[\zeta_1, \zeta_1^{-1}, \zeta_2, \zeta_2^{-1}, (\zeta_1 - \zeta_2)^{-1}]$. Let $(s_1, s_2) = \mathbf{s}(v_1, v_2; w)$ be as defined in Theorem 3.26. Let $C_i(r)$ be a circle in the ζ_i-plane of radius r centered at the origin, and let $C_1(\zeta_2, \epsilon)$ be a circle in the ζ_1-plane of radius ϵ centered at $\zeta_1 = \zeta_2$. Then for $0 < r < \rho < R$ and $0 < \epsilon < \min\{R - \rho, \rho - r\}$, we have*

$$\int_{C_2(\rho)} \int_{C_1(R)} \zeta_1^{s_1} \zeta_2^{s_2} (Y(v_1, \zeta_1) Y(v_2, \zeta_2) w, w') f(\zeta_1, \zeta_2) d\zeta_1 d\zeta_2$$

$$- \int_{C_2(\rho)} \int_{C_1(r)} (-1)^{|v_1||v_2|} \zeta_1^{s_1} \zeta_2^{s_2} (Y(v_2, \zeta_2) Y(v_1, \zeta_1) w, w') f(\zeta_1, \zeta_2) d\zeta_1 d\zeta_2$$

$$= \int_{C_2(\rho)} \int_{C_1(\zeta_2, \epsilon)} \zeta_2^{s_1 + s_2} (1 + (\zeta_1 - \zeta_2)/\zeta_2)^{s_1}$$

$$\cdot (Y(Y(v_1, \zeta_1 - \zeta_2) v_2, \zeta_2) w, w') f(\zeta_1, \zeta_2) d\zeta_1 d\zeta_2$$

where the function $f(\zeta_1, \zeta_2)$ is to be expanded in the three integrands as in (0.48)-(0.50), respectively, with $z = \zeta_1$ and $z_0 = \zeta_2$.

PROOF. From the assumption on $f(\zeta_1, \zeta_2)$, Theorems 3.29 and 3.30 say that all three integrands represent the same rational function of ζ_1 and ζ_2 having possible poles only at $\zeta_1 = 0$, $\zeta_2 = 0$ and $\zeta_1 = \zeta_2$. With ϵ chosen such that $0 < \epsilon < R - \rho$ and $0 < \epsilon < \rho - r$, Cauchy's theorem applied to the integration with respect to ζ_1 gives the result. ∎

DEFINITION. *For operators L_1 and L_2 on $\mathbf{V}(\frac{1}{2}\mathbb{Z})$ let $[L_1, L_2]_\pm = L_1 L_2 \pm L_2 L_1$.*

COROLLARY 3.34. *Let $v_i \in \mathbf{CM}(\mathbb{Z}+\frac{1}{2})^{\alpha_i} = \mathbf{V}_{\alpha_i}$ for $i = 1, 2$. Then for any $r \in \mathbb{Z}$, for $m, n \in \mathbb{Z}$ on $\mathbf{CM}(\mathbb{Z}+\frac{1}{2})$, and for $m \in \mathbb{Z}+\frac{1}{2}\alpha_1$, $n \in \mathbb{Z}+\frac{1}{2}\alpha_2$ on $\mathbf{CM}(\mathbb{Z})$, we have*

$$\sum_{0 \le i \in \mathbb{Z}} (-1)^i \binom{r}{i} (\{v_1\}_{m+r-i} \{v_2\}_{n+i} - (-1)^{|v_1||v_2|+r} \{v_2\}_{n+r-i} \{v_1\}_{m+i})$$

$$= \sum_{0 \le k \in \mathbb{Z}} \binom{m}{k} \{\{v_1\}_{r+k} v_2\}_{m+n-k} \cdot$$

$$\tag{3.137}$$

The sum over k is finite, and if $k_i = wt(v_i)$ then $0 \le k \le k_1 + k_2 - r - 1$.

PROOF. Let $\kappa = 2\pi\mathbf{i}$ and let $v_i \in \mathbf{CM}(\mathbb{Z}+\frac{1}{2})^{\alpha_i}$, $w, w' \in \mathbf{CM}(Z)$, $n \in \mathbb{Z}+s_i$, where $(s_1, s_2) = \mathbf{s}(v_1, v_2; w)$, that is, $n \in \mathbb{Z}$ if $Z = \mathbb{Z}+\frac{1}{2}$, $n \in \mathbb{Z}+\frac{1}{2}\alpha_i$ if $Z = \mathbb{Z}$. From (3.59) and the discussion after (3.25), Cauchy's theorem gives

$$\kappa^{-1} \int_{C_i(r)} (Y(v_i, \zeta_i) w, w') \zeta_i^n d\zeta_i = (\{v_i\}_n w, w') \tag{3.138}$$

where $C_i(r)$ is a circle in the ζ_i-plane of radius $r > 0$ centered at 0. Use Theorem 3.33 with $f(\zeta_1, \zeta_2) = \zeta_1^{m-s_1} \zeta_2^{n-s_2} (\zeta_1 - \zeta_2)^r$. In the first integral on the left side we have $|\zeta_1| > |\zeta_2|$, so we should expand $(\zeta_1 - \zeta_2)^r$ as $\zeta_1^r (1 - \zeta_2/\zeta_1)^r$, but in the second integral we have $|\zeta_2| > |\zeta_1|$, so we should expand $(\zeta_1 - \zeta_2)^r$ as $(-\zeta_2)^r(1 - \zeta_1/\zeta_2)^r$. Then (3.138) gives κ^2 times the left side of (3.137) applied to w and paired with w'. With the substitution $\zeta = \zeta_1 - \zeta_2$ in the integral on the right side of Theorem 3.33, the inside integral is over $C(\epsilon)$, a circle around 0 of radius ϵ in the ζ-plane. This gives

$$\int_{C_2(\rho)} \int_{C(\epsilon)} \sum_{\substack{p \in \mathbb{Z} \\ q \in \mathbb{Z}+s_1+s_2}} (\{\{v_1\}_p v_2\}_q w, w') \zeta^{r-p-1} \zeta_2^{m+n-q-1} (1 + \zeta/\zeta_2)^m d\zeta d\zeta_2$$

$$= \int_{C_2(\rho)} \int_{C(\epsilon)} \sum_{\substack{p \in \mathbb{Z} \\ q \in \mathbb{Z}+s_1+s_2}} \sum_{0 \le k \in \mathbb{Z}} \binom{m}{k} (\{\{v_1\}_p v_2\}_q w, w') \zeta^{r-p-1+k} \zeta_2^{m+n-q-1-k} d\zeta d\zeta_2$$

$$= \kappa^2 \sum_{0 \le k \in \mathbb{Z}} \binom{m}{k} (\{\{v_1\}_{r+k} v_2\}_{m+n-k} w, w').$$

$$(3.139)$$

For $k_i = \mathrm{wt}(v_i)$, $\mathrm{wt}(\{v_1\}_p v_2) = k_1 + k_2 - p - 1$, so $\{v_1\}_p v_2 = 0$ if $p \ge k_1 + k_2$. Since any v_1, v_2 are finite sums of such homogeneous vectors, the sum is finite. Since the Hermitian form on $\mathbf{CM}(Z)$ is nondegenerate, we have the assertion of the corollary. ∎

COROLLARY 3.35. *Under the assumptions of Corollary 3.34, for $m, n \in \mathbb{Z}$ on* $\mathbf{CM}(\mathbb{Z} + \frac{1}{2})$*, and for $m \in \mathbb{Z} + \frac{1}{2}\alpha_1$, $n \in \mathbb{Z} + \frac{1}{2}\alpha_2$ on $\mathbf{CM}(\mathbb{Z})$, we have*

$$[\{v_1\}_m, \{v_2\}_n]_\pm = \sum_{0 \le k \in \mathbb{Z}} \binom{m}{k} \{\{v_1\}_k v_2\}_{m+n-k} \qquad (3.140)$$

where $\pm$ is $-(-1)^{|v_1||v_2|}$. This sum is finite, and if $k_i = \mathrm{wt}(v_i)$ then $0 \le k \le k_1 + k_2 - 1$.

PROOF. Take $r = 0$ in Corollary 3.34. ∎

Corollary 3.35 and Proposition 3.17 (a) show that, representing the coset of $v \otimes t^n$ by the operator $\{v\}_n$, we have superalgebra structures on $\hat{\mathbf{V}}/D(-1)\hat{\mathbf{V}}$ and on $\hat{\mathbf{V}}'/D(-1)\hat{\mathbf{V}}'$ (recall (3.7) and (3.8)) represented on $\mathbf{V} = \mathbf{V}_0 \oplus \mathbf{V}_1$ and on $\mathbf{V}_2 \oplus \mathbf{V}_3$, respectively.

Define the isomorphisms

$$\Upsilon : \mathbf{A} \to (\mathbf{V}_0)_{1/2}, \qquad \Upsilon' : \mathbf{o}(2l) \to (\mathbf{V}_0)_1 \qquad (3.141)$$

by

$$\Upsilon(a_1) = a_1(-\tfrac{1}{2})\mathbf{vac}(\mathbb{Z} + \tfrac{1}{2}), \qquad \Upsilon'(\substack{\circ\\\circ}a_1a_2\substack{\circ\\\circ}) = a_1(-\tfrac{1}{2})a_2(-\tfrac{1}{2})\mathbf{vac}(\mathbb{Z} + \tfrac{1}{2}) \quad (3.142)$$

for $a_1, a_2 \in \mathbf{A}$. Recall from (3.9) and (3.59) that

$$a(n) = Y_n(\Upsilon(a)) = \{\Upsilon(a)\}_{n-\frac{1}{2}} \qquad (3.143)$$

for $a \in \mathbf{A}$, $n \in \frac{1}{2}\mathbb{Z}$. For $x = \substack{\circ\\\circ}a_1a_2\substack{\circ\\\circ} \in \mathbf{o}(2l)$, $m \in \mathbb{Z}$, define the notation

$$x(m) = Y_m(\Upsilon'(x)) = \{\Upsilon'(x)\}_m \qquad (3.144)$$

and note that

$$x(m) = \sum_{k \in \frac{1}{2}\mathbb{Z}} {}^{\circ}_{\circ} a_1(k) a_2(m-k) {}^{\circ}_{\circ}. \tag{3.145}$$

As was stated in Theorem 2.1, these are the operators on $\mathbf{V}(\frac{1}{2}\mathbb{Z})$ representing $\hat{\mathbf{o}}(2l)$. In Corollary 3.37 we will see how this result follows from Corollary 3.35 with $v_1, v_2 \in (\mathbf{V}_0)_1$. Note that we have

$$x(0)\Upsilon(a) = \Upsilon(x \cdot a) \tag{3.146}$$

$$x(0)\Upsilon'(y) = \Upsilon'([x,y]), \tag{3.147}$$

$$x(1)\Upsilon'(y) = \langle x, y \rangle \mathbf{vac}(\mathbb{Z} + \tfrac{1}{2}) \tag{3.148}$$

for $x, y \in \mathbf{o}(2l)$, $a \in \mathbf{A}$, which follow from (2.34)-(2.36) and (3.145).

COROLLARY 3.36. *For $v \in \mathbf{V}$, $m \in \mathbb{Z}$, on $\mathbf{V}(\frac{1}{2}\mathbb{Z})$ we have*

(a) $[D(m), Y(v,z)] = \displaystyle\sum_{0 \leq k \in \mathbb{Z}} \binom{m+1}{k} z^{m+1-k} Y(D(k-1)v, z)$

and the sum is finite,

(b) *If $D(n)v = 0$ for all $n > 0$, then we have*

$$[D(m), Y(v,z)] = -z^{m+1}\frac{d}{dz}Y(v,z) + (m+1)z^m Y(D(0)v, z).$$

PROOF. Part (a) follows from Corollary 3.35 using (3.61). The sum is finite since $\mathrm{wt}(D(k-1)v) = \mathrm{wt}(v) - k + 1$ for homogeneous vectors v. Part (b) follows from (a). ∎

Note that (2.58) - (2.62) follow from Corollary 3.35.

COROLLARY 3.37. *For $x \in \mathbf{o}(2l)$, $m \in \mathbb{Z}$, the operators $x(m) = Y_m(\Upsilon'(x))$ represent $\hat{\mathbf{o}}(2l)$ on $\mathbf{V}(\frac{1}{2}\mathbb{Z})$. For any $v \in \mathbf{V}$ we have*

$$[x(m), Y(v, \zeta)] = \zeta^m \sum_{0 \leq k \in \mathbb{Z}} \binom{m}{k} \zeta^{-k} Y(x(k)v, \zeta) \tag{3.149}$$

and for $v \in (\mathbf{V}_0)_0$ or $v \in (\mathbf{V}_1)_{1/2}$, we have

$$[x(m), Y(v, \zeta)] = \zeta^m Y(x(0)u, \zeta). \tag{3.150}$$

PROOF. For $x, y \in \mathbf{o}(2l)$, $m, n \in \mathbb{Z}$, from Corollary 3.35 with $v = \Upsilon'(y)$, (3.147) and (3.148), we get

$$[x(m), y(n)] = \{x(0)\Upsilon'(y)\}_{m+n} + m\{x(1)\Upsilon'(y)\}_{m+n-1}$$
$$= \{\Upsilon'([x,y])\}_{m+n} + m\{\langle x, y \rangle \mathbf{vac}(\mathbb{Z} + \tfrac{1}{2})\}_{m+n-1}$$
$$= [x,y](m+n) + m\langle x, y \rangle \delta_{m,-n}$$

so these operators represent $\hat{\mathbf{o}}(2l)$ on $\mathbf{V}(\frac{1}{2}\mathbb{Z})$. The rest also follows from Corollary 3.35. ∎

COROLLARY 3.38. *For $x \in (\mathbf{V}_0)_1$, $m \in \mathbb{Z}$, $v \in \mathbf{V}$, on $\mathbf{V}(\frac{1}{2}\mathbb{Z})$ we have*

$$Y(x(-m)v, \zeta) =$$
$$\sum_{0 \leq i \in \mathbb{Z}} \binom{m+i-1}{i} [\zeta^i x(-m-i)Y(v,\zeta) - (-1)^m \zeta^{-m-i} Y(v,\zeta)x(i)] \qquad (3.151)$$

and

$$Y(D(-m-1)v, \zeta) =$$
$$\sum_{0 \leq i \in \mathbb{Z}} \binom{m+i-1}{i} [\zeta^i D(-m-i-1)Y(v,\zeta) - (-1)^m \zeta^{-m-i} Y(v,\zeta)D(i-1)].$$
$$(3.152)$$

PROOF. The first formula follows from Corollary 3.34 with $m = 0$, r replaced by $-m$, v_1 replaced by $x \in (\mathbf{V}_0)_1$, v_2 replaced by v, after multiplying both sides by ζ^{-n-1} and summing over $n \in \mathbb{Z} + \frac{1}{2}\alpha_2$. The second formula similarly follows with v_1 replaced by $D(-2)\mathbf{vac}$. ∎

THEOREM 3.39. *With $\mathbf{V} = \mathbf{CM}(\mathbb{Z} + \frac{1}{2})$, $\mathbf{1} = \mathbf{vac}(\mathbb{Z} + \frac{1}{2})$ and $\omega = D(-2)\mathbf{1}$, $(\mathbf{V}, Y(\ , z), \mathbf{1}, \omega)$ is a vertex operator superalgebra as defined in the Introduction. With $\mathbf{W} = \mathbf{CM}(\mathbb{Z})$, $(\mathbf{W}, Y(\ , z))$ is a canonically $\mathbb{Z}_2$-twisted $\mathbf{V}$-module.*

CHAPTER 4

Spinor Construction of the Chevalley Algebra and Triality for D_4

The principle of triality for $\mathbf{o}(8)$ is more than just an S_3 acting as Lie algebra automorphisms on $\mathbf{o}(8)$. It also involves an action of S_3 on the direct sum of the three 8-dimensional $\mathbf{o}(8)$-modules intertwining the action of $\mathbf{o}(8)$ on the modules. In this chapter we present Chevalley's purely spinor approach to triality [C], which involves putting a commutative nonassociative algebra structure on the direct sum of the three 8-dimensional $\mathbf{o}(8)$-modules. Later we will see how this "Chevalley algebra" gives a spinor construction of E_8.

Let $\mathbf{A} \cong \mathbb{C}^8$ be a complex vector space with nondegenerate symmetric bilinear form $\langle\,,\,\rangle$ such that $\mathbf{A} = \mathbf{A}^+ \oplus \mathbf{A}^-$ is a polarization of $\mathbf{A}$ into maximally isotropic subspaces $\mathbf{A}^\pm \cong \mathbb{C}^4$. Let $\vartheta : \mathbf{A} \to \mathbf{A}$ be a $\mathbb{C}$-antilinear involution such that $\vartheta(\mathbf{A}^+) = \mathbf{A}^-$ and for $a, b \in \mathbf{A}$ we have

$$\langle \vartheta a, \vartheta b \rangle = \overline{\langle a, b \rangle} \tag{4.1}$$

and

$$0 < \langle a, \vartheta a \rangle \in \mathbb{R} \quad \text{if} \quad a \neq 0. \tag{4.2}$$

Choose a nonzero vector $\mathbf{vac}$ in the one-dimensional space $\wedge^4 \mathbf{A}^+$. Let $\mathbf{Cliff}$ be the Clifford algebra generated by $\mathbf{A}$ and $\langle\,,\,\rangle$, and let

$$\mathbf{CM} = \mathbf{Cliff} \cdot \mathbf{vac} = (\wedge \mathbf{A}^-) \cdot \mathbf{vac} \tag{4.3}$$

be the Clifford module determined by the polarization of $\mathbf{A}$. We have

$$\mathbf{CM} = \mathbf{CM}^0 \oplus \mathbf{CM}^1 \tag{4.4}$$

where

$$\mathbf{CM}^0 = (\wedge^{even} \mathbf{A}^-) \cdot \mathbf{vac}, \quad \mathbf{CM}^1 = (\wedge^{odd} \mathbf{A}^-) \cdot \mathbf{vac}. \tag{4.5}$$

Define the "main antiautomorphism" $\alpha : \mathbf{Cliff} \to \mathbf{Cliff}$ by

$$\alpha(a_1 \ldots a_k) = a_k \ldots a_1 \quad \text{for} \quad a_1, \ldots, a_k \in \mathbf{A}. \tag{4.6}$$

Note that $\alpha(\mathbf{vac}) = \mathbf{vac}$ since $\dim(\mathbf{A}^+) = 4$. It is easy to see that for any $u \in \mathbf{Cliff}$, $(\mathbf{vac})u(\mathbf{vac}) \in \wedge^4 \mathbf{A}^+$, so $(\mathbf{vac})u(\mathbf{vac})$ is some multiple of $\mathbf{vac}$. This allows us to define a bilinear form on $\mathbf{CM}$ by setting

$$\langle b, c \rangle \mathbf{vac} = \alpha(b)c \quad \text{for} \quad b, c \in \mathbf{CM}. \tag{4.7}$$

74

It is easy to see that for $b \in (\wedge^r \mathbf{A}^-) \cdot \mathbf{vac}$ and $c \in (\wedge^s \mathbf{A}^-) \cdot \mathbf{vac}$, $\langle b, c \rangle = 0$ if $r + s \neq 4$.

PROPOSITION 4.1. *The form* $\langle \, , \, \rangle$ *on* $\mathbf{CM}$ *defined by (4.7) is symmetric, nondegenerate, and for* $a \in \mathbf{A}$, $b, c \in \mathbf{CM}$ *we have*

(a) $\langle ab, c \rangle = \langle b, ac \rangle$,

(b) $\langle ab, ac \rangle = \frac{1}{2} \langle a, a \rangle \langle b, c \rangle$,

(c) $\langle b, c \rangle = 0$ *if* $b \in \mathbf{CM}^0$, $c \in \mathbf{CM}^1$.

PROOF. Clear. ■

Let

$$\mathbf{C} = \mathbf{A} \oplus \mathbf{CM} \tag{4.8}$$

and define

$$\mathbf{C}^{(1)} = \mathbf{A}, \quad \mathbf{C}^{(2)} = \mathbf{CM}^0, \quad \mathbf{C}^{(3)} = \mathbf{CM}^1. \tag{4.9}$$

With the understanding that $\langle a, b \rangle = 0$ for $a \in \mathbf{A}$, $b \in \mathbf{CM}$, we have defined a nondegenerate symmetric bilinear form on $\mathbf{C}$ whose restriction to each subspace $\mathbf{C}^{(k)}$, $1 \leq k \leq 3$, is nondegenerate. Each $u \in \mathbf{C}$ can be uniquely written $u = a + b + c$ for $a \in \mathbf{A}$, $b \in \mathbf{CM}^0$, $c \in \mathbf{CM}^1$. Define the cubic form $F : \mathbf{C} \to \mathbb{C}$ by

$$F(u) = \langle ab, c \rangle. \tag{4.10}$$

If $a = 0$ or $b = 0$ or $c = 0$ then $F(u) = 0$. By polarization of F we define the symmetric trilinear form $\Phi : \mathbf{C} \times \mathbf{C} \times \mathbf{C} \to \mathbb{C}$ by

$$\begin{aligned} \Phi(u_1, u_2, u_3) = \; &F(u_1 + u_2 + u_3) - F(u_2 + u_3) - F(u_1 + u_3) \\ &- F(u_1 + u_2) + F(u_1) + F(u_2) + F(u_3). \end{aligned} \tag{4.11}$$

Since the form $\langle \, , \, \rangle$ is nondegenerate on $\mathbf{C}$, we can define an operation $\circ$ on $\mathbf{C}$ as follows. If $u_1, u_2 \in \mathbf{C}$, let $u_1 \circ u_2$ be the unique element of $\mathbf{C}$ such that for all $u \in \mathbf{C}$,

$$\Phi(u_1, u_2, u) = \langle u_1 \circ u_2, u \rangle. \tag{4.12}$$

PROPOSITION 4.2. *Let* (k, k', k'') *be a cyclic permutation of* $(1, 2, 3)$. *We have*

(a) *The operation* $\circ$ *on* $\mathbf{C}$ *is commutative,*

(b) $a \circ b = ab$ *and* $a \circ c = ac$ *for* $a \in \mathbf{C}^{(1)}$, $b \in \mathbf{C}^{(2)}$, $c \in \mathbf{C}^{(3)}$,

(c) $\mathbf{C}^{(k)} \circ \mathbf{C}^{(k)} = \{0\}$ *and* $\mathbf{C}^{(k)} \circ \mathbf{C}^{(k')} \subseteq \mathbf{C}^{(k'')}$ *for* $1 \leq k \leq 3$,

(d) $\langle u_1 \circ u_2, u_3 \rangle = \langle u_1, u_2 \circ u_3 \rangle$ *for* $u_1, u_2, u_3 \in \mathbf{C}$.

PROOF. (a) Commutativity of $\circ$ follows from the symmetry of Φ.

(b) We have $\Phi(a, b, c) = F(a + b + c) = \langle ab, c \rangle$, since the other terms in $\Phi(a, b, c)$ are zero. If $u \in \mathbf{C}^{(1)}$ or $u \in \mathbf{C}^{(2)}$ then $\Phi(a, b, u) = 0 = \langle ab, u \rangle$, so $\Phi(a, b, u) = \langle ab, u \rangle$ for all $u \in \mathbf{C}$, and from (4.12), $a \circ b = ab$. Also, $\Phi(a, c, b) = F(a + b + c) = \langle ab, c \rangle = \langle c, ab \rangle = \langle ac, b \rangle$ from Proposition 4.1 (a). If $u \in \mathbf{C}^{(1)}$ or $u \in \mathbf{C}^{(3)}$ then $\Phi(a, c, u) = 0 = \langle ac, u \rangle$, so $\Phi(a, c, u) = \langle ac, u \rangle$ for all $u \in \mathbf{C}$.

(c) These follow easily from the fact that $\Phi(a, b, c) = 0$ if $a, b \in \mathbf{C}^{(k)}$, $c \in \mathbf{C}^{(k')}$ or $c \in \mathbf{C}^{(k'')}$.

(d) This follows from the symmetry of Φ. ∎

PROPOSITION 4.3. *For $a \in \mathbf{C}^{(1)}$, $b \in \mathbf{C}^{(2)}$, $c \in \mathbf{C}^{(3)}$ we have*

(a) $b \circ (b \circ a) = \frac{1}{2}\langle b, b\rangle a,$
(b) $c \circ (c \circ a) = \frac{1}{2}\langle c, c\rangle a.$

PROOF. (a) For any $y \in \mathbf{C}^{(1)}$ we have $\langle b \circ (b \circ a), y\rangle = \langle b \circ a, b \circ y\rangle = \frac{1}{2}(\langle a \circ b, y \circ b\rangle + \langle y \circ b, a \circ b\rangle) = \frac{1}{2}\langle y \circ (a \circ b) + a \circ (y \circ b), b\rangle = \frac{1}{2}\langle a, y\rangle\langle b, b\rangle = \langle \frac{1}{2}\langle b, b\rangle a, y\rangle$. The nondegeneracy of $\langle\ ,\ \rangle$ on $\mathbf{C}^{(1)}$ then gives (a). The proof for (b) is similar. ∎

For $a \in \mathbf{C}^{(k)}$, $1 \le k \le 3$, with $\langle a, a\rangle \neq 0$ define $r_a : \mathbf{C}^{(k)} \to \mathbf{C}^{(k)}$ to be the reflection with respect to a,

$$r_a(b) = b - \frac{2\langle b, a\rangle}{\langle a, a\rangle} a. \tag{4.13}$$

Of course, r_a preserves $\langle\ ,\ \rangle$ and r_a^2 is the identity operator on $\mathbf{C}^{(k)}$.

Let $\mathbf{e}_1 \in \mathbf{A}$ satisfy $\langle \mathbf{e}_1, \mathbf{e}_1\rangle = 2$, so $\mathbf{e}_1^2 = 1$ in **Cliff**. Then

$$\mathbf{e}_1 \circ (\mathbf{e}_1 \circ b) = \mathbf{e}_1\mathbf{e}_1 b = b \quad \text{for} \quad b \in \mathbf{CM}. \tag{4.14}$$

Define $\rho(\mathbf{e}_1) : \mathbf{C} \to \mathbf{C}$ by

$$\rho(\mathbf{e}_1)b = \mathbf{e}_1 \circ b \quad \text{for} \quad b \in \mathbf{CM},$$
$$\rho(\mathbf{e}_1)a = \mathbf{e}_1 a \mathbf{e}_1 = \langle a, \mathbf{e}_1\rangle \mathbf{e}_1 - a = -r_{\mathbf{e}_1}(a) \quad \text{for} \quad a \in \mathbf{A}. \tag{4.15}$$

Let $\mathbf{e}_2 \in \mathbf{CM}^0$ satisfy $\langle \mathbf{e}_2, \mathbf{e}_2\rangle = 2$ and define $\rho(\mathbf{e}_2) : \mathbf{C} \to \mathbf{C}$ by

$$\rho(\mathbf{e}_2)a = \mathbf{e}_2 \circ a \quad \text{for} \quad a \in \mathbf{A},$$
$$\rho(\mathbf{e}_2)b = -r_{\mathbf{e}_2}(b) = \langle b, \mathbf{e}_2\rangle \mathbf{e}_2 - b \quad \text{for} \quad b \in \mathbf{CM}^0, \tag{4.16}$$
$$\rho(\mathbf{e}_2)c = \mathbf{e}_2 \circ c \quad \text{for} \quad c \in \mathbf{CM}^1.$$

PROPOSITION 4.4. *Let $\mathbf{e}_1 \in \mathbf{A}$ and $\mathbf{e}_2 \in \mathbf{CM}^0$ satisfy $\langle \mathbf{e}_1, \mathbf{e}_1\rangle = \langle \mathbf{e}_2, \mathbf{e}_2\rangle = 2$, and let $\rho(\mathbf{e}_1)$ and $\rho(\mathbf{e}_2)$ be the operators on $\mathbf{C}$ defined by (4.15) and (4.16). Then $\rho(\mathbf{e}_1)^2$ and $\rho(\mathbf{e}_2)^2$ are both the identity operator on $\mathbf{C}$.*

PROOF. It is clear from Proposition 4.2 (b) that $\rho(\mathbf{e}_1)^2 = Id$ on $\mathbf{C}$. From Proposition 4.3 (a), $\rho(\mathbf{e}_2)^2 a = \mathbf{e}_2 \circ (\mathbf{e}_2 \circ a) = a$, which also means that $\rho(\mathbf{e}_2) : \mathbf{A} \to \mathbf{CM}^1$ is an isomorphism. For any $c \in \mathbf{CM}^1$, $c = \mathbf{e}_2 \circ a$ for some $a \in \mathbf{A}$, so $\rho(\mathbf{e}_2)^2 c = \mathbf{e}_2 \circ (\mathbf{e}_2 \circ c) = \mathbf{e}_2 \circ (\mathbf{e}_2 \circ (\mathbf{e}_2 \circ a)) = \mathbf{e}_2 \circ a = c$. ∎

PROPOSITION 4.5. *The operators $\rho(\mathbf{e}_1)$ and $\rho(\mathbf{e}_2)$ on $\mathbf{C}$ preserve the forms $\langle\ ,\ \rangle$ and F.*

PROOF. It is easy to check that $\rho(\mathbf{e}_1)$ preserves $\langle\ ,\ \rangle$ and F, and that $\rho(\mathbf{e}_2)$ preserves $\langle\ ,\ \rangle$. Let $z = a + b + c \in \mathbf{C}$, with $a \in \mathbf{A}$, $b \in \mathbf{CM}^0$, $c \in \mathbf{CM}^1$. Write

$c = y\mathbf{e}_2 = \mathbf{e}_2 \circ y$ for some $y \in \mathbf{A}$, and from Proposition 4.3 (a), $y = \mathbf{e}_2 \circ c$. Then we have

$$\begin{aligned}
F(\rho(\mathbf{e}_2)z) &= F(a\mathbf{e}_2 - r_{\mathbf{e}_2}(b) + \mathbf{e}_2 \circ c) = F(y - r_{\mathbf{e}_2}(b) + a\mathbf{e}_2) \\
&= -\langle yr_{\mathbf{e}_2}(b), a\mathbf{e}_2 \rangle = \langle\langle b, \mathbf{e}_2 \rangle y\mathbf{e}_2 - yb, a\mathbf{e}_2 \rangle = \langle b, \mathbf{e}_2 \rangle\langle y\mathbf{e}_2, a\mathbf{e}_2 \rangle - \langle yb, a\mathbf{e}_2 \rangle \\
&= \langle b, \mathbf{e}_2 \rangle\langle y, a \rangle - \langle b, ya\mathbf{e}_2 \rangle = \langle b, \mathbf{e}_2 \rangle\langle y, a \rangle - \langle b, \langle a, y \rangle\mathbf{e}_2 - ay\mathbf{e}_2 \rangle \\
&= \langle b, ay\mathbf{e}_2 \rangle = \langle ab, c \rangle = F(z).
\end{aligned}$$
∎

COROLLARY 4.6. *The operators $\rho(\mathbf{e}_1)$ and $\rho(\mathbf{e}_2)$ are automorphisms of the algebra $(\mathbf{C}, \circ)$.*

PROOF. Since the product $\circ$ on $\mathbf{C}$ is defined from the forms $\langle\ ,\ \rangle$ and F, any vector space automorphism of $\mathbf{C}$ which preserves these forms is an automorphism of the algebra $(\mathbf{C}, \circ)$. ∎

Let $\mathbf{e}_3 = \mathbf{e}_1\mathbf{e}_2 \in \mathbf{CM}^1$. Then $\langle \mathbf{e}_3, \mathbf{e}_3 \rangle = 2$ and we define $\rho(\mathbf{e}_3)$ on $\mathbf{C}$ by

$$\begin{aligned}
\rho(\mathbf{e}_3)a &= \mathbf{e}_3 \circ a \quad \text{for}\quad a \in \mathbf{A}, \\
\rho(\mathbf{e}_3)b &= \mathbf{e}_3 \circ b \quad \text{for}\quad b \in \mathbf{CM}^0, \\
\rho(\mathbf{e}_3)c &= -r_{\mathbf{e}_3}(c) = \langle c, \mathbf{e}_3 \rangle\mathbf{e}_3 - c \quad \text{for}\quad c \in \mathbf{CM}^1.
\end{aligned} \tag{4.17}$$

PROPOSITION 4.7. *As operators on $\mathbf{C}$ we have*

$$\rho(\mathbf{e}_1)\rho(\mathbf{e}_2)\rho(\mathbf{e}_1) = \rho(\mathbf{e}_2)\rho(\mathbf{e}_1)\rho(\mathbf{e}_2) = \rho(\mathbf{e}_1\mathbf{e}_2).$$

PROOF. It is easy to see that these operators agree when applied to $a \in \mathbf{A}$ or to $b \in \mathbf{CM}^0$. Let $c = a\mathbf{e}_2 = \mathbf{e}_1 b \in \mathbf{CM}^1$ for some $a \in \mathbf{A}$, $b \in \mathbf{CM}^0$. Then

$$\begin{aligned}
\rho(\mathbf{e}_1)\rho(\mathbf{e}_2)\rho(\mathbf{e}_1)c &= \rho(\mathbf{e}_1)\rho(\mathbf{e}_2)b = \rho(\mathbf{e}_1)(\langle b, \mathbf{e}_2 \rangle\mathbf{e}_2 - b) \\
&= \langle b, \mathbf{e}_2 \rangle\mathbf{e}_1\mathbf{e}_2 - \mathbf{e}_1 b = \langle \mathbf{e}_1 b, \mathbf{e}_1\mathbf{e}_2 \rangle\mathbf{e}_1\mathbf{e}_2 - \mathbf{e}_1 b = -r_{\mathbf{e}_1\mathbf{e}_2}(c) = \rho(\mathbf{e}_1\mathbf{e}_2)c
\end{aligned}$$

and

$$\begin{aligned}
\rho(\mathbf{e}_2)\rho(\mathbf{e}_1)\rho(\mathbf{e}_2)c &= \rho(\mathbf{e}_2)\rho(\mathbf{e}_1)a = \rho(\mathbf{e}_2)(\langle a, \mathbf{e}_1 \rangle\mathbf{e}_1 - a) \\
&= \langle a, \mathbf{e}_1 \rangle\mathbf{e}_1\mathbf{e}_2 - a\mathbf{e}_2 = \langle a\mathbf{e}_2, \mathbf{e}_1\mathbf{e}_2 \rangle\mathbf{e}_1\mathbf{e}_2 - a\mathbf{e}_2 = -r_{\mathbf{e}_1\mathbf{e}_2}(c) = \rho(\mathbf{e}_1\mathbf{e}_2)c.
\end{aligned}$$
∎

Defining

$$\sigma = \rho(\mathbf{e}_1)\rho(\mathbf{e}_2), \quad \tau = \rho(\mathbf{e}_1), \tag{4.18}$$

we have

$$\sigma^3 = 1 = \tau^2, \ \tau\sigma\tau = \sigma^{-1} \tag{4.19}$$

and

$$\begin{array}{cccc}
& \sigma & & \acute{\tau} \\
\mathbf{C}^{(1)} & \longrightarrow & \mathbf{C}^{(2)} \qquad \mathbf{C}^{(1)} & \longrightarrow \quad \mathbf{C}^{(1)} \\
\sigma\nwarrow & & \nearrow\sigma \qquad \tau & \\
& \mathbf{C}^{(3)} & \qquad \mathbf{C}^{(2)} & \leftrightarrow \quad \mathbf{C}^{(3)}
\end{array} \tag{4.20}$$

The "triality" group of automorphisms of $(\mathbf{C}, \circ)$ generated by σ and τ is denoted by G and it is isomorphic to the symmetric group S_3. Note that

$$\sigma(\mathbf{e}_1) = \mathbf{e}_2, \quad \sigma(\mathbf{e}_2) = \mathbf{e}_3, \quad \sigma(\mathbf{e}_3) = \mathbf{e}_1,$$
$$\tau(\mathbf{e}_1) = \mathbf{e}_1, \quad \tau(\mathbf{e}_2) = \mathbf{e}_3, \quad \tau(\mathbf{e}_3) = \mathbf{e}_2. \tag{4.21}$$

PROPOSITION 4.8. *For $1 \le k \le 3$, let $a \in \mathbf{C}^{(k)}$ and let $b \in \mathbf{C}^{(k')} \oplus \mathbf{C}^{(k'')}$. Then*

$$a \circ (a \circ b) = \tfrac{1}{2} \langle a, a \rangle b.$$

PROOF. Apply σ and σ^2 to the formulas in Proposition 4.3. ∎

For $1 \le k \le 3$, let $\mathbf{Cliff}^{(k)}$ be the Clifford algebra generated by $\mathbf{C}^{(k)}$ and $\langle\,,\,\rangle$. Let $\mathbf{C}^{(k)} = \mathbf{C}^{(k)+} \oplus \mathbf{C}^{(k)-}$ be the polarization into maximally isotropic subspaces transported from $\mathbf{C}^{(1)} = \mathbf{A}$ by σ,

$$\mathbf{C}^{(k)+} = \sigma^{k-1}(\mathbf{A}^+), \quad \mathbf{C}^{(k)-} = \sigma^{k-1}(\mathbf{A}^-). \tag{4.22}$$

Since σ is orthogonal with respect to $\langle\,,\,\rangle$ σ induces Clifford algebra isomorphisms

$$\sigma^{(k)} : \mathbf{Cliff}^{(k)} \to \mathbf{Cliff}^{(k')}, \quad 1 \le k \le 3. \tag{4.23}$$

Let

$$\mathbf{vac}^{(1)} = \mathbf{vac} \in \wedge^4 \mathbf{C}^{(1)+}, \tag{4.24}$$

$$\mathbf{vac}^{(2)} = \sigma^{(1)}(\mathbf{vac}^{(1)}) \in \wedge^4 \mathbf{C}^{(2)+}, \tag{4.25}$$

$$\mathbf{vac}^{(3)} = \sigma^{(2)}(\mathbf{vac}^{(2)}) \in \wedge^4 \mathbf{C}^{(3)+}, \tag{4.26}$$

and define $\mathbf{Cliff}^{(k)}$-module isomorphisms

$$\Psi_k : (\wedge \mathbf{C}^{(k)-}) \cdot \mathbf{vac}^{(k)} \to \mathbf{C}^{(k')} \oplus \mathbf{C}^{(k'')}, \quad 1 \le k \le 3, \tag{4.27}$$

by setting

$$\Psi_k(\mathbf{vac}^{(k)}) = \sigma^{k-1}(\mathbf{vac}) \in \mathbf{C}^{(k')},$$
$$\Psi_k(ab) = a \circ \Psi_k(b) \quad \text{for} \quad a \in \mathbf{C}^{(k)}, \ b \in (\wedge \mathbf{C}^{(k)-}) \cdot \mathbf{vac}^{(k)}. \tag{4.28}$$

Dropping the Ψ_k notation, we identify

$$\mathbf{CM}^{(k)} = \mathbf{C}^{(k')} \oplus \mathbf{C}^{(k'')} \tag{4.29}$$

as the $\mathbf{Cliff}^{(k)}$-module with vacuum vector

$$\mathbf{vac}^{(k)} = \sigma^{k-1}(\mathbf{vac}) \in \mathbf{C}^{(k')}. \tag{4.30}$$

From (4.28) the even and odd parity subspaces of $\mathbf{CM}^{(k)}$ are, respectively,

$$\mathbf{CM}^{(k)0} = \mathbf{C}^{(k')} \quad \text{and} \quad \mathbf{CM}^{(k)1} = \mathbf{C}^{(k'')}. \tag{4.31}$$

DEFINITION. *Let $\{c_1, \ldots, c_4\}$ be a basis of $\mathbf{A}^+$ such that*

$$\langle c_r, \vartheta c_s \rangle = \delta_{rs}, \quad 1 \le r, s \le 4. \tag{4.32}$$

Then we say $\{c_1, \ldots, c_4, \vartheta c_1, \ldots, \vartheta c_4\}$ is a canonical basis of $\mathbf{A}$ adapted to ϑ.

From (4.32) and the anticommutation relations in **Cliff**, we see that

$$c_1 c_2 c_3 c_4 \vartheta c_1 \vartheta c_2 \vartheta c_3 \vartheta c_4 c_1 c_2 c_3 c_4 = c_1 c_2 c_3 c_4. \tag{4.33}$$

Let $0 \ne \vartheta(\mathbf{vac}) \in (\wedge^4 \mathbf{A}^-) \cdot \mathbf{vac}$. Then there are scalars $\xi, \lambda \in \mathbb{C}^*$ such that

$$\mathbf{vac} = \xi c_1 c_2 c_3 c_4, \quad \vartheta(\mathbf{vac}) = \lambda \vartheta c_1 \vartheta c_2 \vartheta c_3 \vartheta c_4 \mathbf{vac}. \tag{4.34}$$

Note that $\mathbf{A}^- \cdot \vartheta(\mathbf{vac}) = \{0\}$. Then a $\mathbb{C}$-antilinear transformation $\vartheta : \mathbf{CM} \to \mathbf{CM}$ is determined by

$$\vartheta(ab) = \vartheta(a)\vartheta(b), \quad a \in \mathbf{A}, \quad b \in \mathbf{CM}. \tag{4.35}$$

From (4.33) and (4.34) we get

$$\langle \mathbf{vac}, \vartheta(\mathbf{vac}) \rangle = \xi \lambda, \quad \vartheta(\vartheta(\mathbf{vac})) = |\lambda|^2 \mathbf{vac}. \tag{4.36}$$

For arbitrary choice of $\mathbf{vac}$, $\vartheta(\mathbf{vac})$ can be chosen so that either $\langle \mathbf{vac}, \vartheta(\mathbf{vac}) \rangle = 1$ or $\vartheta(\vartheta(\mathbf{vac})) = \mathbf{vac}$, but not both unless $|\xi| = 1$. Since we do want both conditions, $\mathbf{vac}$ and $\vartheta(\mathbf{vac})$ in (4.34) must be chosen with $|\xi| = 1$ and $\lambda = \xi^{-1}$.

PROPOSITION 4.9. *Let $\mathbf{vac}$ and $\vartheta(\mathbf{vac})$ have the form in (4.34) with $|\xi| = 1$ and $\lambda = \xi^{-1}$, where $\{c_1, \ldots, c_4, \vartheta c_1, \ldots, \vartheta c_4\}$ is a canonical basis of $\mathbf{A}$ adapted to ϑ. Let $\vartheta : \mathbf{CM} \to \mathbf{CM}$ be the $\mathbb{C}$-antilinear transformation determined by (4.35). Then for $b, c \in \mathbf{CM}$ we have*

   (a) *ϑ is an involution on all of $\mathbf{C}$ preserving each $\mathbf{C}^{(k)}$, $1 \le k \le 3$,*
   (b) *$\langle \vartheta(b), \vartheta(c) \rangle = \overline{\langle b, c \rangle}$,*
   (c) *$0 < \langle b, \vartheta(b) \rangle \in \mathbb{R}$ if $b \ne 0$.*

PROOF. (a) We are given that ϑ is an involution on $\mathbf{A}$. From (4.35) and $\vartheta(\vartheta(\mathbf{vac})) = \mathbf{vac}$ we see that ϑ is an involution on $\mathbf{CM}$ preserving $\mathbf{C}^{(1)}$ and $\mathbf{C}^{(2)}$.

(b) Let $b = a_1 \ldots a_k \mathbf{vac} \in \mathbf{CM}$ with $a_1, \ldots, a_k \in \mathbf{A}^-$. Then $a_k \ldots a_1 c$ equals $t\vartheta(\mathbf{vac})$ plus a sum of terms in $(\wedge^r \mathbf{A}^-) \cdot \mathbf{vac}$ for $0 \le r \le 3$, where the scalar $t = \langle a_k \ldots a_1 c, \mathbf{vac} \rangle$. So $\vartheta(a_k \ldots a_1 c)$ equals $\bar{t}\mathbf{vac}$ plus a sum of terms in $(\wedge^s \mathbf{A}^-) \cdot \mathbf{vac}$ for $1 \le s \le 4$. Then

$$\langle \vartheta(b), \vartheta(c) \rangle = \langle \vartheta(a_1) \ldots \vartheta(a_k)\vartheta(\mathbf{vac}), \vartheta(c) \rangle = \langle \vartheta(\mathbf{vac}), \vartheta(a_k \ldots a_1 c) \rangle$$

$$= \langle \vartheta(\mathbf{vac}), \bar{t}\mathbf{vac} \rangle = \bar{t} = \overline{\langle \mathbf{vac}, a_k \ldots a_1 c \rangle} = \overline{\langle b, c \rangle}.$$

(c) For $a_1, \ldots, a_r, x_1, \ldots, x_s \in \mathbf{A}^-$, one can show that

$$\langle a_1 \ldots a_r \mathbf{vac}, \; x_1 \ldots x_s \mathbf{vac} \rangle = \delta_{rs} \det(\langle a_i, \vartheta x_j \rangle).$$

Along with (4.2), this gives (c). $\blacksquare$

PROPOSITION 4.10. ϑ *is a* $\mathbb{C}$-*antilinear automorphism of the algebra* $(\mathbf{C}, \circ)$.

PROOF. From (4.35), $\vartheta(a \circ b) = \vartheta(a) \circ \vartheta(b)$ for $a \in \mathbf{A}$, $b \in \mathbf{CM}$. It only remains to check that $\vartheta(b \circ c) = \vartheta(b) \circ \vartheta(c)$ for $b \in \mathbf{CM}^0$, $c \in \mathbf{CM}^1$. For any $a \in \mathbf{A}$, we have $\langle \vartheta(a), \vartheta(b) \circ \vartheta(c) \rangle = \langle \vartheta(ab), \vartheta(c) \rangle = \overline{\langle ab, c \rangle} = \overline{\langle a, b \circ c \rangle} = \langle \vartheta(a), \vartheta(b \circ c) \rangle$. ∎

PROPOSITION 4.11. (a) ϑ *commutes with* $\tau = \rho(\mathbf{e}_1)$ *if and only if* $\vartheta(\mathbf{e}_1) = \mathbf{e}_1$.
(b) ϑ *commutes with* $\tau\sigma = \rho(\mathbf{e}_2)$ *if and only if* $\vartheta(\mathbf{e}_2) = \mathbf{e}_2$.

PROOF. (a) For $a \in \mathbf{A}$, $\vartheta\rho(\mathbf{e}_1)a = \vartheta(\langle a, \mathbf{e}_1 \rangle \mathbf{e}_1 - a) = \overline{\langle a, \mathbf{e}_1 \rangle}\vartheta\mathbf{e}_1 - \vartheta a = \langle \vartheta a, \vartheta \mathbf{e}_1 \rangle \vartheta \mathbf{e}_1 - \vartheta a = \rho(\vartheta \mathbf{e}_1)\vartheta a$, while for $b \in \mathbf{CM}$, $\vartheta\rho(\mathbf{e}_1)b = \vartheta(\mathbf{e}_1 \circ b) = \vartheta(\mathbf{e}_1) \circ \vartheta(b) = \rho(\vartheta(\mathbf{e}_1))\vartheta b$. The proof for (b) is similar. ∎

Let $\mathbf{e}_1 \in \mathbf{A}$ and $\mathbf{e}_2 \in \mathbf{CM}^0$ be chosen to satisfy $\langle \mathbf{e}_1, \mathbf{e}_1 \rangle = 2 = \langle \mathbf{e}_2, \mathbf{e}_2 \rangle$, $\vartheta(\mathbf{e}_1) = \mathbf{e}_1$ and $\vartheta(\mathbf{e}_2) = \mathbf{e}_2$. Then from Proposition 4.11 and (4.18) we get that ϑ commutes with the group G.

For $1 \leq k \leq 3$, let $\mathbf{g}^{(k)}$ be the subspace of $\mathbf{Cliff}^{(k)}$ spanned by

$$\{x = {}_\circ^\circ ab_\circ^\circ = \tfrac{1}{2}(ab - ba) \mid a, b \in \mathbf{C}^{(k)}\}. \tag{4.37}$$

Then $\mathbf{g}^{(k)}$ acts on $\mathbf{C}$ by

$$\begin{aligned}
x \cdot c &= \langle b, c \rangle a - \langle a, c \rangle b, \quad c \in \mathbf{C}^{(k)}, \\
x \cdot u &= \tfrac{1}{2}(a \circ (b \circ u) - b \circ (a \circ u)), \quad u \in \mathbf{C}^{(k')} \oplus \mathbf{C}^{(k'')} = \mathbf{CM}^{(k)}.
\end{aligned} \tag{4.38}$$

This is just the type D_4 spinor construction given in Chapter 2 applied to $\mathbf{Cliff}^{(k)}$, so $\mathbf{g}^{(k)}$ is a Lie algebra isomorphic to $\mathbf{o}(8)$ and $\mathbf{C} = \mathbf{C}^{(k)} \oplus \mathbf{C}^{(k')} \oplus \mathbf{C}^{(k'')}$ is the decomposition of $\mathbf{C}$ into irreducible $\mathbf{g}^{(k)}$-modules. Under the action

$$g \cdot {}_\circ^\circ ab_\circ^\circ = {}_\circ^\circ (g \cdot a)(g \cdot b)_\circ^\circ, \quad g \in G, \tag{4.39}$$

the triality group G permutes the three algebras $\mathbf{g}^{(k)}$, $1 \leq k \leq 3$, as it permutes the modules $\mathbf{C}^{(k)}$, the Clifford algebras $\mathbf{Cliff}^{(k)}$ and the Clifford modules $\mathbf{CM}^{(k)}$.

PROPOSITION 4.12. *For* $g \in G$, $x \in \mathbf{g}^{(k)}$, $u \in \mathbf{C}$, *we have*

$$g \cdot (x \cdot u) = (g \cdot x) \cdot (g \cdot u).$$

PROOF. This follows from (4.38) since G acts as automorphisms of $(\mathbf{C}, \circ)$ and as isomorphisms among the Clifford algebras. ∎

PROPOSITION 4.13. *For* $x \in \mathbf{g}^{(k)}$, $u, v \in \mathbf{C}$, *we have*

$$x \cdot (u \circ v) = (x \cdot u) \circ v + u \circ (x \cdot v).$$

PROOF. Using the action of G, we may assume $k = 1$. Since $\mathbf{g}^{(1)}$ preserves each $\mathbf{C}^{(r)}$, $1 \leq r \leq 3$, and $\mathbf{C}^{(r)} \circ \mathbf{C}^{(r)} = \{0\}$, both sides of the equation are zero if $u, v \in \mathbf{C}^{(r)}$. Let $x = {}_\circ^\circ ab_\circ^\circ$ for $a, b \in \mathbf{C}^{(1)}$.
Case 1. Suppose that $u \in \mathbf{C}^{(1)}$ and $v \in \mathbf{C}^{(2)} \oplus \mathbf{C}^{(3)}$. From (4.38) we have

$$x \cdot u = \langle b, u \rangle a - \langle a, u \rangle b \tag{4.40}$$

and

$$(x{\cdot}u)\circ v + u\circ(x{\cdot}v) = \langle b, u\rangle a\circ v - \langle a, u\rangle b\circ v + u\circ\tfrac{1}{2}(a\circ(b\circ v)) - u\circ\tfrac{1}{2}(b\circ(a\circ v)). \quad (4.41)$$

Since $a \circ v,\, b \circ v \in \mathbf{CM}^{(1)}$, we have

$$\langle b, u\rangle a\circ v - \langle a, u\rangle b\circ v = b\circ(u\circ(a\circ v)) + u\circ(b\circ(a\circ v)) - a\circ(u\circ(b\circ v)) - u\circ(a\circ(b\circ v)).$$

Combining this with (4.41) we get

$$
\begin{aligned}
(x \cdot u) &\circ v + u \circ (x \cdot v)\\
&= \tfrac{1}{2}\langle b, u\rangle a \circ v - \tfrac{1}{2}\langle a, u\rangle b \circ v + \tfrac{1}{2}b \circ (u \circ (a \circ v)) - \tfrac{1}{2}a \circ (u \circ (b \circ v))\\
&= \tfrac{1}{2}a \circ (\langle b, u\rangle v - u \circ (b \circ v)) + \tfrac{1}{2}b \circ (u \circ (a \circ v) - \langle a, u\rangle v)\\
&= \tfrac{1}{2}a \circ (b \circ (u \circ v)) - \tfrac{1}{2}b \circ (a \circ (u \circ v))\\
&= {}_\circ^\circ a b_\circ^\circ \cdot (u \circ v) = x \cdot (u \circ v).
\end{aligned}
$$

Case 2. Suppose that $u \in \mathbf{C}^{(2)}$ and $v \in \mathbf{C}^{(3)}$. Then we have

$$
\begin{aligned}
x \cdot (u \circ v) &= \langle b, u \circ v\rangle a - \langle a, u \circ v\rangle b\\
&= \langle b \circ u, v\rangle a - \langle v \circ a, u\rangle b\\
&= (b \circ u) \circ (v \circ a) + v \circ ((b \circ u) \circ a) - u \circ ((v \circ a) \circ b) - (v \circ a) \circ (u \circ b)\\
&= v \circ ((b \circ u) \circ a) - u \circ ((v \circ a) \circ b)\\
&= (a \circ (b \circ u)) \circ v - \tfrac{1}{2}\langle a, b\rangle u \circ v - u \circ (b \circ (a \circ v)) + \tfrac{1}{2}\langle a, b\rangle u \circ v\\
&= (x \cdot u) \circ v + u \circ (x \cdot v). \qquad\blacksquare
\end{aligned}
$$

PROPOSITION 4.14. *For $a_k \in \mathbf{C}^{(k)}$, $1 \le k \le 3$, as operators on $\mathbf{C}$ we have*

$$
{}_\circ^\circ a_1(a_2 \circ a_3)_\circ^\circ + {}_\circ^\circ a_2(a_3 \circ a_1)_\circ^\circ + {}_\circ^\circ a_3(a_1 \circ a_2)_\circ^\circ = 0.
$$

PROOF. For any $b_1 \in \mathbf{C}^{(1)}$ we have

$$
\begin{aligned}
{}_\circ^\circ a_1&(a_2 \circ a_3)_\circ^\circ \cdot b_1 + {}_\circ^\circ a_2(a_3 \circ a_1)_\circ^\circ \cdot b_1 + {}_\circ^\circ a_3(a_1 \circ a_2)_\circ^\circ \cdot b_1\\
&= \langle a_2 \circ a_3, b_1\rangle a_1 - \langle a_1, b_1\rangle a_2 \circ a_3 + a_2 \circ ((a_3 \circ a_1) \circ b_1) - \tfrac{1}{2}\langle a_2, a_3 \circ a_1\rangle b_1\\
&\quad + a_3 \circ ((a_1 \circ a_2) \circ b_1) - \tfrac{1}{2}\langle a_3, a_1 \circ a_2\rangle b_1\\
&= {}_\circ^\circ a_1 b_1{}_\circ^\circ \cdot (a_2 \circ a_3) + a_2 \circ (b_1 \circ (a_1 \circ a_3)) + (b_1 \circ (a_1 \circ a_2)) \circ a_3 - \langle a_1, b_1\rangle a_2 \circ a_3\\
&= {}_\circ^\circ a_1 b_1{}_\circ^\circ \cdot (a_2 \circ a_3) - a_2 \circ ({}_\circ^\circ a_1 b_1{}_\circ^\circ \cdot a_3) - ({}_\circ^\circ a_1 b_1{}_\circ^\circ \cdot a_2) \circ a_3 = 0
\end{aligned}
$$

by Proposition 4.13. Similar calculations give the same result on $\mathbf{C}^{(2)}$ and $\mathbf{C}^{(3)}$. $\blacksquare$

PROPOSITION 4.15. *As Lie algebras of operators on $\mathbf{C}$, $\mathbf{g}^{(1)} = \mathbf{g}^{(2)} = \mathbf{g}^{(3)}$.*

PROOF. Let $0 \ne a \in \mathbf{C}^{(1)}$ with $\langle a, a\rangle = 0$. Let $L = L_a$ denote left multiplication by a on $\mathbf{CM} = \mathbf{CM}^{(1)} = \mathbf{C}^{(2)} \oplus \mathbf{C}^{(3)}$, that is, $L(u) = au = a \circ u$ for $u \in \mathbf{CM}$. Note that L switches $\mathbf{C}^{(2)}$ and $\mathbf{C}^{(3)}$, and $L^2 = 0$ since $\langle a, a\rangle = 0$, so $\mathrm{Image}(L) \subseteq \mathrm{Ker}(L)$. $\mathrm{Ker}(L)$ is an isotropic subspace of $\mathbf{CM}$ because if $au = 0$

then $0 = (a \circ u) \circ u = \frac{1}{2}\langle u, u \rangle a$, so $\langle u, u \rangle = 0$. In fact, $\mathrm{Ker}(L)$ is a maximal isotropic subspace of $\mathbf{CM}$ because

$$\dim(\mathrm{Image}(L)) \leq \dim(\mathrm{Ker}(L)) \leq \tfrac{1}{2}\dim(\mathbf{CM}).$$

Since

$$\dim(\mathrm{Image}(L)) + \dim(\mathrm{Ker}(L)) = \dim(\mathbf{CM}),$$

we get

$$\dim(\mathrm{Ker}(L)) = \tfrac{1}{2}\dim(\mathbf{CM}) = 8.$$

Since L switches $\mathbf{C}^{(2)}$ and $\mathbf{C}^{(3)}$, we have

$$\mathrm{Ker}(L) = (\mathrm{Ker}(L) \cap \mathbf{C}^{(2)}) \oplus (\mathrm{Ker}(L) \cap \mathbf{C}^{(3)}),$$

and $\mathrm{Ker}(L) \cap \mathbf{C}^{(k)}$ is a maximal isotropic subspace of $\mathbf{C}^{(k)}$ for $k = 2, 3$.

We can see as follows that $\mathrm{Ker}(L) \cap \mathbf{C}^{(3)}$ determines $a \in \mathbf{C}^{(1)}$ up to a scalar multiple. Recall that $\mathbf{CM}^{(3)} = \mathbf{C}^{(1)} \oplus \mathbf{C}^{(2)}$ is a $\mathbf{Cliff}^{(3)}$-module. If P is a maximal isotropic subspace of $\mathbf{C}^{(3)}$ then there is a unique one-dimensional subspace $K \subset \mathbf{CM}^{(3)}$ such that $P \circ K = \{0\}$. This is true because P can be taken as the positive half of a polarization of $\mathbf{C}^{(3)}$, and $K = \wedge^4 P$ is the one-dimensional subspace of $\mathbf{CM}^{(3)}$ containing all vacuum vectors for that polarization. Therefore, with $P = \mathrm{Ker}(L) \cap \mathbf{C}^{(3)}$, $a \in K$ so $K = \mathbb{C}a$. (Remark: Chevalley proves that either $K \subset \mathbf{CM}^{(3)0}$ or $K \subset \mathbf{CM}^{(3)1}$, that is, a vacuum vector for a new polarization is either even or odd with respect to the old polarization.) A similar argument shows that $\mathrm{Ker}(L) \cap \mathbf{C}^{(2)}$ also determines $a \in \mathbf{C}^{(1)}$ up to a scalar multiple.

Now suppose that $a, b \in \mathbf{C}^{(1)}$ with $a \neq 0$, $b \neq 0$, $\langle a, a \rangle = \langle b, b \rangle = \langle a, b \rangle = 0$ and a is not a multiple of b. Then

$$L_a L_b = -L_b L_a$$

on $\mathbf{CM}^{(1)}$. It is easy to see that

$$\mathrm{Ker}(L_a L_b) = \mathrm{Ker}(L_a) + \mathrm{Ker}(L_b),$$

not a direct sum, and that

$$\dim(\mathrm{Ker}(L_a) \cap \mathrm{Ker}(L_b)) = \tfrac{1}{4}\dim(\mathbf{CM}) = 4.$$

Just take a polarization $\mathbf{A} = \mathbf{A}^+ \oplus \mathbf{A}^-$ such that $a, b \in \mathbf{A}^-$, and use $\mathbf{CM} \cong (\wedge \mathbf{A}^-) \cdot \mathbf{vac}$ for any vector $0 \neq \mathbf{vac} \in \wedge \mathbf{A}^+$. Then the claim is clear in the exterior algebra $\wedge \mathbf{A}^-$.

From above we have that

$$P_a = \mathrm{Ker}(L_a) \cap \mathbf{C}^{(3)} \quad \text{and} \quad P_b = \mathrm{Ker}(L_b) \cap \mathbf{C}^{(3)}$$

are distinct maximal isotropic subspaces of $\mathbf{C}^{(3)}$. Therefore,

$$\dim((\mathrm{Ker}(L_a) + \mathrm{Ker}(L_b)) \cap \mathbf{C}^{(3)}) \geq 5$$

(and, in fact, it equals 6), so $(\mathrm{Ker}(L_a) + \mathrm{Ker}(L_b)) \cap \mathbf{C}^{(3)}$ is not isotropic. Let

$$v \in (\mathrm{Ker}(L_a) + \mathrm{Ker}(L_b)) \cap \mathbf{C}^{(3)}$$

be a vector such that $\langle v, v \rangle = 2$, so $(b \circ v) \circ v = b$. It also follows that

$$L_a L_b(v) = a \circ (b \circ v) = 0.$$

In Proposition 4.14, if we take $a_1 = a$, $a_2 = b \circ v$ and $a_3 = v$ then we get

$$_\circ^\circ a((b \circ v) \circ v)_\circ^\circ + {}_\circ^\circ(b \circ v)(v \circ a)_\circ^\circ + {}_\circ^\circ v(a \circ (b \circ v))_3^\circ = 0,$$

which reduces to

$$_\circ^\circ a b_\circ^\circ = {}_\circ^\circ(v \circ a)(b \circ v)_\circ^\circ.$$

Since $_\circ^\circ(v \circ a)(b \circ v)_\circ^\circ \in \mathbf{g}^{(2)}$ and $\mathbf{g}^{(1)}$ is generated by brackets of such $_\circ^\circ a b_\circ^\circ$, we get $\mathbf{g}^{(1)} \subseteq \mathbf{g}^{(2)}$, and so they are equal. The same arguments apply to $\mathbf{C}^{(2)}$ and $\mathbf{C}^{(3)}$, so we get $\mathbf{g}^{(1)} = \mathbf{g}^{(2)} = \mathbf{g}^{(3)}$. ∎

We can now drop the $\mathbf{g}^{(k)}$ notation and just refer to one $\mathbf{o}(8)$ acting on $\mathbf{C}$, constructible in three ways. We can conveniently restate some of the above results as follows.

PROPOSITION 4.16. *For $1 \le k \le 3$, let $a_k \in \mathbf{C}^{(k)}$. We have*

(a) $\mathbf{e}_k \circ \mathbf{e}_{k'} = \mathbf{e}_{k''}$, $\vartheta \mathbf{e}_k = \mathbf{e}_k$,

(b) $\tau \sigma^k(a_k) = a_k \circ \mathbf{e}_{k'}$,

(c) $\tau \sigma^k(a_{k'}) = -r_{\mathbf{e}_{k'}}(a_{k'})$,

(d) $\tau \sigma^k(a_{k''}) = \mathbf{e}_{k'} \circ a_{k''}$,

(e) $\sigma(a_k) = (a_k \circ \mathbf{e}_{k'}) \circ \mathbf{e}_k = -r_{\mathbf{e}_{k'}}(\mathbf{e}_{k''} \circ a_k) = -\mathbf{e}_{k''} \circ r_{\mathbf{e}_k}(a_k)$,

(f) $\sigma^2(a_k) = \mathbf{e}_k \circ (\mathbf{e}_{k''} \circ a_k) = -r_{\mathbf{e}_{k''}}(a_k \circ \mathbf{e}_{k'}) = -r_{\mathbf{e}_k}(a_k) \circ \mathbf{e}_{k'}$,

(g) ϑ *commutes with σ and τ on* $\mathbf{C}$,

(h) *The triality group G acts as Lie algebra automorphisms of* $\mathbf{o}(8)$,

(i) *For $g \in G$, $x \in \mathbf{o}(8)$, $u \in \mathbf{C}$, we have $g \cdot (x \cdot u) = (g \cdot x) \cdot (g \cdot u)$.*

PROOF. (a) We have chosen $\mathbf{e}_3 = \mathbf{e}_1 \circ \mathbf{e}_2$, so applying the automorphism σ we get $\mathbf{e}_k \circ \mathbf{e}_{k'} = \mathbf{e}_{k''}$. We have chosen $\mathbf{e}_1$ and $\mathbf{e}_2$ such that ϑ fixes them, so $\vartheta \mathbf{e}_3 = \vartheta \mathbf{e}_1 \circ \vartheta \mathbf{e}_2 = \mathbf{e}_3$. Parts (b)-(f) are restatements of (4.15)-(4.17). Although (g) results from Proposition 4.11 and our choice of $\mathbf{e}_1$ and $\mathbf{e}_2$, note that it also follows from (a) - (f) and Proposition 4.10. (h) follows from (4.39), Proposition 4.5 and the bracket formula (2.66). (i) is the statement of Proposition 4.12. ∎

We will now show that, with a new restriction on $\mathbf{e}_2$, we can choose a canonical basis of $\mathbf{C}^{(1)}$ adapted to ϑ and a Cartan subalgebra $\mathbf{h}$ in $\mathbf{o}(8)$ such that the action of G permutes the weight spaces of $\mathbf{C}$ with respect to $\mathbf{h}$. If $\{c_1, \ldots, c_4, \vartheta c_1, \ldots, \vartheta c_4\}$ is a canonical basis of $\mathbf{C}^{(1)}$ adapted to ϑ, define

$$h_i = {}_\circ^\circ c_i \vartheta c_i{}_\circ^\circ \in \mathbf{o}(8), \quad 1 \le i \le 4. \tag{4.42}$$

Since

$$\langle c_r, \vartheta c_s \rangle = \delta_{rs}, \quad \langle c_r, c_s \rangle = \langle \vartheta c_r, \vartheta c_s \rangle = 0, \tag{4.43}$$

the brackets (2.66) show that $\{h_i \mid 1 \le i \le 4\}$ spans a Cartan subalgebra $\mathbf{h}$, and $\langle h_r, h_s \rangle = \delta_{rs}$ for the invariant form $\langle \ , \ \rangle$ on $\mathbf{o}(8)$ given in (2.67). In the dual space $\mathbf{h}^*$ let $\{\epsilon_i \mid 1 \leq i \leq 4\}$ be the dual basis, so $\epsilon_i(h_r) = \delta_{ir}$. With the standard choice of simple roots

$$\alpha_1 = \epsilon_1 - \epsilon_2, \quad \alpha_2 = \epsilon_2 - \epsilon_3, \quad \alpha_3 = \epsilon_3 - \epsilon_4, \quad \alpha_4 = \epsilon_3 + \epsilon_4, \tag{4.44}$$

the root vectors ${}^{\circ}_{\circ}c_r \vartheta c_s {}^{\circ}_{\circ}$ and ${}^{\circ}_{\circ}c_r c_s {}^{\circ}_{\circ}$, $1 \leq r < s \leq 4$, correspond to the positive roots $\epsilon_r - \epsilon_s$ and $\epsilon_r + \epsilon_s$, respectively. Then under the action (2.65) it is easy to see that c_i is a weight vector with weight ϵ_i, ϑc_i is a weight vector with weight $-\epsilon_i$, and that c_1 is a highest weight vector in $\mathbf{C}^{(1)}$.

Since $h_i = -\vartheta c_i c_i + \frac{1}{2}$ in $\mathbf{Cliff}$, $h_i \mathbf{vac} = \frac{1}{2}\mathbf{vac}$, so the weight of $\mathbf{vac}$ is $\frac{1}{2}(\epsilon_1 + \epsilon_2 + \epsilon_3 + \epsilon_4)$. One may check that $\mathbf{vac}$ is a highest weight vector in $\mathbf{CM}^0 = \mathbf{C}^{(2)}$. One also finds $h_i \vartheta(\mathbf{vac}) = -\frac{1}{2}\vartheta(\mathbf{vac})$, so the weight of $\vartheta(\mathbf{vac})$ is $-\frac{1}{2}(\epsilon_1 + \epsilon_2 + \epsilon_3 + \epsilon_4)$. More generally, a basis for $\mathbf{CM} = (\wedge \mathbf{C}^{(1)-}) \cdot \mathbf{vac}$ consists of vectors of the form

$$\vartheta c_{i_1} \ldots \vartheta c_{i_n} \mathbf{vac} \tag{4.45}$$

for $i_1 < i_2 < \ldots < i_n$, $0 \leq n \leq 4$, and the weight of such a vector is $\frac{1}{2}(\epsilon_1 + \epsilon_2 + \epsilon_3 + \epsilon_4) - \epsilon_{i_1} - \ldots - \epsilon_{i_n}$. In $\mathbf{CM}^1 = \mathbf{C}^{(3)}$ one finds a highest weight vector $\vartheta c_4 \mathbf{vac}$ of weight $\frac{1}{2}(\epsilon_1 + \epsilon_2 + \epsilon_3 - \epsilon_4)$.

Given any $\mathbf{e}_1 \in \mathbf{C}^{(1)}$ satisfying $\langle \mathbf{e}_1, \mathbf{e}_1 \rangle = 2$ and $\vartheta \mathbf{e}_1 = \mathbf{e}_1$, we can uniquely write

$$\mathbf{e}_1 = \mathbf{e}_1^+ + \mathbf{e}_1^- \tag{4.46}$$

where $\mathbf{e}_1^{\pm} \in \mathbf{C}^{(1)\pm}$, $\vartheta \mathbf{e}_1^+ = \mathbf{e}_1^-$ and

$$\langle \mathbf{e}_1^+, \mathbf{e}_1^- \rangle = (\mathbf{e}_1^+, \mathbf{e}_1^+) = (\mathbf{e}_1^-, \mathbf{e}_1^-) = 1. \tag{4.47}$$

We choose

$$c_4 = \mathbf{e}_1^+, \tag{4.48}$$

so that

$$\mathbf{e}_1 = c_4 + \vartheta c_4. \tag{4.49}$$

We wish to know when $\sigma = \rho(\mathbf{e}_1)\rho(\mathbf{e}_2)$ satisfies

$$\sigma^2(\mathbf{vac}) \in \mathbf{C}^{(1)+}. \tag{4.50}$$

This means that $\sigma^2(\mathbf{vac}) \circ \mathbf{vac} = 0$, which is equivalent to

$$\sigma(\mathbf{vac}) \circ \mathbf{vac} = 0. \tag{4.51}$$

From (4.15)-(4.16) and Proposition 4.8, we have

$$\sigma(\mathbf{vac}) \circ \mathbf{vac} = (\mathbf{e}_1 \circ (\langle \mathbf{e}_2, \mathbf{vac}\rangle \mathbf{e}_2 - \mathbf{vac})) \circ \mathbf{vac}$$
$$= \langle \mathbf{e}_2, \mathbf{vac}\rangle \mathbf{e}_3 \circ \mathbf{vac}. \tag{4.52}$$

If $\mathbf{e}_3 \circ \mathbf{vac} = 0$ then $0 = \mathbf{e}_3 \circ (\mathbf{e}_3 \circ \mathbf{vac}) = \frac{1}{2}\langle \mathbf{e}_3, \mathbf{e}_3\rangle \mathbf{vac} = \mathbf{vac}$ contradicts our choice of $\mathbf{vac}$, so (4.50) is equivalent to the condition

$$\langle \mathbf{e}_2, \mathbf{vac}\rangle = 0. \tag{4.53}$$

Under that condition, we get

$$\sigma(\mathbf{vac}) = -(\mathbf{e}_1 \circ \mathbf{vac}) = -\vartheta c_4 \mathbf{vac}. \tag{4.54}$$

For σ determined by such an $\mathbf{e}_2$ we claim that

$$\langle \vartheta c_4, \sigma^2(\mathbf{vac}) \rangle = 0. \tag{4.55}$$

Since $c_4 \circ (c_4 \circ \mathbf{e}_2) = 0$ and $\vartheta c_4 \circ \vartheta(\mathbf{vac}) = 0$, we have

$$\begin{aligned}
\overline{\langle \vartheta c_4, \sigma^2(\mathbf{vac}) \rangle} &= \langle \sigma c_4, \vartheta(\mathbf{vac}) \rangle = \langle \mathbf{e}_1 \circ (\mathbf{e}_2 \circ c_4), \vartheta(\mathbf{vac}) \rangle \\
&= \langle (c_4 + \vartheta c_4) \circ (c_4 \circ \mathbf{e}_2), \vartheta(\mathbf{vac}) \rangle = \langle \vartheta c_4 \circ (c_4 \circ \mathbf{e}_2), \vartheta(\mathbf{vac}) \rangle \\
&= \langle (c_4 \circ \mathbf{e}_2), \vartheta c_4 \circ \vartheta(\mathbf{vac}) \rangle = 0.
\end{aligned} \tag{4.56}$$

From (4.50) and (4.55), for any $d \in \mathbb{C}$ with $|d| = 1$, we may choose

$$c_1 = d\sigma^2(\mathbf{vac}) \in \mathbf{C}^{(1)+} \tag{4.57}$$

so that

$$\sigma(c_1) = d\mathbf{vac}. \tag{4.58}$$

Let $c_2, c_3 \in \mathbf{C}^{(1)+}$ be any vectors such that $\{c_1, \ldots, c_4, \vartheta c_1, \ldots, \vartheta c_4\}$ is a canonical basis of $\mathbf{C}^{(1)}$ adapted to ϑ. The most general expression for $\mathbf{e}_2 \in \mathbf{C}^{(2)} = (\wedge^{even}\mathbf{C}^{(1)-}) \cdot \mathbf{vac}$ satisfying $\vartheta \mathbf{e}_2 = \mathbf{e}_2$ is

$$\mathbf{e}_2 = d_0\mathbf{vac} + \sum_{1 \leq i < j \leq 4} d_{ij}\vartheta c_i \vartheta c_j \mathbf{vac} + d_0\vartheta(\mathbf{vac}) \tag{4.59}$$

with the following constraints on the coefficients $d_{ij} \in \mathbb{C}$. If $1 \leq i < j \leq 4$, let $1 \leq i' < j' \leq 4$ with $\{i, j, i', j'\} = \{1, 2, 3, 4\}$. Then the constraint is that

$$d_{i'j'} = (-1)^{i+j}\xi^{-1}\bar{d}_{ij}. \tag{4.60}$$

The condition (4.53) is just that $d_0 = 0$, but the condition $\langle \mathbf{e}_2, \mathbf{e}_2 \rangle = 2$ is a further constraint on the coefficients. We have

$$\begin{aligned}
\sigma(c_1) &= \mathbf{e}_1 \circ (\mathbf{e}_2 \circ c_1) \\
&= (c_4 + \vartheta c_4) \circ (c_1 \circ \mathbf{e}_2),
\end{aligned} \tag{4.61}$$

and

$$c_1 \circ \mathbf{e}_2 = \sum_{2 \leq j \leq 4} d_{1j}\vartheta c_j \mathbf{vac} \tag{4.62}$$

so

$$\sigma(c_1) = d_{14}\mathbf{vac} - d_{12}\vartheta c_2 \vartheta c_4 \mathbf{vac} - d_{13}\vartheta c_3 \vartheta c_4 \mathbf{vac}. \tag{4.63}$$

Therefore, $d_{14} = d$, $d_{12} = d_{13} = 0$ and (4.60) gives $d_{34} = d_{24} = 0$. Then

$$\mathbf{e}_2 = d\vartheta c_1 \vartheta c_4 \mathbf{vac} + \vartheta(d\vartheta c_1 \vartheta c_4 \mathbf{vac}) \tag{4.64}$$

and the constraint $\langle \mathbf{e}_2, \mathbf{e}_2 \rangle = 2$ is just the condition $|d| = 1$.

PROPOSITION 4.17. *Let $\mathbf{e}_1 \in \mathbf{C}^{(1)}$ and $\mathbf{e}_2 \in \mathbf{C}^{(2)}$ satisfying $\vartheta\mathbf{e}_1 = \mathbf{e}_1$, $\vartheta\mathbf{e}_2 = \mathbf{e}_2$, $\langle\mathbf{e}_1,\mathbf{e}_1\rangle = 2$, $\langle\mathbf{e}_2,\mathbf{e}_2\rangle = 2$, and $\langle\mathbf{e}_2,\mathbf{vac}\rangle = 0$ determine $\sigma = \rho(\mathbf{e}_1)\rho(\mathbf{e}_2)$ on $\mathbf{C}$. Let $\{c_1,\ldots,c_4,\vartheta c_1,\ldots,\vartheta c_4\}$ be a canonical basis of $\mathbf{C}^{(1)}$ adapted to ϑ with $c_1 = d\sigma^2(\mathbf{vac})$, $|d| = 1$, and $c_4 = \mathbf{e}_1^+$. Then $\mathbf{e}_2$ is given in (4.64), and we have*

(a) $\sigma(c_1) = d\mathbf{vac},$

$\quad \sigma(c_2) = \xi^{-1}d^{-1}\vartheta c_3\vartheta c_4\mathbf{vac},$

$\quad \sigma(c_3) = -\xi^{-1}d^{-1}\vartheta c_2\vartheta c_4\mathbf{vac},$

$\quad \sigma(c_4) = d\vartheta c_1\vartheta c_4\mathbf{vac},$

(b) $\sigma(\vartheta c_1) = \xi^{-1}d^{-1}\vartheta c_1\vartheta c_2\vartheta c_3\vartheta c_4\mathbf{vac},$

$\quad \sigma(\vartheta c_2) = -d\vartheta c_1\vartheta c_2\mathbf{vac},$

$\quad \sigma(\vartheta c_3) = -d\vartheta c_1\vartheta c_3\mathbf{vac},$

$\quad \sigma(\vartheta c_4) = -\xi^{-1}d^{-1}\vartheta c_2\vartheta c_3\mathbf{vac},$

(c) $\sigma^2(c_1) = -d\vartheta c_4\mathbf{vac},$

$\quad \sigma^2(c_2) = \xi^{-1}d^{-1}\vartheta c_3\mathbf{vac},$

$\quad \sigma^2(c_3) = -\xi^{-1}d^{-1}\vartheta c_2\mathbf{vac},$

$\quad \sigma^2(c_4) = -\xi^{-1}d^{-1}\vartheta c_2\vartheta c_3\vartheta c_4\mathbf{vac},$

(d) $\sigma^2(\vartheta c_1) = \xi^{-1}d^{-1}\vartheta c_1\vartheta c_2\vartheta c_3\mathbf{vac},$

$\quad \sigma^2(\vartheta c_2) = d\vartheta c_1\vartheta c_2\vartheta c_4\mathbf{vac},$

$\quad \sigma^2(\vartheta c_3) = d\vartheta c_1\vartheta c_3\vartheta c_4\mathbf{vac},$

$\quad \sigma^2(\vartheta c_4) = -d\vartheta c_1\mathbf{vac},$

(e) $\sigma(\mathbf{vac}) = -\vartheta c_4\mathbf{vac}.$

PROOF. (a) These follow from $\sigma(c_i) = \mathbf{e}_1 \circ (c_i \circ \mathbf{e}_2)$ and the relations in **Cliff**.

(b) These follow either from part (a), $\sigma(\vartheta c_i) = \vartheta\sigma(c_i)$ and (4.34), or directly from $\sigma(\vartheta c_i) = \mathbf{e}_1 \circ (\vartheta c_i \circ \mathbf{e}_2)$.

(c) These follow from

$$\sigma^2(c_i) = \sigma^{-1}(c_i) = \rho(\mathbf{e}_2)\rho(\mathbf{e}_1)c_i = \mathbf{e}_2 \circ (-r_{\mathbf{e}_1}(c_i))$$

$$= \mathbf{e}_2 \circ (\langle\mathbf{e}_1, c_i\rangle\mathbf{e}_1 - c_i) = \begin{cases} -c_i \circ \mathbf{e}_2 & \text{if } 1 \leq i \leq 3 \\ \\ \vartheta c_i \circ \mathbf{e}_2 & \text{if } i = 4. \end{cases}$$

(d) These follow either from part (c), $\sigma^2(\vartheta c_i) = \vartheta\sigma^2(c_i)$ and (4.34), or directly from $\sigma^2(\vartheta c_i) = \mathbf{e}_2 \circ (\langle\mathbf{e}_1, \vartheta c_i\rangle\mathbf{e}_1 - \vartheta c_i)$.

(e) This is just (4.54). $\blacksquare$

We have seen how σ lifts to isomorphisms $\sigma^{(k)}$ (4.23) cyclically permuting the Clifford algebras **Cliff**$^{(k)}$ and their modules **CM**$^{(k)}$, $1 \leq k \leq 3$. The other elements of the triality group G also lift to isomorphisms permuting these Clifford algebras and their modules since the spaces $\mathbf{C}^{(k)}$, $1 \leq k \leq 3$, are permuted and the form $\langle\ ,\ \rangle$ on $\mathbf{C}$ is invariant under G. Let $g \in G$ permute $\{1,2,3\}$ as it permutes $\{\mathbf{C}^{(1)}, \mathbf{C}^{(2)}, \mathbf{C}^{(3)}\}$, and for $1 \leq k \leq 3$, let

$$g^{(k)} : \mathbf{Cliff}^{(k)} \to \mathbf{Cliff}^{(g(k))} \tag{4.65}$$

be the Clifford algebra isomorphism induced by $g : \mathbf{C}^{(k)} \to \mathbf{C}^{(g(k))}$. Since each $g \in G$ is an automorphism of $(\mathbf{C}, \circ)$, we have

$$g \cdot (a \circ b) = (g \cdot a) \circ (g \cdot b) \tag{4.66}$$

for $a \in \mathbf{C}^{(k)}$, $b \in \mathbf{CM}^{(k)}$. This formula extends to

$$g \cdot (u \cdot b) = (g^{(k)} \cdot u) \cdot (g \cdot b) \tag{4.67}$$

for $u \in \mathbf{Cliff}^{(k)}$, $b \in \mathbf{CM}^{(k)}$ so g intertwines the action of $\mathbf{Cliff}^{(k)}$ on $\mathbf{CM}^{(k)}$.

We can define a positive Hermitian form $(\ ,\ )$ on $\mathbf{C}$ by setting

$$(a, b) = \langle a, \vartheta b \rangle \tag{4.68}$$

for $a, b \in \mathbf{C}$. Positivity of this form is given by Proposition 4.9 (b). From Proposition 4.10 we have

$$(a, b \circ c) = ((\vartheta b) \circ a, c) \tag{4.69}$$

for $a, b, c \in \mathbf{C}$, so the operation on $\mathbf{C}$ of multiplication by ϑb is the adjoint with respect to $(\ ,\ )$ of the operation of multiplication by b. If L is any linear operator on $\mathbf{C}$ we let L^* denote the adjoint of L with respect to $(\ ,\ )$, so that $b^* = \vartheta b$ for $b \in \mathbf{C}$.

PROPOSITION 4.18.

(a) *G acts as unitary transformations on $\mathbf{C}$ with respect to the form $(\ ,\ )$,*

(b) *For $x = {}^{\circ}_{\circ}ab^{\circ}_{\circ} \in \mathbf{o}(8)$, $a, b \in \mathbf{C}^{(k)}$, $1 \leq k \leq 3$, we have*

$$x^* = {}^{\circ}_{\circ}(\vartheta b)(\vartheta a)^{\circ}_{\circ}, \tag{4.70}$$

(c) *For $g \in G$, $a \in \mathbf{C}$, $x \in \mathbf{o}(8)$, we have*

$$g \cdot a^* = (g \cdot a)^* \quad and \quad g \cdot x^* = (g \cdot x)^*, \tag{4.71}$$

(d) *For $1 \leq k \leq 3$ the operator ϑ on $\mathbf{C}^{(k)}$ induces a $\mathbb{C}$-antilinear antiautomorphism of $\mathbf{Cliff}^{(k)}$ giving the adjoints of its elements as operators on $\mathbf{CM}^{(k)}$, consistent with part (b).*

PROOF. (a) For $a, b \in \mathbf{C}$, $g \in G$, $(ga, gb) = \langle ga, \vartheta gb \rangle = \langle ga, g \vartheta b \rangle = \langle a, \vartheta b \rangle = (a, b)$ since G commutes with ϑ.

(b) Let $x = {}^{\circ}_{\circ}ab^{\circ}_{\circ} \in \mathbf{o}(8)$, $a, b \in \mathbf{C}^{(k)}$. We wish to show that for $u, v \in \mathbf{C}$,

$$(u, {}^{\circ}_{\circ}ab^{\circ}_{\circ} \cdot v) = ({}^{\circ}_{\circ}(\vartheta b)(\vartheta a)^{\circ}_{\circ} \cdot u, v). \tag{4.72}$$

Since ϑ preserves each $\mathbf{C}^{(s)}$, $1 \leq s \leq 3$, $(\mathbf{C}^{(r)}, \mathbf{C}^{(s)}) = \langle \mathbf{C}^{(r)}, \vartheta(\mathbf{C}^{(s)}) \rangle = 0$ for $r \neq s$. Since $\mathbf{o}(8)$ preserves each $\mathbf{C}^{(s)}$, it is enough to prove (4.72) for $u, v \in \mathbf{C}^{(s)}$, $1 \leq s \leq 3$. If $u, v \in \mathbf{C}^{(k)}$ then

$$\begin{aligned}
(u, {}^{\circ}_{\circ}ab^{\circ}_{\circ} \cdot v) &= (u, \langle b, v \rangle a - \langle a, v \rangle b) \\
&= \langle u, \vartheta a \rangle \langle \vartheta b, \vartheta v \rangle - \langle u, \vartheta b \rangle \langle \vartheta a, \vartheta v \rangle \\
&= \langle \langle \vartheta a, u \rangle \vartheta b - \langle \vartheta b, u \rangle \vartheta a, \vartheta v \rangle \\
&= ({}^{\circ}_{\circ}(\vartheta b)(\vartheta a)^{\circ}_{\circ} \cdot u, v).
\end{aligned}$$

If $u, v \in \mathbf{C}^{(k')} \oplus \mathbf{C}^{(k'')}$ then

$$
\begin{aligned}
(u, {}_\circ^\circ ab{}_\circ^\circ \cdot v) &= (u, \tfrac{1}{2}(a \circ (b \circ v) - b \circ (a \circ v))) \\
&= \langle u, \tfrac{1}{2}(\vartheta a \circ (\vartheta b \circ \vartheta v) - \vartheta b \circ (\vartheta a \circ \vartheta v)) \rangle \\
&= \langle \tfrac{1}{2}(\vartheta b \circ (\vartheta a \circ u) - \vartheta a \circ (\vartheta b \circ u)), \vartheta v \rangle \\
&= ({}_\circ^\circ(\vartheta b)(\vartheta a){}_\circ^\circ \cdot u, v).
\end{aligned}
$$

(c) The first formula just means ϑ commutes with G, and the second is clear from (4.39) and part (b).

(d) This follows from (4.69). ∎

We conclude this chapter with Chevalley's construction of a product $*$ on $\mathbf{C}^{(1)}$ making it the complex octonians, and with a proposition giving certain tensor product decompositions which will be needed later. For $v, w \in \mathbf{C}^{(1)}$ define

$$
v * w = \sigma\tau(v) \circ \sigma^2\tau(w) = (\mathbf{e}_3 \circ v) \circ (w \circ \mathbf{e}_2). \tag{4.73}
$$

Then $\mathbf{e}_1 \in \mathbf{C}^{(1)}$ is the unit element for this product and

$$
\tau(v * w) = \tau(w) * \tau(v). \tag{4.74}
$$

DEFINITION. *We write $\tau(v) = \bar{v}$ for $v \in \mathbf{C}^{(1)}$.*

PROPOSITION 4.19. *For $x, y, z \in \mathbf{C}^{(1)}$ we have*

(a) $\langle x * y, z \rangle = \langle x, z * \bar{y} \rangle = \langle y, \bar{x} * z \rangle = \langle y * \bar{z}, \bar{x} \rangle,$
(b) $x * \bar{x} = \tfrac{1}{2}\langle x, x \rangle \mathbf{e}_1,$
(c) $(x * \bar{y}) + (y * \bar{x}) = \langle x, y \rangle \mathbf{e}_1.$

PROOF. (a) We have

$$
\begin{aligned}
\langle x * y, z \rangle &= \langle \sigma\tau(x) \circ \sigma^2\tau(y), z \rangle = \langle \sigma\tau(x), \sigma^2\tau(y) \circ z \rangle \\
&= \langle x, \sigma\tau(z) \circ \sigma^2\tau(\tau y) \rangle = \langle x, z * \bar{y} \rangle.
\end{aligned}
$$

Since $\langle \, , \, \rangle$ is τ-invariant,

$$
\begin{aligned}
\langle x * y, z \rangle &= \langle \tau(x * y), \tau(z) \rangle = \langle \bar{y} * \bar{x}, \bar{z} \rangle \\
&= \langle \bar{y}, \bar{z} * x \rangle = \langle \tau(\bar{y}), \tau(\bar{z} * x) \rangle = \langle y, \bar{x} * z \rangle,
\end{aligned}
$$

and

$$
\langle x, z * \bar{y} \rangle = \langle \bar{x}, y * \bar{z} \rangle = \langle y * \bar{z}, \bar{x} \rangle
$$

by symmetry of the form.

(b) Since $\tau(x * \bar{x}) = x * \bar{x}$, $x * \bar{x} = c\mathbf{e}_1$ for some scalar c. We find that $2c = \langle c\mathbf{e}_1, \mathbf{e}_1 \rangle = \langle x * \bar{x}, \mathbf{e}_1 \rangle = \langle x, \mathbf{e}_1 * x \rangle = \langle x, x \rangle.$

(c) Polarize part (b). ∎

PROPOSITION 4.20. *For $v_1, v_2, v_3, v_4 \in \mathbf{C}^{(1)}$ we have*

$$
\langle v_1 * v_2, v_3 * v_4 \rangle + \langle v_1 * v_4, v_3 * v_2 \rangle = \langle v_1, v_3 \rangle \langle v_2, v_4 \rangle.
$$

PROOF. This follows from Proposition 4.8 and (4.73). ∎

COROLLARY 4.21. *For $x, y \in \mathbf{C}^{(1)}$ we have*

$$2\langle x * y, x * y \rangle = \langle x, x \rangle \langle y, y \rangle.$$

DEFINITION. *Let N be the quadratic form on $\mathbf{C}^{(1)}$ given by $N(x) = \frac{1}{2}\langle x, x \rangle$.*

THEOREM 4.22. *For $x, y \in \mathbf{C}^{(1)}$ we have*

(a) $N(x * y) = N(x)N(y)$,
(b) $(y * x) * x = y * (x * x)$,
(c) $x * (x * y) = (x * x) * y$.

PROOF. (a) Clear from Corollary 4.21.

(b) Since (a) means that $\mathbf{C}^{(1)}$ is a composition algebra, the alternative laws follow. We present the following proof to keep our presentation self contained. It is enough to show that for $x_1, x_2 \in \mathbf{C}^{(1)}$,

$$(y * x_1) * x_2 + (y * x_2) * x_1 = y * (x_1 * x_2 + x_2 * x_1). \tag{4.75}$$

This is clearly true if x_1 or x_2 is a multiple of the unit element $\mathbf{e}_1$. Since (4.75) is linear in x_1 and x_2, it remains only to be checked when x_1 and x_2 are in the orthogonal complement of $\mathbf{e}_1$, so that $\bar{x}_1 = -x_1$, $\bar{x}_2 = -x_2$. Then for any $y, z \in \mathbf{C}^{(1)}$ we have

$$\langle (y * x_1) * x_2 + (y * x_2) * x_1, z \rangle = -\langle y * x_1, z * x_2 \rangle - \langle y * x_2, z * x_1 \rangle$$
$$= -\langle x_1, x_2 \rangle \langle y, z \rangle = \langle -\langle x_1, x_2 \rangle y, z \rangle$$

so that

$$(y * x_1) * x_2 + (y * x_2) * x_1 = -\langle x_1, x_2 \rangle y$$
$$= -y * (x_1 * \bar{x}_2 + x_2 * \bar{x}_1) = y * (x_1 * x_2 + x_2 * x_1).$$

Setting $x_1 = x_2$ in (4.75) gives (b), and (c) follows from (a) by applying τ. ∎

We have shown that $\mathbf{C}^{(1)}$ under $*$ is an alternative composition algebra of dimension 8 over $\mathbb{C}$, so it is a complex octonian (Cayley) algebra [**S**]. (See also [**GNORS**].)

THEOREM 4.23. *(The Principle of Local Triality)*
For $x \in \mathbf{o}(8)$, $v, w \in \mathbf{C}^{(1)}$, we have

$$x \cdot (v * w) = (\sigma\tau(x) \cdot v) * w + v * (\sigma^2 \tau(x) \cdot w).$$

PROOF. This follows immediately from Proposition 4.13. ∎

In later chapters we will need to know certain tensor product decompositions of $\mathbf{o}(8)$-modules, and how to express the $\mathbf{o}(8)$-module maps from the tensor products to certain components. Let 8_1, 8_2, 8_3 denote the three eight-dimensional $\mathbf{o}(8)$-modules $\mathbf{C}^{(1)}$, $\mathbf{C}^{(2)}$, $\mathbf{C}^{(3)}$. Let $\mathbf{C}^{(1)} = \mathbf{C}^{(1)+} \oplus \mathbf{C}^{(1)-}$ be a polarization, $\vartheta : \mathbf{C}^{(1)} \to \mathbf{C}^{(1)}$ an antilinear involution and let $\{c_1, \dots, c_4, \vartheta c_1, \dots, \vartheta c_4\}$ be a canonical basis of $\mathbf{C}^{(1)}$ adapted to ϑ, chosen as in Proposition 4.17. Let h_i, ϵ_i

and α_i, $1 \leq i \leq 4$, and the Cartan subalgebra $\mathbf{h}$, be as described in (4.42) - (4.44). Then the fundamental weights of $\mathbf{o}(8)$ are

$$\omega_1 = \epsilon_1, \quad \omega_2 = \epsilon_1 + \epsilon_2, \quad \omega_3 = \tfrac{1}{2}(\epsilon_1 + \epsilon_2 + \epsilon_3 - \epsilon_4), \quad \omega_4 = \tfrac{1}{2}(\epsilon_1 + \epsilon_2 + \epsilon_3 + \epsilon_4) \quad (4.76)$$

and we denote the irreducible $\mathbf{o}(8)$-module with highest weight $n_1\omega_1 + n_2\omega_2 + n_3\omega_3 + n_4\omega_4$ by (n_1, n_2, n_3, n_4). We use the abbreviations

$$
\begin{aligned}
1 &= (0,0,0,0), & 8_1 &= (1,0,0,0), & 8_2 &= (0,0,0,1), & 8_3 &= (0,0,1,0), \\
28 &= (0,1,0,0), & 35_1 &= (2,0,0,0), & 35_2 &= (0,0,0,2), & 35_3 &= (0,0,2,0), \\
56_1 &= (0,0,1,1), & 56_2 &= (1,0,1,0), & 56_3 &= (1,0,0,1), \\
160_1 &= (1,1,0,0), & 160_2 &= (0,1,0,1), & 160_3 &= (0,1,1,0), \\
224_1 &= (2,0,0,1), & 224_2 &= (0,0,1,2), & 224_3 &= (1,0,2,0), & 350 &= (1,0,1,1)
\end{aligned}
$$

which indicate the dimension of the module. Then standard Lie algebra techniques give the following proposition.

PROPOSITION 4.24. *For $1 \leq k \leq 3$ and (k, k', k'') a permutation of (1,2,3), we have the tensor product decompositions*

    (a) $8_k \otimes 8_k = 35_k \oplus 1 \oplus 28$, *where $35_k \oplus 1$ consists of the symmetric tensors and 28 consists of the antisymmetric tensors,*

    (b) $8_k \otimes 8_{k'} = 56_{k''} \oplus 8_{k''}$,

    (c) $35_k \otimes 8_{k'} = 224_k \oplus 56_{k'}$,

    (d) $28 \otimes 8_{k'} = 160_{k'} \oplus 56_{k'} \oplus 8_{k'}$,

    (e) $56_k \otimes 8_k = 350 \oplus 35_{k'} \oplus 35_{k''} \oplus 28$.

In $8_k \otimes 8_k$ the trivial representation 1 is explicitly provided by the symmetric pairing $a \otimes a_1 \to \langle a, a_1 \rangle$, and the adjoint representation 28 is provided by the normal ordering $a \otimes a_1 \to {}^\circ_\circ a a_1 {}^\circ_\circ$. In $8_k \otimes 8_{k'}$ the $8_{k''}$ representation is provided by the Chevalley operation $a \otimes b \to a \circ b$. In

$$8_k \otimes 8_k \otimes 8_{k'} = (35_k \otimes 8_{k'}) \oplus (1 \otimes 8_{k'}) \oplus (28 \otimes 8_{k'})$$

$$= 224_k \oplus 56_{k'} \oplus 8_{k'} \oplus 160_{k'} \oplus 56_{k'} \oplus 8_{k'}$$

one $8_{k'}$ representation is provided by $a \otimes a_1 \otimes b \to \langle a, a_1 \rangle b$ and the other is provided by $a \otimes a_1 \otimes b \to {}^\circ_\circ a a_1 {}^\circ_\circ \cdot b$. In

$$8_1 \otimes 8_2 \otimes 8_3 = (56_3 \otimes 8_3) \oplus (8_3 \otimes 8_3)$$

$$= 350 \oplus 35_1 \oplus 35_2 \oplus 28 \oplus 35_3 \oplus 1 \oplus 28$$

the trivial representation 1 is provided by $a \otimes b \otimes c \to \langle a \circ b, c \rangle$.

CHAPTER 5

Spinor Construction of Triality for $D_4^{(1)}$

We begin with a nondegenerate symmetric bilinear form $\langle\,,\,\rangle$ on $\mathbf{A} \cong \mathbb{C}^8$, a polarization $\mathbf{A} = \mathbf{A}^+ \oplus \mathbf{A}^-$ into maximally isotropic subspaces, and a $\mathbb{C}$-antilinear involution $\vartheta : \mathbf{A} \to \mathbf{A}$ satisfying (2.63) - (2.65). Let $\mathbf{A}$ and $\langle\,,\,\rangle$ generate the Clifford algebra **Cliff** and construct the **Cliff**-module $\mathbf{CM} = (\wedge \mathbf{A}^-) \cdot \mathbf{vac}$, $0 \neq \mathbf{vac} \in \wedge^4 \mathbf{A}^+$, from the polarization of $\mathbf{A}$. Let

$$\mathbf{C} = \mathbf{A} \oplus \mathbf{CM} = \mathbf{C}^{(1)} \oplus \mathbf{C}^{(2)} \oplus \mathbf{C}^{(3)} \tag{5.1}$$

where $\mathbf{C}^{(1)} = \mathbf{A}$, $\mathbf{C}^{(2)} = \mathbf{CM}^0 = (\wedge^{even} \mathbf{A}^-) \cdot \mathbf{vac}$, $\mathbf{C}^{(3)} = (\wedge^{odd} \mathbf{A}^-) \cdot \mathbf{vac}$. In Chapter 4 we gave Chevalley's extension of the form $\langle\,,\,\rangle$ to all of $\mathbf{C}$, his construction of a commutative nonassociative product $\circ$ on $\mathbf{C}$, and the action of the triality group $G \cong S_3$ as automorphisms of the algebra $(\mathbf{C}, \circ)$ commuting with ϑ. Under left $\circ$ multiplication, $\mathbf{CM}^{(k)} = \mathbf{C}^{(k')} \oplus \mathbf{C}^{(k'')}$, $1 \leq k \leq 3$, is an irreducible module for the Clifford algebra $\mathbf{Cliff}^{(k)}$ generated by $\mathbf{C}^{(k)}$ and $\langle\,,\,\rangle$. Recall that $\sigma \in G$ cyclically permutes $\mathbf{C}^{(1)}, \mathbf{C}^{(2)}, \mathbf{C}^{(3)}$, and $\tau \in G$ preserves $\mathbf{C}^{(1)}$ and switches $\mathbf{C}^{(2)}$ and $\mathbf{C}^{(3)}$.

We have also shown that for each k, $1 \leq k \leq 3$, the Lie algebra $\mathbf{o}(8)$ is represented on $\mathbf{C}$ by the span of normally ordered quadratic elements $x = {}^{\circ}_{\circ} ab {}^{\circ}_{\circ}$, where $a, b \in \mathbf{C}^{(k)}$, and $g \in G$ acts on $\mathbf{o}(8)$ by $g \cdot {}^{\circ}_{\circ} ab {}^{\circ}_{\circ} = {}^{\circ}_{\circ} (g \cdot a)(g \cdot b) {}^{\circ}_{\circ}$ so as to intertwine the action of $\mathbf{o}(8)$ on $\mathbf{C}$, $g \cdot (x \cdot a) = (g \cdot x) \cdot (g \cdot a)$ for $a \in \mathbf{C}$.

Now we proceed with the affine construction. Let $Z = \mathbb{Z}$ or $\mathbb{Z} + \frac{1}{2}$ and let $Z' = \frac{1}{2}\mathbb{Z} - Z$. Let $\mathbf{A}(Z)$, $\mathbf{Cliff}(Z)$ and $\mathbf{CM}(Z)$ be constructed from $\mathbf{A}$ and the form $\langle\,,\,\rangle$ as in Chapter 2. Write the four irreducible $\hat{\mathbf{o}}(8)$-modules

$$\mathbf{U}_0 = \mathbf{CM}(\mathbb{Z}+\tfrac{1}{2})^0, \quad \mathbf{U}_1 = \mathbf{CM}(\mathbb{Z}+\tfrac{1}{2})^1, \quad \mathbf{U}_2 = \mathbf{CM}(\mathbb{Z})^0, \quad \mathbf{U}_3 = \mathbf{CM}(\mathbb{Z})^1 \tag{5.2}$$

and let

$$\mathbf{U} = \mathbf{U}_0 \oplus \mathbf{U}_1 \oplus \mathbf{U}_2 \oplus \mathbf{U}_3. \tag{5.3}$$

We let $a(n)$, $n \in Z$, $a \in \mathbf{A}$, act on $\mathbf{U}$ by having it act as usual on $\mathbf{CM}(Z)$ and as zero on $\mathbf{CM}(Z')$. We take all generating functions to be summed over $\frac{1}{2}\mathbb{Z}$, so for $a, b \in \mathbf{A}$,

$$a(\zeta) = \sum_{n \in \frac{1}{2}\mathbb{Z}} a(n)\zeta^{-n} \tag{5.4}$$

and

$$\,_\circ^\circ a(\zeta)b(\zeta)\,_\circ^\circ = \sum_{m\in\frac{1}{2}\mathbb{Z}}\ \sum_{n\in\frac{1}{2}\mathbb{Z}}\,_\circ^\circ a(n)b(m-n)\,_\circ^\circ \zeta^{-m}. \tag{5.5}$$

Then for $m \in \frac{1}{2}\mathbb{Z}$

$$\,_\circ^\circ a(\zeta)b(\zeta)\,_\circ^\circ {}_m = \sum_{n\in\frac{1}{2}\mathbb{Z}}\,_\circ^\circ a(n)b(m-n)\,_\circ^\circ \tag{5.6}$$

is a well-defined operator on $\mathbf{U}$, which is zero for $m \in \mathbb{Z} + \frac{1}{2}$.

In Chapter 2 we defined the operators $D^Z(n)$, $n \in \mathbb{Z}$, which represent the Virasoro algebra $\mathbf{Vir}$ on $\mathbf{CM}(Z)$. Then the operators

$$D(n) = D^Z(n) + D^{Z+\frac{1}{2}}(n), \quad n \in \mathbb{Z}, \tag{5.7}$$

represent $\mathbf{Vir}$ on $\mathbf{U}$, preserving each summand in (5.3). The span of the identity operator, $D(0)$, and the operators $\,_\circ^\circ a(\zeta)b(\zeta)\,_\circ^\circ {}_m$ for $m \in \mathbb{Z}$, $a,b \in \mathbf{A}$, represent $\hat{\mathbf{o}}(8)$ on $\mathbf{U}$.

Recall from Chapter 2, (2.63) - (2.65), that the $\mathbb{C}$-antilinear involution $\vartheta : \mathbf{A} \to \mathbf{A}$ gives a positive Hermitian form $(\ ,\ )$ on $\mathbf{CM}(Z)$, and therefore on $\mathbf{U}$. Also recall that the adjoint of $D(n)$ is $D(-n)$, and for $a,b \in \mathbf{A}$ the adjoint of $\,_\circ^\circ a(\zeta)b(\zeta)\,_\circ^\circ {}_n$ is $\,_\circ^\circ (\vartheta b)(\zeta)(\vartheta a)(\zeta)\,_\circ^\circ {}_{-n}$.

From (2.70), with $l = 4$, we have

$$D(0)\mathbf{vac}(Z) = -\tfrac{1}{4}(1 + \iota)\mathbf{vac}(Z). \tag{5.8}$$

Combining this with (2.60) we get the gradings

$$\mathbf{U}_0 = (\mathbf{U}_0)_0 \oplus (\mathbf{U}_0)_1 \oplus (\mathbf{U}_0)_2 \oplus \ldots, \tag{5.9}$$

$$\mathbf{U}_k = (\mathbf{U}_k)_{1/2} \oplus (\mathbf{U}_k)_{3/2} \oplus (\mathbf{U}_k)_{5/2} \oplus \ldots, \quad \text{for } 1 \le k \le 3, \tag{5.10}$$

where

$$(\mathbf{U}_k)_n = \{u \in \mathbf{U}_k \mid D(0)u = -nu\} \tag{5.11}$$

for $0 \le k \le 3$, $n \in \frac{1}{2}\mathbb{Z}$. It is clear that $(\mathbf{U}_0)_0$ is one dimensional with basis $\{\mathbf{vac}(\mathbb{Z} + \frac{1}{2})\}$, and that $(\mathbf{U}_0)_1$ is 28-dimensional, spanned by

$$\{a(-\tfrac{1}{2})b(-\tfrac{1}{2})\mathbf{vac}(\mathbb{Z} + \tfrac{1}{2}) \mid a, b \in \mathbf{A}\}. \tag{5.12}$$

For $1 \le k \le 3$, $(\mathbf{U}_k)_{1/2}$ is 8-dimensional, $(\mathbf{U}_1)_{1/2}$ is spanned by

$$\{a(-\tfrac{1}{2})\mathbf{vac}(\mathbb{Z} + \tfrac{1}{2}) \mid a \in \mathbf{A}\}, \tag{5.13}$$

$(\mathbf{U}_2)_{1/2}$ is spanned by

$$\{a_1(0)\ldots a_r(0)\mathbf{vac}(\mathbb{Z}) \mid a_1,\ldots,a_r \in \mathbf{A}^-, \ r = 0, 2, 4\}, \tag{5.14}$$

and $(\mathbf{U}_3)_{1/2}$ is spanned by

$$\{a_1(0)\ldots a_r(0)\mathbf{vac}(\mathbb{Z}) \mid a_1,\ldots,a_r \in \mathbf{A}^-, \ r = 1, 3\}. \tag{5.15}$$

From (5.9) and (5.10) we see that

$$(\mathbf{U})_{1/2} = (\mathbf{U}_1)_{1/2} \oplus (\mathbf{U}_2)_{1/2} \oplus (\mathbf{U}_3)_{1/2}. \tag{5.16}$$

For $a, b \in \mathbf{A}$, $x = {}^\circ_\circ ab{}^\circ_\circ \in \mathbf{o}(8)$, $n \in \mathbb{Z}$, let us write $x(n) = {}^\circ_\circ a(\zeta)b(\zeta){}^\circ_{\circ n}$ for the corresponding operator on $\mathbf{U}$ representing $\hat{\mathbf{o}}(8)$. We have seen that the operator $x(0)$ represents $x \in \mathbf{o}(8)$ on $(\mathbf{U})_{1/2}$. If $x(n)^*$ denotes the adjoint of $x(n)$ on $\mathbf{U}$ then

$$x(n)^* = (\vartheta x)(-n), \tag{5.17}$$

where

$$\vartheta x = {}^\circ_\circ (\vartheta b)(\vartheta a){}^\circ_\circ \tag{5.18}$$

is a $\mathbb{C}$-antilinear antiautomorphism of $\mathbf{o}(8)$.

Define the unitary isomorphism

$$\Upsilon : \mathbf{C} \to (\mathbf{U})_{1/2} \tag{5.19}$$

by

$$\Upsilon(a) = a(-\tfrac{1}{2})\mathbf{vac}(\mathbb{Z} + \tfrac{1}{2}) \quad \text{for} \quad a \in \mathbf{A},$$
$$\Upsilon(\mathbf{vac}) = \mathbf{vac}(\mathbb{Z}), \tag{5.20}$$
$$\Upsilon(a \circ v) = a(0)\Upsilon(v) \quad \text{for} \quad a \in \mathbf{A}, \quad v \in \mathbf{CM}.$$

From (2.41) this is a well defined linear isomorphism such that $\Upsilon(\mathbf{A}) = (\mathbf{U}_1)_{1/2}$, $\Upsilon(\mathbf{CM}^0) = (\mathbf{U}_2)_{1/2}$ and $\Upsilon(\mathbf{CM}^1) = (\mathbf{U}_3)_{1/2}$. From (2.34), (2.49) and (5.20) it is easy to check that for $x \in \mathbf{o}(8)$, $u \in \mathbf{C}$, we have

$$\Upsilon(x \cdot u) = x(0)\Upsilon(u). \tag{5.21}$$

Note that the unitarity of Υ,

$$(u, v) = (\Upsilon u, \Upsilon v) \quad \text{for} \quad u, v \in \mathbf{C} \tag{5.22}$$

follows from (2.68) and (4.43). Using Υ to transport the action of G from $\mathbf{C}$ to $(\mathbf{U})_{1/2}$, that is,

$$g\Upsilon(u) = \Upsilon(gu) \quad \text{for} \quad g \in G, \ u \in \mathbf{C}, \tag{5.23}$$

we see that each $g \in G$ is unitary on $(\mathbf{U})_{1/2}$.

The isomorphism

$$\Upsilon' : \mathbf{o}(8) \to (\mathbf{U})_1 \tag{5.24}$$

defined by

$$\Upsilon'({}^\circ_\circ ab{}^\circ_\circ) = a(-\tfrac{1}{2})b(-\tfrac{1}{2})\mathbf{vac}(\mathbb{Z} + \tfrac{1}{2}) \quad \text{for} \quad a, b \in \mathbf{A} \tag{5.25}$$

satisfies

$$x(0)\Upsilon'(y) = \Upsilon'([x, y]), \quad x(1)\Upsilon'(y) = \langle x, y \rangle \mathbf{vac}(\mathbb{Z} + \tfrac{1}{2}) \quad \text{for} \quad x, y \in \mathbf{o}(8). \tag{5.26}$$

We can use the invariant form $\langle \, , \, \rangle$ on $\mathbf{o}(8)$ given in (2.36) and (5.18) to define a positive Hermitian form $(\, , \,)$ on $\mathbf{o}(8)$ by

$$(x, y) = \langle x, \vartheta y \rangle. \tag{5.27}$$

This gives

$$({}^\circ_\circ ab{}^\circ_\circ, {}^\circ_\circ cd{}^\circ_\circ) = (a, c)(b, d) - (a, d)(b, c) \tag{5.28}$$

and (2.68) then says Υ' is a unitary isomorphism,

$$(x, y) = (\Upsilon'x, \Upsilon'y). \tag{5.29}$$

The invariance of the form $\langle \ , \ \rangle$ on $\mathbf{o}(8)$ says that ad_x^*, the adjoint of ad_x with respect to the Hermitian form (5.27), is $ad_{\vartheta x}$.

Suppose $\nu : \mathbf{o}(8) \to \mathbf{o}(8)$ is any Lie algebra automorphism, so the invariant symmetric bilinear form on $\mathbf{o}(8)$ satisfies $\langle \nu x, \nu y \rangle = \langle x, y \rangle$. Then ν extends to an automorphism $\hat{\nu}$ of $\hat{\mathbf{o}}(8)$ by

$$\hat{\nu}(x(n)) = (\nu x)(n), \quad \hat{\nu}(c) = c, \quad \hat{\nu}(d) = d. \tag{5.30}$$

Since $(\mathbf{U})_{1/2}$ is the direct sum of the three 8-dimensional representations of $\mathbf{o}(8)$, there must be an isomorphism $\nu : (\mathbf{U})_{1/2} \to (\mathbf{U})_{1/2}$ such that

$$\nu x(0)\nu^{-1} = (\nu x)(0) \quad \text{on} \quad (\mathbf{U})_{1/2}. \tag{5.31}$$

With the help of the following lemma, we can define $\hat{\nu} : \mathbf{U} \to \mathbf{U}$ extending ν uniquely.

LEMMA 5.1. *Suppose that $u \in (\mathbf{U})_n$ for $1 \leq n \in \frac{1}{2}\mathbb{Z}$, and $z(k)u = 0$ for all $z(k) \in \hat{\mathbf{o}}(8)$ with $k > 0$. Then we have $u = 0$.*

PROOF. For any $z_1(-k)u_1 \in (\mathbf{U})_n$ with $k > 0$, using the Hermitian form we have $(u, z_1(-k)u_1) = ((\vartheta z_1)(k)u, u_1) = 0$. Since $(\mathbf{U})_n$ is spanned by all vectors of the form $z_1(-k)u_1$, and the form is positive definite, we get $u = 0$. ∎

PROPOSITION 5.2. *Let $\nu : \mathbf{o}(8) \to \mathbf{o}(8)$ be a Lie algebra automorphism, and suppose $\nu : (\mathbf{U})_{1/2} \to (\mathbf{U})_{1/2}$ is an isomorphism such that for $x \in \mathbf{o}(8)$, $\nu x(0)\nu^{-1} = (\nu x)(0)$ on $(\mathbf{U})_{1/2}$. Then there is a unique isomorphism $\hat{\nu} : \mathbf{U} \to \mathbf{U}$ such that $\hat{\nu}$ restricts to ν on $(\mathbf{U})_{1/2}$, $\hat{\nu}$ restricts to the identity operator on $(\mathbf{U})_0$ and $\hat{\nu}x(n)\hat{\nu}^{-1} = (\nu x)(n)$ on $\mathbf{U}$ for $x(n) \in \hat{\mathbf{o}}(8)$.*

PROOF. We define $\hat{\nu}$ on $(\mathbf{U})_0$ to be the identity operator, and we define $\hat{\nu}$ on $(\mathbf{U})_{1/2}$ to be ν. We proceed to define $\hat{\nu}$ on higher levels by induction. Suppose we have defined isomorphisms $\hat{\nu} : (\mathbf{U})_n \to (\mathbf{U})_n$ for $0 \leq n \leq N$ such that $\hat{\nu}(x(-p)u) = \hat{\nu}(x(-p))\hat{\nu}(u)$ for $u \in (\mathbf{U})_n$ with $n \leq N$ and $n + p \leq N$. Note that for any $1 \leq n \in \frac{1}{2}\mathbb{Z}$, $(\mathbf{U})_n$ is spanned by vectors of the form $x(-p)u$ for $u \in (\mathbf{U})_t$ with $p > 0$ and $p + t = n$. Suppose $x(-p)u \in (\mathbf{U})_{N+1}$ and let $y(k) \in \hat{\mathbf{o}}(8)$. Then for some $z(k) \in \hat{\mathbf{o}}(8)$ we have $y(k) = \hat{\nu}(z(k))$. If $k > 0$ then

$$y(k)\hat{\nu}(x(-p))\hat{\nu}(u) = \hat{\nu}(z(k))\hat{\nu}(x(-p))\hat{\nu}(u)$$
$$= \hat{\nu}(x(-p))\hat{\nu}(z(k))\hat{\nu}(u) + [\hat{\nu}(z(k)), \hat{\nu}(x(-p))]\hat{\nu}(u).$$

Since $u \in (\mathbf{U})_{N+1-p}$, $k > 0$, and $z(k)u \in (\mathbf{U})_{N+1-p-k}$, our inductive assumption says

$$\hat{\nu}(x(-p))\hat{\nu}(z(k))\hat{\nu}(u) = \hat{\nu}(x(-p))\hat{\nu}(z(k)u) = \hat{\nu}(x(-p)z(k)u).$$

Since $\hat{\nu}$ is a Lie algebra automorphism of $\hat{\mathfrak{o}}(8)$, and because of the induction assumption, we have

$$[\hat{\nu}(z(k)), \hat{\nu}(x(-p))]\hat{\nu}(u) = \hat{\nu}([z(k), x(-p)]u).$$

So we get

$$\begin{aligned}
y(k)\hat{\nu}(x(-p))\hat{\nu}(u) &= \hat{\nu}(z(k))\hat{\nu}(x(-p))\hat{\nu}(u) \\
&= \hat{\nu}(x(-p)z(k)u + [z(k), x(-p)]u) \\
&= \hat{\nu}(z(k)x(-p)u).
\end{aligned}$$

Any dependence relation in $(\mathbf{U})_{N+1}$, $0 = \displaystyle\sum_{1 \leq r \leq N+1} x_r(-r)u_r$, gives

$$y(k) \sum_{1 \leq r \leq N+1} \hat{\nu}(x_r(-r))\hat{\nu}(u_r) = \hat{\nu}\left(z(k) \sum_{1 \leq r \leq N+1} x_r(-r)u_r \right) = 0.$$

By Lemma 5.1 we have

$$0 = \sum_{1 \leq r \leq N+1} \hat{\nu}(x_r(-r))\hat{\nu}(u_r);$$

so $\hat{\nu} : (\mathbf{U})_{N+1} \to (\mathbf{U})_{N+1}$ defined by

$$\hat{\nu}(x(-p)u) = \hat{\nu}(x(-p))\hat{\nu}(u) \quad \text{for} \quad p > 0, \; x(-p)u \in (\mathbf{U})_{N+1},$$

is a well-defined isomorphism.

For $k > 0$ and $u \in (\mathbf{U})_{N+1}$ we have seen that $\hat{\nu}(z(k)u) = \hat{\nu}(z(k))\hat{\nu}(u)$. With $\hat{\nu}$ defined on $(\mathbf{U})_{N+1}$, we can repeat the above argument with $p = 0$ to get $y(k)\hat{\nu}(x(0))\hat{\nu}(u) = y(k)\hat{\nu}(x(0)u)$ for $k > 0$, so Lemma 5.1 gives $\hat{\nu}(x(0))\hat{\nu}(u) = \hat{\nu}(x(0)u)$. Thus, $\hat{\nu}$ satisfies the inductive hypothesis. Lemma 5.1 also implies that $\hat{\nu}$ is unique on $\mathbf{U}$, since the difference of two extensions would be zero on $(\mathbf{U})_0$ and $(\mathbf{U})_{1/2}$, and by induction would be zero on all $(\mathbf{U})_n$. ∎

PROPOSITION 5.3. *Let ν be as in Proposition 5.2, suppose ν and the $\mathbb{C}$-antilinear antiautomorphism ϑ of $\mathfrak{o}(8)$ commute, and suppose ν is unitary on $(\mathbf{U})_{1/2}$. Then $\hat{\nu}$ commutes with adjoint * on $\hat{\mathfrak{o}}(8)$ and the extension $\hat{\nu} : \mathbf{U} \to \mathbf{U}$ from Proposition 5.2 is unitary on all of $\mathbf{U}$. Furthermore, we have $\hat{\nu}D(n)\hat{\nu}^{-1} = D(n)$ for $n \in \mathbb{Z}$.*

PROOF. For any $x(n) \in \hat{\mathfrak{o}}(8)$, from (5.17) we have $\hat{\nu}(x(n)^*) = \hat{\nu}((\vartheta x)(-n)) = (\nu\vartheta x)(-n) = (\vartheta\nu x)(-n) = (\nu x)(n)^* = \hat{\nu}(x(n))^*$. Since $\hat{\nu}$ is the identity operator on $(\mathbf{U})_0$, it is unitary on $(\mathbf{U})_0 \oplus (\mathbf{U})_{1/2}$. Assume $\hat{\nu}$ is unitary on $(\mathbf{U})_n$ for $0 \leq n \leq N$, and let $p > 0$, $x(-p)u_1, u_2 \in (\mathbf{U})_{N+1}$. Then we have

$$\begin{aligned}
(\hat{\nu}(x(-p)u_1), \hat{\nu}(u_2)) &= (\hat{\nu}(x(-p))\hat{\nu}(u_1), \hat{\nu}(u_2)) = (\hat{\nu}(u_1), \hat{\nu}(x(-p)^*)\hat{\nu}(u_2)) \\
&= (\hat{\nu}(u_1), \hat{\nu}(x(-p)^*u_2)) = (u_1, x(-p)^*u_2) = (x(-p)u_1, u_2).
\end{aligned}$$

By the construction of $\hat{\nu}$ from Proposition 5.2, $\hat{\nu}$ preserves each subspace $(\mathbf{U})_n$, so $\hat{\nu}$ commutes with $D(0)$. An induction argument similar to that given in

Proposition 5.2 shows that $\hat{\nu}$ commutes with $D(n)$ for $n > 0$. Since $\hat{\nu}$ is unitary, $[\hat{\nu}, D(-n)] = [D(n), \hat{\nu}^{-1}]^* = 0$. ∎

THEOREM 5.4. *Suppose G is any group of automorphisms of $\mathbf{o}(8)$ commuting with the $\mathbb{C}$-antilinear antiautomorphism ϑ of $\mathbf{o}(8)$. Suppose G also acts as unitary isomorphisms on $(\mathbf{U})_{1/2}$ intertwining the action of $\mathbf{o}(8)$. Then G extends to an action on $\hat{\mathbf{o}}(8)$ commuting with adjoint * and G extends uniquely to unitary isomorphisms of $\mathbf{U}$ intertwining the action of $\hat{\mathbf{o}}(8)$.*

PROOF. Propositions 5.2 and 5.3 are valid for groups of automorphisms satisfying the assumptions because the uniqueness of the extensions implies that relations also lift. ∎

THEOREM 5.5. *Let the triality group G act on $\mathbf{C}$ as described in Chapter 4. Let Υ, defined in (5.20), transfer that action to a unitary action of G on $(\mathbf{U})_{1/2}$ as in (5.23). Let G act on $\mathbf{o}(8)$ by $g \cdot {}^{\circ}_{\circ}ab{}^{\circ}_{\circ} = {}^{\circ}_{\circ}(g \cdot a)(g \cdot b){}^{\circ}_{\circ}$ for $g \in G$, $a, b \in \mathbf{C}^{(k)}$, $1 \le k \le 3$, so that G intertwines the action of $\mathbf{o}(8)$ on $\mathbf{C}$, $g \cdot (x \cdot a) = (g \cdot x) \cdot (g \cdot a)$ for $g \in G$, $x \in \mathbf{o}(8)$, $a \in \mathbf{C}$, and G commutes with ϑ on $\mathbf{o}(8)$. Let G act on $\hat{\mathbf{o}}(8)$ by $\hat{g}(x(n)) = (gx)(n)$, $\hat{g}(c) = c$, $\hat{g}(d) = d$ for $x \in \mathbf{o}(8)$, $n \in \mathbb{Z}$. Then G lifts uniquely to a group of unitary isomorphisms $\hat{g} : \mathbf{U} \to \mathbf{U}$ such that $\hat{g}$ restricts to the identity operator on $(\mathbf{U})_0$, $\hat{g}$ restricts to g on $(\mathbf{U})_{1/2}$, and $\hat{g}x(n)\hat{g}^{-1} = (gx)(n)$ on $\mathbf{U}$ for $x(n) \in \hat{\mathbf{o}}(8)$. Furthermore, G commutes with the adjoint * on $\hat{\mathbf{o}}(8)$ and G commutes with the Virasoro operators $D(n)$, $n \in \mathbb{Z}$, on $\mathbf{U}$.*

THEOREM 5.6. *For any $g \in G$, $u \in \mathbf{U}_0$, on $\mathbf{U}$ we have*

$$\hat{g}Y(u, \zeta)\hat{g}^{-1} = Y(\hat{g}u, \zeta). \tag{5.32}$$

PROOF. We will prove this by induction on $\mathrm{wt}(u)$ as in the proof of Proposition 5.2. The base case, when $\mathrm{wt}(u) = 0$ and we may take $u = \mathbf{vac}(\mathbb{Z}+\frac{1}{2})$, is valid since $\hat{g}\mathbf{vac}(\mathbb{Z}+\frac{1}{2}) = \mathbf{vac}(\mathbb{Z}+\frac{1}{2})$ and $Y(\mathbf{vac}(\mathbb{Z}+\frac{1}{2}), \zeta)$ is the identity operator on $\mathbf{U}$. If $\mathrm{wt}(u) > 0$ then u is a linear combination of vectors of the form $x(-m)u_0$ for $1 \le m \in \mathbb{Z}$, $x \in (\mathbf{U}_0)_1 \cong \mathbf{o}(8)$, $\mathrm{wt}(u_0) < \mathrm{wt}(u)$. Since (5.32) is linear in u, it suffices to do the inductive step with $u = x(-m)u_0$. From (3.154) in Corollary 3.39, Theorem 5.5 and the induction assumption $\hat{g}Y(u_0, \zeta)\hat{g}^{-1} = Y(\hat{g}u_0, \zeta)$, we have

$$\hat{g}Y(u, \zeta)\hat{g}^{-1} = \hat{g}Y(x(-m)u_0, \zeta)\hat{g}^{-1}$$

$$= \hat{g} \sum_{0 \le k \in \mathbb{Z}} \binom{m+k-1}{k} [\zeta^k x(-m-k)Y(u_0, \zeta) - (-1)^m \zeta^{-m-k}Y(u_0, \zeta)x(k)]\hat{g}^{-1}$$

$$= \sum_{0 \le k \in \mathbb{Z}} \binom{m+k-1}{k} [\zeta^k (gx)(-m-k)Y(\hat{g}u_0, \zeta)$$

$$- (-1)^m \zeta^{-m-k}Y(\hat{g}u_0, \zeta)(gx)(k)]$$

$$= Y((gx)(-m)\hat{g}u_0, \zeta) = Y(\hat{g}u, \zeta). \quad ∎$$

For $u \in \mathbf{U}_1$ define the vertex operators

$$Y(\hat{\sigma}u, \zeta) = \hat{\sigma}Y(u, \zeta)\hat{\sigma}^{-1} \tag{5.33}$$

and

$$Y(\hat{\sigma}^2 u, \zeta) = \hat{\sigma}^2 Y(u, \zeta)\hat{\sigma}^{-2}. \tag{5.34}$$

For all homogeneous $u \in \mathbf{U}$ this defines

$$Y(u, \zeta) = \sum_{n \in \frac{1}{2}\mathbb{Z}} Y_{n+1-\mathrm{wt}(u)}(u)\zeta^{-n-1} = \sum_{n \in \frac{1}{2}\mathbb{Z}} \{u\}_n \zeta^{-n-1} \tag{5.35}$$

and on $\mathbf{U}$ we have

$$\hat{\sigma}Y(u, \zeta)\hat{\sigma}^{-1} = Y(\hat{\sigma}u, \zeta). \tag{5.36}$$

For $a \in \mathbf{C}^{(1)}$, $u = a(-\frac{1}{2})\mathbf{vac}(\mathbb{Z} + \frac{1}{2}) \in (\mathbf{U}_1)_{1/2}$ and we have the operators on $\mathbf{U}$

$$a(m) = Y_m(u) = \{u\}_{m-\frac{1}{2}} \tag{5.37}$$

for $m \in \frac{1}{2}\mathbb{Z}$. Recall that for $m \in \mathbb{Z}$, $a(m)$ is zero on $\mathbf{CM}(Z')$, but $a(m)$ represents $\mathbf{Cliff}(Z)$ on $\mathbf{CM}(Z)$, where $Z' = \frac{1}{2}\mathbb{Z} - Z$. Formulas (5.33) - (5.34) then define on $\mathbf{U}$ the operators

$$(\sigma a)(m) = Y_m(\hat{\sigma}u) = \{\hat{\sigma}u\}_{m-\frac{1}{2}} = \hat{\sigma}a(m)\hat{\sigma}^{-1}, \tag{5.38}$$

and

$$(\sigma^2 a)(m) = Y_m(\hat{\sigma}^2 u) = \{\hat{\sigma}^2 u\}_{m-\frac{1}{2}} = \hat{\sigma}^2 a(m)\hat{\sigma}^{-2} \tag{5.39}$$

for $m \in \frac{1}{2}\mathbb{Z}$. Conjugation of the Clifford relations in $\mathbf{Cliff}(Z)$ by $\hat{\sigma}$ makes it clear that for $1 \leq k \leq 3$, $Z = \mathbb{Z} + \frac{1}{2}$ or $\mathbb{Z}$, we have an irreducible representation of the Clifford algebra $\mathbf{Cliff}^{(k)}(Z)$ generated by

$$\mathbf{C}^{(k)}(Z) = \{a(m) \mid a \in \mathbf{C}^{(k)}, m \in Z\} \tag{5.40}$$

and the form (2.39) on

$$\mathbf{CM}^{(k)}(Z) = \hat{\sigma}^{k-1}\mathbf{CM}(Z) = \begin{cases} \mathbf{U}_0 \oplus \mathbf{U}_k & \text{if } Z = \mathbb{Z} + \frac{1}{2} \\[2mm] \mathbf{U}_{k'} \oplus \mathbf{U}_{k''} & \text{if } Z = \mathbb{Z}. \end{cases} \tag{5.41}$$

This is the irreducible $\mathbf{Cliff}^{(k)}(Z)$-module constructed in Chapter 2, where $\hat{\sigma}$ transfers the polarization of $\mathbf{C}^{(1)}(Z)$ to $\mathbf{C}^{(k)}(Z)$ and distinguishes a vacuum vector

$$\mathbf{vac}^{(k)}(Z) = \hat{\sigma}^{k-1}\mathbf{vac}(Z). \tag{5.42}$$

Since $\hat{g}$ is the identity operator on $(\mathbf{U})_0$ for all $g \in G$, we have

$$\mathbf{vac}(\mathbb{Z} + \tfrac{1}{2}) = \mathbf{vac}^{(1)}(\mathbb{Z} + \tfrac{1}{2}) = \mathbf{vac}^{(2)}(\mathbb{Z} + \tfrac{1}{2}) = \mathbf{vac}^{(3)}(\mathbb{Z} + \tfrac{1}{2}) \in (\mathbf{U})_0. \tag{5.43}$$

For (k, k', k'') a cyclic permutation of $(1,2,3)$ we have the even and odd parity subspaces

$$\begin{aligned} \mathbf{U}_0 &= \mathbf{CM}^{(k)}(\mathbb{Z} + \tfrac{1}{2})^0, & \mathbf{U}_k &= \mathbf{CM}^{(k)}(\mathbb{Z} + \tfrac{1}{2})^1, \\ \mathbf{U}_{k'} &= \mathbf{CM}^{(k)}(\mathbb{Z})^0, & \mathbf{U}_{k''} &= \mathbf{CM}^{(k)}(\mathbb{Z})^1, \end{aligned} \tag{5.44}$$

If $u \in \mathbf{U}_m$ and $v \in \mathbf{U}_n$ then from (5.44) it follows that

$$Y_r(u)v \in \mathbf{U}_{m+n} \ for \ r \in \tfrac{1}{2}\mathbb{Z}, \tag{5.45}$$

where we have put the additive $\mathbb{Z}_2 \times \mathbb{Z}_2$ group structure on $\{0, 1, 2, 3\}$. We also define the action of $G \cong Aut(\mathbb{Z}_2 \times \mathbb{Z}_2)$ on $\{0, 1, 2, 3\}$ such that $\hat{g}\mathbf{U}_k = \mathbf{U}_{g(k)}$ for $0 \le k \le 3$.

Since the action (5.23) of G on $(\mathbf{U})_{1/2}$ is transported by Υ from its action on $\mathbf{C}$, we have $\hat{g}\Upsilon(a) = \Upsilon(ga)$ for $g \in G$, $a \in \mathbf{C}$. Let $b \in \mathbf{CM}^0 = \mathbf{C}^{(2)}$, then $b = \sigma a$ for some $a \in \mathbf{A} = \mathbf{C}^{(1)}$ and

$$\begin{aligned}
\Upsilon(b) = \Upsilon(\sigma a) = \hat{\sigma}\Upsilon(a) &= \hat{\sigma}(a(-\tfrac{1}{2})\mathbf{vac}(\mathbb{Z} + \tfrac{i}{2})) \\
&= (\sigma a)(-\tfrac{1}{2})\mathbf{vac}(\mathbb{Z} + \tfrac{1}{2}) = b(-\tfrac{1}{2})\mathbf{vac}(\mathbb{Z} + \tfrac{1}{2}).
\end{aligned} \tag{5.46}$$

Similarly, for $c = \sigma^2 a \in \mathbf{CM}^1 = \mathbf{C}^{(3)}$, we get

$$\Upsilon(c) = c(-\tfrac{1}{2})\mathbf{vac}(\mathbb{Z} + \tfrac{1}{2}). \tag{5.47}$$

By applying $\hat{\sigma}$ to the third formula in (5.20) we get

$$\Upsilon(a \circ b) = a(0)\Upsilon(b) \tag{5.48}$$

for all $a, b \in \mathbf{C}$, taking into account where the operator $a(0)$ is zero. Then for $a, b \in \mathbf{C}$, we have

$$\begin{aligned}
a(0)b(-\tfrac{1}{2})\mathbf{vac}(\mathbb{Z} + \tfrac{1}{2}) &= a(0)\Upsilon(b) = \Upsilon(a \circ b) \\
&= (a \circ b)(-\tfrac{1}{2})\mathbf{vac}(\mathbb{Z} + \tfrac{1}{2}).
\end{aligned} \tag{5.49}$$

For $a(m), b(n) \in \mathbf{C}^{(1)}(Z)$, $i = 1, 2$, we have

$${}^{\circ}_{\circ}(\sigma^i a)(m)(\sigma^i b)(n){}^{\circ}_{\circ} = \hat{\sigma}^i {}^{\circ}_{\circ} a(m)b(n){}^{\circ}_{\circ}\hat{\sigma}^{-i}, \tag{5.50}$$

so that for $r \in \mathbb{Z}$,

$$\begin{aligned}
{}^{\circ}_{\circ}(\sigma^i a)(\zeta)(\sigma^i b)(\zeta){}^{\circ}_{\circ r} &= \hat{\sigma}^i {}^{\circ}_{\circ} a(\zeta)b(\zeta){}^{\circ}_{\circ r}\hat{\sigma}^{-i} \\
&= \hat{\sigma}^i({}^{\circ}_{\circ}ab{}^{\circ}_{\circ})(r)\hat{\sigma}^{-i} = (\sigma^i \cdot {}^{\circ}_{\circ}ab{}^{\circ}_{\circ})(r) = ({}^{\circ}_{\circ}(\sigma^i a)(\sigma^i b){}^{\circ}_{\circ})(r)
\end{aligned} \tag{5.51}$$

where the third equality comes from the part of Theorem 5.5 which says that conjugation by $\hat{\sigma}$ preserves $\hat{\mathbf{o}}(8)$. Hence, (5.51) implies that the spinor constructions of $\hat{\mathbf{o}}(8)$ on $\mathbf{U}$ from each $\mathbf{Cliff}^{(k)}(Z)$, $1 \le k \le 3$, coincide. Therefore, from (2.58) applied to $\mathbf{Cliff}^{(k)}(Z)$,

$$[x(m), a(r)] = (x \cdot a)(m + r) \tag{5.52}$$

for $x \in \mathbf{o}(8)$, $a \in \mathbf{C}$, $m \in \mathbb{Z}$, $r \in \tfrac{1}{2}\mathbb{Z}$. Let $D^{(k)}(m)$, $m \in \mathbb{Z}$, be the Virasoro operators constructed from $\mathbf{Cliff}^{(k)}(Z)$. Since σ takes a "canonical" basis of $\mathbf{C}^{(k)}$ to a canonical basis of $\mathbf{C}^{(k')}$, we see that

$$D^{(k')}(m) = \hat{\sigma}D^{(k)}(m)\hat{\sigma}^{-1}. \tag{5.53}$$

But in Proposition 5.3 we have seen that $\hat{g}D^{(1)}(m)\hat{g}^{-1} = D^{(1)}(m)$ for all $g \in G$, so

$$D^{(1)}(m) = D^{(2)}(m) = D^{(3)}(m) \quad \text{for} \quad m \in \mathbb{Z}, \tag{5.54}$$

and we may write just $D(m)$ for the Virasoro operators constructed from any $\mathbf{Cliff}^{(k)}(Z)$. It follows that

$$[D(m), a(r)] = (r + \tfrac{1}{2}m)a(r + m) \tag{5.55}$$

for $a \in \mathbf{C}$, $m \in \mathbb{Z}$, $r \in \tfrac{1}{2}\mathbb{Z}$.

All the results of Chapter 3 can now be applied from any of the three points of view, $1 \le k \le 3$, provided by the three Clifford constructions of $\mathbf{U}$, with $\mathbf{V}_0$, $\mathbf{V}_1$, $\mathbf{V}_2$, $\mathbf{V}_3$ and $\mathbf{V}(\tfrac{1}{2}\mathbb{Z})$ replaced by $\mathbf{U}_0$, $\mathbf{U}_k$, $\mathbf{U}_{k'}$, $\mathbf{U}_{k''}$ and $\mathbf{U}$, respectively. In particular, we will be applying the results of Corollaries 3.37 and 3.38 in the proof of the next theorem.

THEOREM 5.7. *For any $g \in G$, $u \in \mathbf{U}$, on $\mathbf{U}$ we have*

$$\hat{g}Y(u, \zeta)\hat{g}^{-1} = Y(\hat{g}u, \zeta). \tag{5.56}$$

PROOF. We have already established the result for any $g \in G$ and $u \in \mathbf{U}_0$ in Theorem 5.6, and for $sgn(g) = 1$ and any $u \in \mathbf{U}$ (5.36). So assume $sgn(g) = -1$ and u is in $\mathbf{U}_1$, $\mathbf{U}_2$ or $\mathbf{U}_3$. We will prove for $v \in \mathbf{U}$ that

$$\hat{g}Y(u, \zeta)v = Y(\hat{g}u, \zeta)\hat{g}v \tag{5.57}$$

by induction on $\mathrm{wt}(u) + \mathrm{wt}(v) \ge \tfrac{1}{2}$. The base cases are when $\mathrm{wt}(u) = \tfrac{1}{2}$ and $\mathrm{wt}(v) = 0$ or $\tfrac{1}{2}$. First we do the inductive step which reduces us to the base cases. If $\mathrm{wt}(u) > \tfrac{1}{2}$ then u is a linear combination of vectors of the form $x(-m)u_0$ for $1 \le m \in \mathbb{Z}$, $x \in (\mathbf{U}_0)_1 \cong \mathbf{o}(8)$, $\mathrm{wt}(u_0) < \mathrm{wt}(u)$. Since (5.57) is linear in u, it suffices to do the inductive step with $u = x(-m)u_0$. From (3.151) in Corollary 3.38, Theorem 5.6 and the induction assumption, we have

$$\hat{g}Y(u, \zeta)v = \hat{g}Y(x(-m)u_0, \zeta)v$$

$$= \hat{g}\sum_{0 \le k \in \mathbb{Z}} \binom{m + k - 1}{k}[\zeta^k x(-m - k)Y(u_0, \zeta) - (-1)^m \zeta^{-m-k} Y(u_0, \zeta)x(k)]v$$

$$= \sum_{0 \le k \in \mathbb{Z}} \binom{m + k - 1}{k}[\zeta^k (gx)(-m - k)Y(\hat{g}u_0, \zeta)$$

$$\qquad\qquad\qquad\qquad - (-1)^m \zeta^{-m-k} Y(\hat{g}u_0, \zeta)(gx)(k)]\hat{g}v$$

$$= Y((gx)(-m)\hat{g}u_0, \zeta)\hat{g}v = Y(\hat{g}u, \zeta)\hat{g}v.$$

If $\mathrm{wt}(v) > \tfrac{1}{2}$ then, as above, it suffices to do the inductive step with $v = x(-m)v_0$ where $1 \le m \in \mathbb{Z}$, $x \in (\mathbf{U}_0)_1$ and $\mathrm{wt}(v_0) < \mathrm{wt}(v)$. From (3.149) in Corollary

3.37, Theorem 5.6 and the induction assumption, we have

$$
\begin{aligned}
\hat{g}Y(u,\zeta)v &= \hat{g}Y(u,\zeta)x(-m)v_0 \\
&= \hat{g}x(-m)Y(u,\zeta)v_0 - \hat{g}[x(-m),Y(u,\zeta)]v_0 \\
&= (gx)(-m)\hat{g}Y(u,\zeta)v_0 - \hat{g}\zeta^{-m}\sum_{0\leq k\in\mathbb{Z}}\binom{-m}{k}\zeta^{-k}Y(x(k)u,\zeta)v_0 \\
&= (gx)(-m)Y(\hat{g}u,\zeta)\hat{g}v_0 - \zeta^{-m}\sum_{0\leq k\in\mathbb{Z}}\binom{-m}{k}\zeta^{-k}Y((gx)(k)\hat{g}u,\zeta)\hat{g}v_0 \\
&= (gx)(-m)Y(\hat{g}u,\zeta)\hat{g}v_0 - [(gx)(-m),Y(\hat{g}u,\zeta)]\hat{g}v_0 \\
&= Y(\hat{g}u,\zeta)(gx)(-m)\hat{g}v_0 \\
&= Y(\hat{g}u,\zeta)\hat{g}v.
\end{aligned}
$$

Having completed the inductive step, we now turn to the base cases. It will be convenient to use the abbreviation $\mathbf{vac} = \mathbf{vac}(\mathbb{Z}+\frac{1}{2})$. Since the theorem is known for $sgn(g) = 1$, it suffices to establish only the base cases where $u = a(-\frac{1}{2})\mathbf{vac} \in (\mathbf{U}_1)_{1/2}$ and $\mathrm{wt}(v) = 0$ or $\frac{1}{2}$. Then

$$
\hat{g}u = \hat{g}a(-\tfrac{1}{2})\mathbf{vac} = \hat{g}\Upsilon(a) = \Upsilon(ga) = (ga)(-\tfrac{1}{2})\mathbf{vac} \tag{5.58}
$$

and from the definitions, (5.38) and (5.39), on $\mathbf{CM}^{(g(1))}(Z)$ we have

$$
Y(\hat{g}u,\zeta) = \sum_{k\in\mathbb{Z}}(ga)(k)\zeta^{-k-\frac{1}{2}}. \tag{5.59}
$$

If $\mathrm{wt}(v) = 0$ then we may take $v = \mathbf{vac} \in (\mathbf{U}_0)_0$, so $\hat{g}v = v$, and we wish to prove for $0 \leq n \in \mathbb{Z}$ that

$$
\hat{g}a(-n-\tfrac{1}{2})\mathbf{vac} = (ga)(-n-\tfrac{1}{2})\mathbf{vac}. \tag{5.60}
$$

The $n = 0$ case is (5.58). If (5.60) is true for some n, $0 \leq n \in \mathbb{Z}$, then from (5.55), $[D(-1),\hat{g}] = 0$ and $D(-1)\mathbf{vac} = 0$, we get

$$
\begin{aligned}
\hat{g}a(-n-\tfrac{3}{2})\mathbf{vac} &= -(n+1)^{-1}\hat{g}D(-1)a(-n-\tfrac{1}{2})\mathbf{vac} \\
&= -(n+1)^{-1}D(-1)\hat{g}a(-n-\tfrac{1}{2})\mathbf{vac} \\
&= -(n+1)^{-1}D(-1)(ga)(-n-\tfrac{1}{2})\mathbf{vac} \tag{5.61} \\
&= -(n+1)^{-1}[D(-1),(ga)(-n-\tfrac{1}{2})]\mathbf{vac} \\
&= (ga)(-n-\tfrac{3}{2})\mathbf{vac}.
\end{aligned}
$$

So, by induction, we have established (5.60) for all $0 \leq n \in \mathbb{Z}$.

If $\mathrm{wt}(v) = \frac{1}{2}$ there are three cases, as $v \in (\mathbf{U}_k)_{1/2}$ for $1 \leq k \leq 3$. First, letting $v = a_1(-\frac{1}{2})\mathbf{vac} \in (\mathbf{U}_1)_{1/2}$, so $\hat{g}v = (ga_1)(-\frac{1}{2})\mathbf{vac}$ as in (5.58), we wish to prove for $0 \leq n \in \mathbb{Z}$ that

$$
\hat{g}a(-n+\tfrac{1}{2})a_1(-\tfrac{1}{2})\mathbf{vac} = (ga)(-n+\tfrac{1}{2})(ga_1)(-\tfrac{1}{2})\mathbf{vac}. \tag{5.62}
$$

The $n = 0$ case is

$$\hat{g}a(\tfrac{1}{2})a_1(-\tfrac{1}{2})\mathbf{vac} = \langle a, a_1 \rangle \mathbf{vac} = \langle ga, ga_1 \rangle \mathbf{vac}$$
$$= (ga)(\tfrac{1}{2})(ga_1)(-\tfrac{1}{2})\mathbf{vac}. \tag{5.63}$$

In the course of the induction argument to prove (5.62) we will need the $n = 1$ case as well. For $x = {}^{\circ}_{\circ}aa_1{}^{\circ}_{\circ} \in \mathbf{o}(8)$, $x(-1) = {}^{\circ}_{\circ}a(\zeta)a_1(\zeta){}^{\circ}_{\circ}{}_{-1} \in \hat{\mathbf{o}}(8)$ and

$$x(-1)\mathbf{vac} = {}^{\circ}_{\circ}a(\zeta)a_1(\zeta){}^{\circ}_{\circ}{}_{-1}\mathbf{vac} = a(-\tfrac{1}{2})a_1(-\tfrac{1}{2})\mathbf{vac}, \tag{5.64}$$

so from (5.51) we have

$$\hat{g}a(-\tfrac{1}{2})a_1(-\tfrac{1}{2})\mathbf{vac} = \hat{g}x(-1)\mathbf{vac} = (gx)(-1)\mathbf{vac}$$
$$= (ga)(-\tfrac{1}{2})(ga_1)(-\tfrac{1}{2})\mathbf{vac}. \tag{5.65}$$

We will give an induction argument which applies to only certain vectors a, a_1, and then show why these imply the result in general.

For $a, a_1 \in \mathbf{C}^{(1)}$ if $x \in \mathbf{o}(8)$ is chosen such that $a_1(-\tfrac{1}{2})x(-1)\mathbf{vac} = 0$, then

$$(x \cdot a)(-n - \tfrac{1}{2})a_1(-\tfrac{1}{2})\mathbf{vac} = [x(-1), a(-n + \tfrac{1}{2})]a_1(-\tfrac{1}{2})\mathbf{vac}$$
$$= x(-1)a(-n + \tfrac{1}{2})a_1(-\tfrac{1}{2})\mathbf{vac} - a(-n + \tfrac{1}{2})x(-1)a_1(-\tfrac{1}{2})\mathbf{vac} \tag{5.66}$$

and

$$a(-n + \tfrac{1}{2})x(-1)a_1(-\tfrac{1}{2})\mathbf{vac} = a(-n + \tfrac{1}{2})[x(-1), a_1(-\tfrac{1}{2})]\mathbf{vac}$$
$$= a(-n + \tfrac{1}{2})(x \cdot a_1)(-\tfrac{3}{2})\mathbf{vac}$$
$$= -a(-n + \tfrac{1}{2})D(-1)(x \cdot a_1)(-\tfrac{1}{2})\mathbf{vac}$$
$$= -D(-1)a(-n + \tfrac{1}{2})(x \cdot a_1)(-\tfrac{1}{2})\mathbf{vac} + [D(-1), a(-n + \tfrac{1}{2})](x \cdot a_1)(-\tfrac{1}{2})\mathbf{vac}$$
$$= -D(-1)a(-n + \tfrac{1}{2})(x \cdot a_1)(-\tfrac{1}{2})\mathbf{vac} - na(-n - \tfrac{1}{2})(x \cdot a_1)(-\tfrac{1}{2})\mathbf{vac}. \tag{5.67}$$

So when $a_1(-\tfrac{1}{2})x(-1)\mathbf{vac} = 0$, we get

$$(x \cdot a)(-n - \tfrac{1}{2})a_1(-\tfrac{1}{2})\mathbf{vac} - na(-n - \tfrac{1}{2})(x \cdot a_1)(-\tfrac{1}{2})\mathbf{vac}$$
$$= x(-1)a(-n + \tfrac{1}{2})a_1(-\tfrac{1}{2})\mathbf{vac} + D(-1)a(-n + \tfrac{1}{2})(x \cdot a_1)(-\tfrac{1}{2})\mathbf{vac}. \tag{5.68}$$

Choosing nonzero vectors a, a_1 and $x = {}^{\circ}_{\circ}aa_1{}^{\circ}_{\circ} \in \mathbf{o}(8)$ such that

$$\langle a, a \rangle = \langle a_1, a_1 \rangle = 0, \quad \langle a, a_1 \rangle = 1 \tag{5.69}$$

then we have

$$x \cdot a = a, \quad x \cdot a_1 = -a_1, \quad a_1(-\tfrac{1}{2})x(-1)\mathbf{vac} = a_1(-\tfrac{1}{2})a(-\tfrac{1}{2})a_1(-\tfrac{1}{2})\mathbf{vac} = 0 \tag{5.70}$$

and (5.68) gives

$$(n + 1)a(-n - \tfrac{1}{2})a_1(-\tfrac{1}{2})\mathbf{vac} = (x(-1) - D(-1))a(-n + \tfrac{1}{2})a_1(-\tfrac{1}{2})\mathbf{vac}. \tag{5.71}$$

Since $a_1(-\frac{1}{2})x(-1)\mathbf{vac} = 0$ and (5.51) imply $(ga_1)(-\frac{1}{2})(gx)(-1)\mathbf{vac} = 0$, (5.66) - (5.71) are valid with a, a_1 and x replaced by ga, ga_1 and gx, giving

$$(n+1)(ga)(-n - \tfrac{1}{2})(ga_1)(-\tfrac{1}{2})\mathbf{vac}$$
$$= ((gx)(-1) - D(-1))(ga)(-n + \tfrac{1}{2})(ga_1)(-\tfrac{1}{2})\mathbf{vac}. \tag{5.72}$$

Assuming that (5.62) is true for some n, $0 \le n \in \mathbb{Z}$, for a, a_1 satisfying (5.69) and $x = {}^{\circ}_{\circ}aa_1{}^{\circ}_{\circ}$, we have

$$(n+1)\hat{g}a(-n - \tfrac{1}{2})a_1(-\tfrac{1}{2})\mathbf{vac}$$
$$= \hat{g}(x(-1) - D(-1))a(-n + \tfrac{1}{2})a_1(-\tfrac{1}{2})\mathbf{vac}$$
$$= ((gx)(-1) - D(-1))\hat{g}a(-n + \tfrac{1}{2})a_1(-\tfrac{1}{2})\mathbf{vac} \tag{5.73}$$
$$= ((gx)(-1) - D(-1))(ga)(-n + \tfrac{1}{2})(ga_1)(-\tfrac{1}{2})\mathbf{vac}$$
$$= (n+1)(ga)(-n - \tfrac{1}{2})(ga_1)(-\tfrac{1}{2})\mathbf{vac}.$$

So, by induction, (5.62) is true for a, a_1 satisfying (5.69) and $0 \le n \in \mathbb{Z}$.

To see why (5.62), with the restrictions (5.69), implies the general case, we argue as follows. For $0 \le n \in \mathbb{Z}$, $1 \le k \le 3$, define the map $F_k : \mathbf{C}^{(k)} \otimes \mathbf{C}^{(k)} \to (\mathbf{U}_0)_n$ by

$$F_k(a \otimes b) = a(-n + \tfrac{1}{2})b(-\tfrac{1}{2})\mathbf{vac}. \tag{5.74}$$

Then F_k is an $\mathbf{o}(8)$-module map, that is,

$$x(0)F_k(a \otimes b) = F_k(x \cdot (a \otimes b)). \tag{5.75}$$

As at the end of Chapter 4, we denote the 8-dimensional $\mathbf{o}(8)$-module $\mathbf{C}^{(k)}$ by 8_k. From Proposition 4.24 (a), we have the tensor product decomposition into irreducible $\mathbf{o}(8)$-modules

$$8_1 \otimes 8_1 = 35_1 \oplus 28 \oplus 1 \tag{5.76}$$

where 28 is the adjoint representation consisting of antisymmetric tensors and $35_1 \oplus 1$ consists of the symmetric tensors. The assertion of (5.62) is that on $\mathbf{C}^{(1)} \otimes \mathbf{C}^{(1)}$,

$$\hat{g}F_1 = F_{g(1)}g, \tag{5.77}$$

where $g(a \otimes a_1) = ga \otimes ga_1$. If we knew, for a particular nonzero vector $w_1 \otimes w_2 \in \mathbf{C}^{(1)} \otimes \mathbf{C}^{(1)}$, that

$$\hat{g}F_1(w_1 \otimes w_2) = F_{g(1)}(g(w_1 \otimes w_2)) \tag{5.78}$$

then for any $x \in \mathbf{o}(8)$ we would have

$$\hat{g}F_1(x \cdot (w_1 \otimes w_2)) = \hat{g}(x(0)F_1(w_1 \otimes w_2)) = (gx)(0)\hat{g}F_1(w_1 \otimes w_2)$$
$$= (gx)(0)F_{g(1)}(g(w_1 \otimes w_2)) = F_{g(1)}((gx) \cdot g(w_1 \otimes w_2)) \tag{5.79}$$
$$= F_{g(1)}(g(x \cdot (w_1 \otimes w_2))).$$

It therefore suffices to establish (5.77) for one nonzero vector in each irreducible component of $8_1 \otimes 8_1$. Let $\{a_1, \ldots, a_4, a_1^*, \ldots, a_4^*\}$ be a canonical basis of $\mathbf{C}^{(1)}$, so

$$\langle a_i, a_j \rangle = \langle a_i^*, a_j^* \rangle = 0, \quad \langle a_i, a_j^* \rangle = \delta_{ij} \text{ for } 1 \leq i, j \leq 4. \tag{5.80}$$

Then (5.78) is true for $w_1 \otimes w_2 = a_i \otimes a_i^*$ and for $w_1 \otimes w_2 = a_i^* \otimes a_i$, $1 \leq i \leq 4$, so it is true for any linear combination of such vectors. It remains only to find a nonzero vector in each component of (5.76) which is such a linear combination. The easiest to find is the nonzero antisymmetric vector $a_i \otimes a_i^* - a_i^* \otimes a_i \in 28$. It is easy to check that the trivial module in (5.76) is spanned by

$$\sum_{1 \leq i \leq 4} (a_i \otimes a_i^* + a_i^* \otimes a_i). \tag{5.81}$$

Let $\{{}^{\circ}_{\circ} a_r a_s {}^{\circ}_{\circ}, {}^{\circ}_{\circ} a_r a_s^* {}^{\circ}_{\circ} \mid 1 \leq r < s \leq 4\}$ span the positive root vectors of $\mathbf{o}(8)$. Then a_1 is a highest weight vector in 8_1 so $a_1 \otimes a_1 \in 35_1$. Then

$$\begin{aligned}
{}^{\circ}_{\circ} a_1^* a_2^* {}^{\circ}_{\circ} \cdot ({}^{\circ}_{\circ} a_2 a_1^* {}^{\circ}_{\circ} \cdot (a_1 \otimes a_1)) &= {}^{\circ}_{\circ} a_1^* a_2^* {}^{\circ}_{\circ} \cdot (a_1 \otimes a_2 + a_2 \otimes a_1) \\
&= (a_1 \otimes a_1^* + a_1^* \otimes a_1) - (a_2 \otimes a_2^* + a_2^* \otimes a_2) \in 35_1.
\end{aligned} \tag{5.82}$$

This completes the proof of (5.62).

If $v = b(-\frac{1}{2})\mathbf{vac} \in (\mathbf{U}_2)_{1/2}$ or $(\mathbf{U}_3)_{1/2}$ then

$$\hat{g}v = \hat{g}\Upsilon(b) = \Upsilon(gb) = (gb)(-\tfrac{1}{2})\mathbf{vac} \tag{5.83}$$

and we wish to prove for $0 \leq n \in \mathbb{Z}$ that

$$\hat{g}a(-n)b(-\tfrac{1}{2})\mathbf{vac} = (ga)(-n)(gb)(-\tfrac{1}{2})\mathbf{vac}. \tag{5.84}$$

The $n = 0$ case is

$$\begin{aligned}
\hat{g}a(0)b(-\tfrac{1}{2})\mathbf{vac} &= \hat{g}(a \circ b)(-\tfrac{1}{2})\mathbf{vac} = \hat{g}\Upsilon(a \circ b) = \Upsilon(g(a \circ b)) \\
&= \Upsilon(ga \circ gb) = (ga \circ gb)(-\tfrac{1}{2})\mathbf{vac} = (ga)(0)(gb)(-\tfrac{1}{2})\mathbf{vac}.
\end{aligned} \tag{5.85}$$

As in the argument given above, we get an inductive formula for certain vectors a, b, and these imply the result in general.

For $a \in \mathbf{C}^{(1)}$, $b \in \mathbf{C}^{(2)}$ or $b \in \mathbf{C}^{(3)}$, if $x \in \mathbf{o}(8)$ is chosen such that $b(-\frac{1}{2})x(-1)\mathbf{vac} = 0$ then calculating as in (5.66) and (5.67) we get

$$\begin{aligned}
(x \cdot a)(-n-1)b(-\tfrac{1}{2})\mathbf{vac} &- (n + \tfrac{1}{2})a(-n-1)(x \cdot b)(-\tfrac{1}{2})\mathbf{vac} \\
&= x(-1)a(-n)b(-\tfrac{1}{2})\mathbf{vac} + D(-1)a(-n)(x \cdot b)(-\tfrac{1}{2})\mathbf{vac}.
\end{aligned} \tag{5.86}$$

Choosing nonzero vectors a, b and $x = {}^{\circ}_{\circ} b \vartheta b {}^{\circ}_{\circ} \in \mathbf{o}(8)$ such that

$$\langle a, a \rangle = \langle b, b \rangle = 0, \quad \langle b, \vartheta b \rangle = 1, \quad b \circ (\vartheta b \circ a) = 0 \tag{5.87}$$

then we have

$$x \cdot a = -\tfrac{1}{2}a, \quad x \cdot b = b, \quad b(-\tfrac{1}{2})x(-1)\mathbf{vac} = 0 \tag{5.88}$$

and (5.86) gives

$$-(n+1)a(-n-1)b(-\tfrac{1}{2})\mathbf{vac} = (x(-1) + D(-1))a(-n)b(-\tfrac{1}{2})\mathbf{vac}. \tag{5.89}$$

As before, $b(-\frac{1}{2})x(-1)\mathbf{vac} = 0$ implies $(gb)(-\frac{1}{2})(gx)(-1)\mathbf{vac} = 0$, we get

$$-(n+1)(ga)(-n-1)(gb)(-\tfrac{1}{2})\mathbf{vac}$$
$$= ((gx)(-1) - D(-1))(ga)(-n)(gb)(-\tfrac{1}{2})\mathbf{vac}, \tag{5.90}$$

and the induction argument goes as it did in (5.73), proving (5.84) for a and b satisfying (5.87).

We now show how these special cases imply the general case. For $0 \le n \in \mathbb{Z}$, $1 \le k \le 3$, define the map $G_k : \mathbf{C}^{(k)} \otimes \mathbf{C}^{(k')} \to (\mathbf{U}_{k''})_{n+\frac{1}{2}}$ by

$$G_k(a \otimes b) = a(-n)\Upsilon(b) \tag{5.91}$$

and define the map $H_k : \mathbf{C}^{(k)} \otimes \mathbf{C}^{(k'')} \to (\mathbf{U}_{k'})_{n+\frac{1}{2}}$ by

$$H_k(a \otimes b) = a(-n)\Upsilon(b). \tag{5.92}$$

Then G_k and H_k are $\mathbf{o}(8)$-module maps, that is,

$$x(0)G_k(a \otimes b) = G_k(x \cdot (a \otimes b)) \tag{5.93}$$

and

$$x(0)H_k(a \otimes b) = H_k(x \cdot (a \otimes b)) \tag{5.94}$$

for $x \in \mathbf{o}(8)$. The assertion of (5.84) is that

$$\hat{g}G_1 = H_{g(1)}g \quad \text{on} \quad \mathbf{C}^{(1)} \otimes \mathbf{C}^{(2)} \tag{5.95}$$

and

$$\hat{g}H_1 = G_{g(1)}g \quad \text{on} \quad \mathbf{C}^{(1)} \otimes \mathbf{C}^{(3)}. \tag{5.96}$$

Working with (5.95), as in (5.79) the equality

$$\hat{g}G_1(w_1 \otimes w_2) = H_{g(1)}g(w_1 \otimes w_2) \tag{5.97}$$

for some vector $w_1 \otimes w_2 \in \mathbf{C}^{(1)} \otimes \mathbf{C}^{(2)}$ implies

$$\hat{g}G_1(x \cdot (w_1 \otimes w_2)) = H_{g(1)}g(x \cdot (w_1 \otimes w_2)) \tag{5.98}$$

for all $x \in \mathbf{o}(8)$, and similarly for (5.96). So, from Proposition 4.24 (b), it suffices to establish (5.97) for one nonzero vector in each irreducible component of

$$8_1 \otimes 8_2 = 56_3 \oplus 8_3. \tag{5.99}$$

Let $\mathbf{h}$ be a Cartan subalgebra of $\mathbf{o}(8)$ spanned by $\{h_i = {}^{\circ}_{\circ}c_i \vartheta c_i {}^{\circ}_{\circ} \mid 1 \le i \le 4\}$, where $\{c_1, \ldots, c_4, \vartheta c_1, \ldots, \vartheta c_4\}$, a canonical basis of $\mathbf{C}^{(1)}$ adapted to ϑ, and $\mathbf{vac} \in \mathbf{CM}$ are chosen as in Proposition 4.17. With respect to the form (2.36) on $\mathbf{o}(8)$, $\{h_i \mid 1 \le i \le 4\}$ is an orthonormal basis of $\mathbf{h}$. Let $\{\epsilon_i \mid 1 \le i \le 4\}$ be the dual orthonormal basis of $\mathbf{h}^*$ and take

$$\alpha_1 = \epsilon_1 - \epsilon_2, \quad \alpha_2 = \epsilon_2 - \epsilon_3, \quad \alpha_3 = \epsilon_3 - \epsilon_4, \quad \alpha_4 = \epsilon_3 + \epsilon_4, \tag{5.100}$$

as simple roots of $\mathbf{o}(8)$. From Proposition 4.17, the triality group has been constructed on $\mathbf{C}$ so that σ cyclically permutes highest weight vectors in 8_1, 8_2 and 8_3. The weights of $8_1 \otimes 8_2$ with respect to the Cartan subalgebra $\mathbf{h}$ are the sums of weights of 8_1 and weights of 8_2. As was discussed in Chapter 4, the weights of weight-vectors in each space are easily determined. The highest weights in 8_1, 8_2 and 8_3 are ϵ_1, $\frac{1}{2}(\epsilon_1 + \epsilon_2 + \epsilon_3 + \epsilon_4)$ and $\frac{1}{2}(\epsilon_1 + \epsilon_2 + \epsilon_3 - \epsilon_4)$, respectively. There are only four ways to express $\frac{1}{2}(\epsilon_1 + \epsilon_2 + \epsilon_3 - \epsilon_4)$ as a sum of two weights, one from 8_1 and the other from 8_2,

$$
\begin{aligned}
\tfrac{1}{2}(\epsilon_1 + \epsilon_2 + \epsilon_3 - \epsilon_4) &= -\epsilon_4 + \tfrac{1}{2}(\epsilon_1 + \epsilon_2 + \epsilon_3 + \epsilon_4) \\
&= \epsilon_3 + \tfrac{1}{2}(\epsilon_1 + \epsilon_2 - \epsilon_3 - \epsilon_4) \\
&= \epsilon_2 + \tfrac{1}{2}(\epsilon_1 - \epsilon_2 + \epsilon_3 - \epsilon_4) \\
&= \epsilon_1 + \tfrac{1}{2}(-\epsilon_1 + \epsilon_2 + \epsilon_3 - \epsilon_4).
\end{aligned}
\tag{5.101}
$$

These correspond to the four tensors which form a basis of the $\frac{1}{2}(\epsilon_1 + \epsilon_2 + \epsilon_3 - \epsilon_4)$ weight-space of $8_1 \otimes 8_2$,

$$
\vartheta c_4 \otimes \mathbf{vac}, \quad c_3 \otimes \vartheta c_3 \vartheta c_4 \mathbf{vac}, \quad c_2 \otimes \vartheta c_2 \vartheta c_4 \mathbf{vac}, \quad c_1 \otimes \vartheta c_1 \vartheta c_4 \mathbf{vac}.
\tag{5.102}
$$

Their span decomposes into a three-dimensional weight space of the 56_3 component of the tensor product, and the one-dimensional highest weight space of the 8_3 component. It is easy to see from (4.34) and (4.35) that each vector in (5.102) satisfies all the conditions in (5.87), so (5.95) has been proven. The proof of (5.96) is similar. This finally completes the proof of the theorem.   ■

CHAPTER 6

Spinor Construction of a Vertex Operator Para-algebra for $D_4^{(1)}$

In Chapter 5 we have seen how the order three automorphism $\hat{\sigma}$ in the triality group G acting on $\mathbf{U} = \mathbf{U}_0 \oplus \mathbf{U}_1 \oplus \mathbf{U}_2 \oplus \mathbf{U}_3$ provides for each k, $1 \leq k \leq 3$, a Clifford algebra $\mathbf{Cliff}^{(k)}(\mathbb{Z} + \frac{1}{2})$ with Clifford module $\mathbf{CM}^{(k)}(\mathbb{Z} + \frac{1}{2}) = \mathbf{U}_0 \oplus \mathbf{U}_k$ and a Clifford algebra $\mathbf{Cliff}^{(k)}(\mathbb{Z})$ with Clifford module $\mathbf{CM}^{(k)}(\mathbb{Z}) = \mathbf{U}_{k'} \oplus \mathbf{U}_{k''}$. These are the affine spinor constructions as given in Chapter 2, giving the same four representations of one affine algebra $\hat{o}(8)$ and of one Virasoro algebra $\mathbf{Vir}$ from three points of view. From the results of Chapter 3, we have a vertex operator superalgebra $(\mathbf{U}_0 \oplus \mathbf{U}_k, Y(\ ,\zeta), \mathbf{vac}(\mathbb{Z} + \frac{1}{2}), D(-2)\mathbf{vac}(\mathbb{Z} + \frac{1}{2}))$ and its canonically $\mathbb{Z}_2$-twisted representation on $\mathbf{U}_{k'} \oplus \mathbf{U}_{k''}$. In particular, we have a representation of the superalgebra $(\hat{\mathbf{U}}_0 \oplus \hat{\mathbf{U}}_k)/D(-1)(\hat{\mathbf{U}}_0 \oplus \hat{\mathbf{U}}_k)$ on $\mathbf{U}_0 \oplus \mathbf{U}_k$ and we have a representation of the superalgebra $(\hat{\mathbf{U}}_0 \oplus \hat{\mathbf{U}}_k')/D(-1)(\hat{\mathbf{U}}_0 \oplus \hat{\mathbf{U}}_k')$ on $\mathbf{U}_{k'} \oplus \mathbf{U}_{k''}$, where

$$
\begin{aligned}
\hat{\mathbf{U}}_0 \oplus \hat{\mathbf{U}}_k &= (\mathbf{U}_0 \oplus \mathbf{U}_k) \otimes \mathbb{C}[t, t^{-1}], \\
\hat{\mathbf{U}}_0 \oplus \hat{\mathbf{U}}_k' &= \mathbf{U}_0 \otimes \mathbb{C}[t, t^{-1}] \oplus \mathbf{U}_k \otimes t^{1/2}\mathbb{C}[t, t^{-1}].
\end{aligned}
\tag{6.1}
$$

These representations are given by the vertex operators $Y(u, \zeta)$, which are defined for all $u \in \mathbf{U}$ and which satisfy the properties established in Chapter 3. In this chapter we will prove appropriately modified versions of the rationality, permutability, and associativity theorems which apply to operators $Y(u_1, \zeta_1)$ and $Y(u_2, \zeta_2)$ for $u_1 \in \mathbf{U}_k$ and $u_2 \in \mathbf{U}_{k'}$. These theorems will play a vital role in the spinor construction of vertex operator algebras for $E_8^{(1)}$ in Chapter 8. They also give a modified version of the Jacobi identity, showing that for appropriate Δ, $\Gamma = \mathbb{Z}_2 \times \mathbb{Z}_2$, and η, $(\mathbf{U}, Y(\ ,\zeta), \mathbf{vac}(\mathbb{Z} + \frac{1}{2}), D(-2)\mathbf{vac}(\mathbb{Z} + \frac{1}{2}), \Gamma, \Delta, \eta)$ is a vertex operator para-algebra as defined in the Introduction.

We first prove relations which will later be used to establish the base cases of inductive proofs of the rationality, permutability and associativity theorems. We emphasize that all of our proofs use only the spinor constructions given in earlier chapters, whose notations we use freely. For example, we write $\mathbf{vac} = \mathbf{vac}(\mathbb{Z} + \frac{1}{2})$.

LEMMA 6.1. *The coefficients $f_k = (-1)^k \binom{-\frac{1}{2}}{k}$ of the power series*

$$(1 - \zeta)^{-1/2} = \sum_{k \geq 0} f_k \zeta^k$$

are recursively determined by $f_0 = 1$, $f_{k+1} = f_k(k + \frac{1}{2})/(k+1)$. The coefficients $g_k = (-1)^k \binom{\frac{1}{2}}{k} = -f_k/(2k - 1)$ of the power series

$$(1 - \zeta)^{1/2} = \sum_{k \geq 0} g_k \zeta^k$$

are recursively determined by $g_0 = 1$, $g_{k+1} = g_k(k - \frac{1}{2})/(k+1)$.

PROPOSITION 6.2. *Let $a, a_1 \in \mathbf{C}^{(1)}$, $b, b_1 \in \mathbf{C}^{(2)}$, $c \in \mathbf{C}^{(3)}$. Then, as generating functions of identities we have*

(a) $\displaystyle\sum_{k \geq 0} \zeta^k a(k)b(-k - \frac{1}{2})\mathbf{vac} = (1 - \zeta)^{-1/2}(a \circ b)(-\frac{1}{2})\mathbf{vac},$

(b) $\displaystyle\sum_{k \geq 0} \zeta^k a(k + \frac{1}{2})b(-k)c(-\frac{1}{2})\mathbf{vac} = (1 - \zeta)^{-1/2}\langle a \circ b, c\rangle \mathbf{vac},$

(c) $\displaystyle\sum_{k \geq 0} \zeta^k a(k)b(-k)a_1(-\frac{1}{2})\mathbf{vac}$

$\qquad = -\zeta(1 - \zeta)^{-1/2}\langle a, a_1\rangle b(-\frac{1}{2})\mathbf{vac} + (1 - \zeta)^{-1/2}(a \circ (b \circ a_1))(-\frac{1}{2})\mathbf{vac},$

(d) $\displaystyle\sum_{k \geq 0} \zeta^k a(k - \frac{1}{2})b(-k + \frac{1}{2})b_1(-\frac{1}{2})\mathbf{vac}$

$\qquad = (1 - \zeta)^{1/2}\langle b, b_1\rangle a(-\frac{1}{2})\mathbf{vac} + \zeta(1 - \zeta)^{-1/2}((a \circ b) \circ b_1)(-\frac{1}{2})\mathbf{vac}.$

PROOF. For convenience we will write $\mathbf{C}^{(k)} = 8_k$, $1 \leq k \leq 3$, for the three eight-dimensional $\mathbf{o}(8)$ modules whose direct sum $\mathbf{C}$ has the $\circ$ algebra structure. We also write $\,^{\circ}_{\circ}aa_1{}^{\circ}_{\circ n}$ for $\,^{\circ}_{\circ}a(\zeta)a_1(\zeta)^{\circ}_{\circ n}$. Recall that Υ (5.20) is used to identify $\mathbf{C}$ with $(\mathbf{U})_{1/2}$.

(a) In Chapter 5 we saw how the Virasoro operators $D(m)$, $m \in \mathbb{Z}$, satisfy (5.55). In particular, using

$$[D(-1), a(k + 1)] = (k + \tfrac{1}{2})a(k) \tag{6.2}$$

and

$$[D(-1), b(-k - \tfrac{1}{2})] = -(k + 1)b(-k - \tfrac{3}{2}) \tag{6.3}$$

we get

$$\begin{aligned}
a(k)b(-k - \tfrac{1}{2})\mathbf{vac} &= (k + \tfrac{1}{2})^{-1}[D(-1), a(k + 1)]b(-k - \tfrac{1}{2})\mathbf{vac} \\
&= -(k + \tfrac{1}{2})^{-1}a(k + 1)[D(-1), b(-k - \tfrac{1}{2})]\mathbf{vac} \\
&= (k + 1)(k + \tfrac{1}{2})^{-1}a(k + 1)b(-k - \tfrac{3}{2})\mathbf{vac}
\end{aligned} \tag{6.4}$$

because $D(-1)\mathbf{vac} = 0$ and $a(k + 1)b(-k - \frac{1}{2})\mathbf{vac} = 0$. From (5.48) we have

$$a(0)b(-\tfrac{1}{2})\mathbf{vac} = (a \circ b)(-\tfrac{1}{2})\mathbf{vac} \tag{6.5}$$

and (6.4) inductively gives

$$a(k)b(-k-\tfrac{1}{2})\mathbf{vac} = f_k(a \circ b)(-\tfrac{1}{2})\mathbf{vac}. \tag{6.6}$$

(b) It is clear that for each $k \geq 0$, $a(k+\tfrac{1}{2})b(-k)c(-\tfrac{1}{2})\mathbf{vac} = \alpha_k\mathbf{vac}$ for some scalar α_k. To find the scalar, compute the Hermitian pairing

$$
\begin{aligned}
(a(k+\tfrac{1}{2})b(-k)c(-\tfrac{1}{2})\mathbf{vac}, \mathbf{vac}) &= (c(-\tfrac{1}{2})\mathbf{vac}, (\vartheta b)(k)(\vartheta a)(-k-\tfrac{1}{2})\mathbf{vac}) \\
&= (c(-\tfrac{1}{2})\mathbf{vac}, f_k(\vartheta b \circ \vartheta a)(-\tfrac{1}{2})\mathbf{vac}) = f_k(c, \vartheta b \circ \vartheta a) \\
&= f_k\langle c, \vartheta(\vartheta b \circ \vartheta a)\rangle = f_k\langle c, a \circ b\rangle = f_k\langle a \circ b, c\rangle.
\end{aligned}
$$

(c) It is clear that for each $k \geq 0$, $a(k)b(-k)a_1(-\tfrac{1}{2})\mathbf{vac}$ is of the form $b_1(-\tfrac{1}{2})\mathbf{vac}$ for some $b_1 \in \mathbf{C}^{(2)}$, and the map sending $a \otimes b \otimes a_1 \in 8_1 \otimes 8_2 \otimes 8_1$ to $a(k)b(-k)a_1(-\tfrac{1}{2})\mathbf{vac} \in (\mathbf{U}_2)_{1/2} \cong 8_2$ is an $\mathbf{o}(8)$-module map. Therefore, for each $k \geq 0$, $a(k)b(-k)a_1(-\tfrac{1}{2})\mathbf{vac}$ is some linear combination of $\langle a, a_1\rangle b(-\tfrac{1}{2})\mathbf{vac}$ and $({}^{\circ}_{\circ}aa_1{}^{\circ}_{\circ} \cdot b)(-\tfrac{1}{2})\mathbf{vac}$. From Proposition 4.8 and (4.38) we have

$$({}^{\circ}_{\circ}aa_1{}^{\circ}_{\circ} \cdot b)(-\tfrac{1}{2})\mathbf{vac} = (a \circ (b \circ a_1))(-\tfrac{1}{2})\mathbf{vac} - \tfrac{1}{2}\langle a, a_1\rangle b(-\tfrac{1}{2})\mathbf{vac}. \tag{6.7}$$

This shows that for each $k \geq 0$, there are scalars α_k and β_k such that

$$a(k)b(-k)a_1(-\tfrac{1}{2})\mathbf{vac} = \alpha_k(a \circ (b \circ a_1))(-\tfrac{1}{2})\mathbf{vac} + \beta_k\langle a, a_1\rangle b(-\tfrac{1}{2})\mathbf{vac}. \tag{6.8}$$

The $k = 0$ case is

$$a(0)b(0)a_1(-\tfrac{1}{2})\mathbf{vac} = (a \circ (b \circ a_1))(-\tfrac{1}{2})\mathbf{vac} \tag{6.9}$$

so that $\alpha_0 = 1$ and $\beta_0 = 0$. If $x \in \mathbf{o}(8)$ is chosen such that $a_1(-\tfrac{1}{2})x(-1)\mathbf{vac} = 0$ then we have

$$
\begin{aligned}
(x \cdot a)(k)b(-k)a_1(-\tfrac{1}{2})\mathbf{vac} &= [x(-1), a(k+1)]b(-k)a_1(-\tfrac{1}{2})\mathbf{vac} \\
&= -a(k+1)x(-1)b(-k)a_1(-\tfrac{1}{2})\mathbf{vac} \\
&= -a(k+1)[x(-1), b(-k)]a_1(-\tfrac{1}{2})\mathbf{vac} - a(k+1)b(-k)x(-1)a_1(-\tfrac{1}{2})\mathbf{vac} \\
&= -a(k+1)(x \cdot b)(-k-1)a_1(-\tfrac{1}{2})\mathbf{vac} - a(k+1)b(-k)(x \cdot a_1)(-\tfrac{3}{2})\mathbf{vac} \\
&\quad - a(k+1)b(-k)a_1(-\tfrac{1}{2})x(-1)\mathbf{vac} \\
&= -a(k+1)(x \cdot b)(-k-1)a_1(-\tfrac{1}{2})\mathbf{vac} + a(k+1)b(-k)D(-1)(x \cdot a_1)(-\tfrac{1}{2})\mathbf{vac} \\
&= -a(k+1)(x \cdot b)(-k-1)a_1(-\tfrac{1}{2})\mathbf{vac} + a(k+1)D(-1)b(-k)(x \cdot a_1)(-\tfrac{1}{2})\mathbf{vac} \\
&\quad - a(k+1)[D(-1), b(-k)](x \cdot a_1)(-\tfrac{1}{2})\mathbf{vac} \\
&= -a(k+1)(x \cdot b)(-k-1)a_1(-\tfrac{1}{2})\mathbf{vac} + D(-1)a(k+1)b(-k)(x \cdot a_1)(-\tfrac{1}{2})\mathbf{vac} \\
&\quad - [D(-1), a(k+1)]b(-k)(x \cdot a_1)(-\tfrac{1}{2})\mathbf{vac} \\
&\quad + (k+\tfrac{1}{2})a(k+1)b(-k-1)(x \cdot a_1)(-\tfrac{1}{2})\mathbf{vac} \\
&= -a(k+1)(x \cdot b)(-k-1)a_1(-\tfrac{1}{2})\mathbf{vac} - (k+\tfrac{1}{2})a(k)b(-k)(x \cdot a_1)(-\tfrac{1}{2})\mathbf{vac} \\
&\quad + (k+\tfrac{1}{2})a(k+1)b(-k-1)(x \cdot a_1)(-\tfrac{1}{2})\mathbf{vac}
\end{aligned}
$$

since $a(k+1)b(-k)(x \cdot a_1)(-\frac{1}{2})\mathbf{vac} = 0$. So, when $a_1(-\frac{1}{2})x(-1)\mathbf{vac} = 0$ we have

$$(x \cdot a)(k)b(-k)a_1(-\tfrac{1}{2})\mathbf{vac} + (k + \tfrac{1}{2})a(k)b(-k)(x \cdot a_1)(-\tfrac{1}{2})\mathbf{vac}$$
$$= -a(k+1)(x \cdot b)(-k-1)a_1(-\tfrac{1}{2})\mathbf{vac} \qquad (6.10)$$
$$+ (k + \tfrac{1}{2})a(k+1)b(-k-1)(x \cdot a_1)(-\tfrac{1}{2})\mathbf{vac}.$$

Choosing nonzero vectors a, b, a_1 and $x = {}^{\circ}_{\circ}a_1 a^{\circ}_{\circ} \in \mathbf{o}(8)$ such that

$$\langle a, a \rangle = \langle a_1, a_1 \rangle = 0, \quad \langle a, a_1 \rangle = 1, \quad x \cdot b = -\tfrac{1}{2}b, \qquad (6.11)$$

we have

$$x \cdot a = -a, \quad x \cdot a_1 = a_1, \quad a_1(-\tfrac{1}{2})x(-1)\mathbf{vac} = 0. \qquad (6.12)$$

Since $x \cdot (a \circ b) = (x \cdot a) \circ b + a \circ (x \cdot b) = -\frac{3}{2}(a \circ b)$ and the only eigenvalues of such Cartan elements made from $\mathbf{C}^{(1)}$ acting on $\mathbf{C}^{(3)}$ are $\pm\frac{1}{2}$, we have $a \circ b = 0$, and therefore

$$a \circ (a_1 \circ b) = a \circ (a_1 \circ b) + a_1 \circ (a \circ b) = \langle a, a_1 \rangle b = b \neq 0. \qquad (6.13)$$

For these choices, we then have

$$a(k)b(-k)a_1(-\tfrac{1}{2})\mathbf{vac} = (\alpha_k + \beta_k)b(-\tfrac{1}{2})\mathbf{vac} \qquad (6.14)$$

and (6.10) gives

$$(k - \tfrac{1}{2})(\alpha_k + \beta_k) = (k+1)(\alpha_{k+1} + \beta_{k+1}) \qquad (6.15)$$

so, using $\alpha_0 + \beta_0 = 1$, we get $\alpha_k + \beta_k = g_k$ for $k \geq 0$ from Lemma 6.1.

If we instead choose nonzero vectors a, b, a_1 and $x = {}^{\circ}_{\circ}a_1 \vartheta a^{\circ}_{1\circ} \in \mathbf{o}(8)$ such that

$$\langle a, a \rangle = \langle a_1, a_1 \rangle = 0, \quad \langle a, a_1 \rangle = \langle a, \vartheta a_1 \rangle = 0, \quad \langle a_1, \vartheta a_1 \rangle = 1, \quad x \cdot b = -\tfrac{1}{2}b, \quad (6.16)$$

then we have

$$x \cdot a = 0, \quad x \cdot a_1 = a_1, \quad x \cdot \vartheta a_1 = -\vartheta a_1, \quad a_1(-\tfrac{1}{2})x(-1)\mathbf{vac} = 0. \qquad (6.17)$$

Since

$$a \circ (\vartheta a \circ b) + \vartheta a \circ (a \circ b) = \langle a, \vartheta a \rangle b \neq 0, \qquad (6.18)$$

at least one of the two terms on the left side is nonzero, and we may assume that a is chosen such that $a \circ b \neq 0$. Since $x \cdot (\vartheta a_1 \circ (a \circ b)) = -\frac{3}{2}(\vartheta a_1 \circ (a \circ b))$, as before $\vartheta a_1 \circ (a \circ b) = 0$. Then we get

$$\vartheta a_1 \circ (a_1 \circ (a \circ b)) = \vartheta a_1 \circ (a_1 \circ (a \circ b)) + a_1 \circ (\vartheta a_1 \circ (a \circ b)) = a \circ b \neq 0 \quad (6.19)$$

and so

$$a \circ (b \circ a_1) = -a_1 \circ (b \circ a) \neq 0. \qquad (6.20)$$

For these choices, we then have

$$a(k)b(-k)a_1(-\tfrac{1}{2})\mathbf{vac} = \alpha_k(a \circ (b \circ a_1))(-\tfrac{1}{2})\mathbf{vac} \qquad (6.21)$$

and (6.10) gives

$$(k + \tfrac{1}{2})\alpha_k = (k+1)\alpha_{k+1} \qquad (6.22)$$

so, using $\alpha_0 = 1$, we get $\alpha_k = f_k$ for $k \geq 0$ from Lemma 6.1, giving (c).

(d) It is clear that for each $k \geq 0$, $a(k - \frac{1}{2})b(-k + \frac{1}{2})b_1(-\frac{1}{2})\mathbf{vac}$ is in $(\mathbf{U}_1)_{1/2}$, so it is of the form $a_1(-\frac{1}{2})\mathbf{vac}$ for some $a_1 \in \mathbf{C}^{(2)}$. The function from $8_1 \otimes 8_2 \otimes 8_2$ to 8_1 defined by sending $a \otimes b \otimes b_1$ to $a(k - \frac{1}{2})b(-k + \frac{1}{2})b_1(-\frac{1}{2})\mathbf{vac}$ is clearly an $\mathbf{o}(8)$-module map. So, as in part (c), tensor product decompositions show that for each $k \geq 0$,

$$a(k - \tfrac{1}{2})b(-k + \tfrac{1}{2})b_1(-\tfrac{1}{2})\mathbf{vac}$$
$$= \alpha_k((a \circ b) \circ b_1)(-\tfrac{1}{2})\mathbf{vac} + \beta_k \langle b, b_1 \rangle a(-\tfrac{1}{2})\mathbf{vac} \tag{6.23}$$

for some scalars α_k, β_k depending only on k. The $k = 0$ case is

$$a(-\tfrac{1}{2})b(\tfrac{1}{2})b_1(-\tfrac{1}{2})\mathbf{vac} = -a(-\tfrac{1}{2})b_1(-\tfrac{1}{2})b(\tfrac{1}{2})\mathbf{vac} + \langle b, b_1 \rangle a(-\tfrac{1}{2})\mathbf{vac}$$
$$= \langle b, b_1 \rangle a(-\tfrac{1}{2})\mathbf{vac}, \tag{6.24}$$

so $\alpha_0 = 0$, $\beta_0 = 1$. As in the proof of part (c), if $x \in \mathbf{o}(8)$ is chosen such that $b_1(-\frac{1}{2})x(-1)\mathbf{vac} = 0$ then we have

$$(x \cdot a)(k - \tfrac{1}{2})b(-k + \tfrac{1}{2})b_1(-\tfrac{1}{2})\mathbf{vac} + ka(k - \tfrac{1}{2})b(-k + \tfrac{1}{2})(x \cdot b_1)(-\tfrac{1}{2})\mathbf{vac}$$
$$= -a(k + \tfrac{1}{2})(x \cdot b)(-k - \tfrac{1}{2})b_1(-\tfrac{1}{2})\mathbf{vac} + ka(k + \tfrac{1}{2})b(-k - \tfrac{1}{2})(x \cdot b_1)(-\tfrac{1}{2})\mathbf{vac}. \tag{6.25}$$

Choosing nonzero vectors a, b, b_1 and $x = {}^{\circ}_{\circ}b_1 b{}^{\circ}_{\circ} \in \mathbf{o}(8)$ such that

$$\langle b, b \rangle = \langle b_1, b_1 \rangle = 0, \quad \langle b, b_1 \rangle = 1, \quad x \cdot a = -\tfrac{1}{2}a, \tag{6.26}$$

we have

$$x \cdot b = -b, \quad x \cdot b_1 = b_1, \quad a \circ b = 0, \quad b_1(-\tfrac{1}{2})x(-1)\mathbf{vac} = 0. \tag{6.27}$$

For these choices we then have

$$a(k - \tfrac{1}{2})b(-k + \tfrac{1}{2})b_1(-\tfrac{1}{2})\mathbf{vac} = \beta_k a(-\tfrac{1}{2})\mathbf{vac} \tag{6.28}$$

and (6.25) gives

$$(k - \tfrac{1}{2})\beta_k = (k + 1)\beta_{k+1} \tag{6.29}$$

so, using $\beta_0 = 1$, we get $\beta_k = g_k$ for $k \geq 0$.

If we instead choose nonzero vectors a, b, b_1 and $x = {}^{\circ}_{\circ}b_1 b{}^{\circ}_{\circ} \in \mathbf{o}(8)$ such that

$$\langle b, b \rangle = \langle b_1, b_1 \rangle = 0, \quad \langle b, b_1 \rangle = 1, \quad x \cdot a = \tfrac{1}{2}a, \tag{6.30}$$

then we have

$$x \cdot b = -b, \quad x \cdot b_1 = b_1, \quad a \circ b_1 = 0, \quad b_1(-\tfrac{1}{2})x(-1)\mathbf{vac} = 0 \tag{6.31}$$

and $(a \circ b) \circ b_1 = a \neq 0$. So we get

$$a(k - \tfrac{1}{2})b(-k + \tfrac{1}{2})b_1(-\tfrac{1}{2})\mathbf{vac} = (\alpha_k + \beta_k)a(-\tfrac{1}{2})\mathbf{vac} \tag{6.32}$$

and (6.25) gives

$$(k + \tfrac{1}{2})(\alpha_k + \beta_k) = (k + 1)(\alpha_{k+1} + \beta_{k+1}). \tag{6.33}$$

Using $\alpha_0 + \beta_0 = 1$, we get $\alpha_k + \beta_k = f_k$ for $k \geq 0$, and Lemma 6.1 gives (d). $\blacksquare$

COROLLARY 6.3. *Let $a, a_1 \in \mathbf{C}^{(1)}$, $b, b_1 \in \mathbf{C}^{(2)}$, $c \in \mathbf{C}^{(3)}$. Then for $0 \leq k \in \mathbb{Z}$ we have*

(a) $\displaystyle\sum_{0 \leq p \leq k} f_p a(k-p) b(-k+p-\tfrac{1}{2})\mathbf{vac} = (a \circ b)(-\tfrac{1}{2})\mathbf{vac},$

(b) $\displaystyle\sum_{0 \leq p \leq k} f_p a(k-p+\tfrac{1}{2}) b(-k+p) c(-\tfrac{1}{2})\mathbf{vac} = \langle a \circ b, c \rangle \mathbf{vac},$

(c) $\displaystyle\sum_{0 \leq p \leq k} f_p a(k-p) b(-k+p) a_1(-\tfrac{1}{2})\mathbf{vac} = (a \circ (b \circ a_1))(-\tfrac{1}{2})\mathbf{vac}$ *if* $k = 0,$

$$= -((a \circ b) \circ a_1)(-\tfrac{1}{2})\mathbf{vac} \text{ if } k \geq 1,$$

(d) $\displaystyle\sum_{0 \leq p \leq k} f_p a(k-p-\tfrac{1}{2}) b(-k+p+\tfrac{1}{2}) b_1(-\tfrac{1}{2})\mathbf{vac} = \langle b, b_1 \rangle a(-\tfrac{1}{2})\mathbf{vac}$ *if* $k = 0,$

$$= ((a \circ b) \circ b_1)(-\tfrac{1}{2})\mathbf{vac} \text{ if } k \geq 1.$$

PROOF. These follow immediately from Proposition 6.2 after multiplying by $(1 - \zeta)^{-1/2}$ and comparing the coefficients of ζ^k on both sides. ∎

Remark: We used $a, a_1 \in \mathbf{C}^{(1)}$, $b, b_1 \in \mathbf{C}^{(2)}$, $c \in \mathbf{C}^{(3)}$ in Proposition 6.2 and Corollary 6.3 only for convenience. The statements are still valid for any permutation of $(1, 2, 3)$, not just cyclic permutations.

LEMMA 6.4. *For $x \in (\mathbf{U}_0)_1 \cong \mathbf{o}(8)$, $m \in \mathbb{Z}$, $x(m) = Y_m(x)$ represents $\hat{\mathbf{o}}(8)$ on $\mathbf{U}$ and for any $u \in \mathbf{U}$ we have*

(a) $[x(m), Y(u, \zeta)] = \zeta^m \displaystyle\sum_{0 \leq k \in \mathbb{Z}} \binom{m}{k} \zeta^{-k} Y(x(k)u, \zeta),$

(b) $Y(x(-m)u, \zeta) = $
$$\sum_{0 \leq k \in \mathbb{Z}} \binom{m+k-1}{k} [\zeta^k x(-m-k)Y(u, \zeta) - (-1)^m \zeta^{-m-k} Y(u, \zeta)x(k)]$$

(c) $[x(0), Y(u, \zeta)] = Y(x(0)u, \zeta),$

(d) $Y(x(-1)u, \zeta) = \displaystyle\sum_{0 \leq k \in \mathbb{Z}} [\zeta^k x(-1-k)Y(u, \zeta) + \zeta^{-1-k} Y(u, \zeta)x(k)],$

(e) $[D(-1), Y_{n+1}(u)] = (n + wt(u))Y_n(u) = Y_n(D(-1)u).$

PROOF. Parts (a) and (b) follow from Corollaries 3.37 and 3.38, which are now valid for all $u \in \mathbf{U}$. Part (c) is obtained from either (a) or (b) with $m = 0$, (d) is (b) with $m = 1$, and (e) follows from Proposition 3.17 (a). ∎

PROPOSITION 6.5. *Let $a, a_1 \in \mathbf{C}^{(1)}$, $b, b_1 \in \mathbf{C}^{(2)}$, $c \in \mathbf{C}^{(3)}$. Then, as generating functions of identities we have*

(a) $\displaystyle\sum_{k \geq 0} \zeta^k Y_{-1/2}(a(-k)b(-\tfrac{1}{2})\mathbf{vac})\mathbf{vac} = (a \circ b)(-\tfrac{1}{2})\mathbf{vac},$

(b) $\displaystyle\sum_{k \geq 0} \zeta^k Y_{1/2}(a(-k)b(-\tfrac{1}{2})\mathbf{vac})c(-\tfrac{1}{2})\mathbf{vac} = (1 + \zeta)^{-1/2} \langle a \circ b, c \rangle \mathbf{vac},$

(c) $\displaystyle\sum_{k \geq 0} \zeta^k Y_0(a(-k)b(-\tfrac{1}{2})\mathbf{vac})a_1(-\tfrac{1}{2})\mathbf{vac}$

$$= -(a \circ (b \circ a_1))(-\tfrac{1}{2})\mathbf{vac} + (1 + \zeta)^{-1} \langle a, a_1 \rangle b(-\tfrac{1}{2})\mathbf{vac},$$

(d) $\displaystyle\sum_{k\geq 0}\zeta^k Y_0(a(-k)b(-\tfrac{1}{2})\mathbf{vac})b_1(-\tfrac{1}{2})\mathbf{vac}$

$$= (1+\zeta)^{1/2}\langle b, b_1\rangle a(-\tfrac{1}{2})\mathbf{vac} - (1+\zeta)^{-1/2}((a\circ b_1)\circ b))(-\tfrac{1}{2})\mathbf{vac}.$$

PROOF. For $x \in \mathbf{o}(8)$, $s \in \tfrac{1}{2}\mathbb{Z}$, $u \in \mathbf{U}$, from Lemma 6.4 (d) we have

$$Y_s((x\cdot a)(-k)b(-\tfrac{1}{2})\mathbf{vac})u = Y_s([x(-1), a(-k+1)]b(-\tfrac{1}{2})\mathbf{vac})u$$

$$= Y_s(x(-1)a(-k+1)b(-\tfrac{1}{2})\mathbf{vac})u - Y_s(a(-k+1)x(-1)b(-\tfrac{1}{2})\mathbf{vac})u$$

$$= \sum_{n\geq 0}[x(-1-n)Y_{s+n+1}(a(-k+1)b(-\tfrac{1}{2})\mathbf{vac})u$$

$$+ Y_{s-n}(a(-k+1)b(-\tfrac{1}{2})\mathbf{vac})x(n)u)]$$

$$- Y_s(a(-k+1)[x(-1), b(-\tfrac{1}{2})]\mathbf{vac})u - Y_s(a(-k+1)b(-\tfrac{1}{2})x(-1)\mathbf{vac})u$$

(6.34)

and from Lemma 6.4 (e) we have

$$Y_s(a(-k+1)[x(-1), b(-\tfrac{1}{2})]\mathbf{vac})u = Y_s(a(-k+1)(x\cdot b)(-\tfrac{3}{2})\mathbf{vac})u$$

$$= -Y_s(a(-k+1)D(-1)(x\cdot b)(-\tfrac{1}{2})\mathbf{vac})u$$

$$= Y_s([D(-1), a(-k+1)](x\cdot b)(-\tfrac{1}{2})\mathbf{vac})u$$

$$- Y_s(D(-1)a(-k+1)(x\cdot b)(-\tfrac{1}{2})\mathbf{vac})u$$

$$= (\tfrac{1}{2} - k)Y_s(a(-k)(x\cdot b)(-\tfrac{1}{2})\mathbf{vac})u$$

$$- (s+k-\tfrac{1}{2})Y_s(a(-k+1)(x\cdot b)(-\tfrac{1}{2})\mathbf{vac})u.$$

(6.35)

Therefore, we have

$$Y_s((x\cdot a)(-k)b(-\tfrac{1}{2})\mathbf{vac})u + (\tfrac{1}{2} - k)Y_s(a(-k)(x\cdot b)(-\tfrac{1}{2})\mathbf{vac})u$$

$$= \sum_{n\geq 0}[x(-1-n)Y_{s+n+1}(a(-k+1)b(-\tfrac{1}{2})\mathbf{vac})u$$

$$+ Y_{s-n}(a(-k+1)b(-\tfrac{1}{2})\mathbf{vac})x(n)u]$$

$$- Y_s(a(-k+1)b(-\tfrac{1}{2})x(-1)\mathbf{vac})u$$

$$+ (s+k-\tfrac{1}{2})Y_s(a(-k+1)(x\cdot b)(-\tfrac{1}{2})\mathbf{vac})u.$$

(6.36)

(a) It is clear from Lemma 6.4 (c) that for each $k \geq 0$, the map sending $a \otimes b \in \mathbf{8}_1 \otimes \mathbf{8}_2$ to $Y_{-1/2}(a(-k)b(-\tfrac{1}{2})\mathbf{vac})\mathbf{vac} \in (\mathbf{U}_3)_{1/2} \cong \mathbf{8}_3$ is an $\mathbf{o}(8)$-module map, so the image must be some multiple of $(a\circ b)(-\tfrac{1}{2})\mathbf{vac}$. The $k = 0$ case

$$Y_{-1/2}(a(0)b(-\tfrac{1}{2})\mathbf{vac})\mathbf{vac} = Y_{-1/2}((a\circ b)(-\tfrac{1}{2})\mathbf{vac})\mathbf{vac}$$

$$= (a\circ b)(-\tfrac{1}{2})\mathbf{vac}$$

(6.37)

follows from (5.49). We may choose nonzero a, b and $x = {}^{\circ}_{\circ}b\vartheta b{}^{\circ}_{\circ} \in \mathbf{o}(8)$ such that

$$x\cdot b = b, \quad x\cdot a = -\tfrac{1}{2}a, \quad a\circ b \neq 0, \quad b(-\tfrac{1}{2})x(-1)\mathbf{vac} = 0.$$

(6.38)

With $u = \mathbf{vac}$, (6.36) then gives

$$- kY_{-1/2}(a(-k)b(-\tfrac{1}{2})\mathbf{vac})\mathbf{vac}$$

$$= (k-1)Y_{-1/2}(a(-k+1)b(-\tfrac{1}{2})\mathbf{vac})\mathbf{vac}$$

(6.39)

which says that $Y_{-1/2}(a(-k)b(-\frac{1}{2})\mathbf{vac})\mathbf{vac} = 0$ for $k \geq 1$.

(b) For each $k \geq 0$, the map sending $a \otimes b \otimes c \in 8_1 \otimes 8_2 \otimes 8_3$ to $Y_{1/2}(a(-k)b(-\frac{1}{2})\mathbf{vac})c(-\frac{1}{2})\mathbf{vac} \in (\mathbf{U}_0)_0$ is an $\mathbf{o}(8)$-module map, so the image must be some multiple of $\langle a \circ b, c\rangle\mathbf{vac}$. The $k = 0$ case is

$$Y_{1/2}(a(0)b(-\tfrac{1}{2})\mathbf{vac})c(-\tfrac{1}{2})\mathbf{vac} = Y_{1/2}((a \circ b)(-\tfrac{1}{2})\mathbf{vac})c(-\tfrac{1}{2})\mathbf{vac}$$
$$= (a \circ b)(\tfrac{1}{2})c(-\tfrac{1}{2})\mathbf{vac} = \langle a \circ b, c\rangle\mathbf{vac} \tag{6.40}$$

using the Clifford relations. Choosing nonzero a, b, c and $x = {}^\circ_\circ b\vartheta b{}^\circ_\circ \in \mathbf{o}(8)$ such that

$$x \cdot b = b, \quad x \cdot a = -\tfrac{1}{2}a, \quad \langle a \circ b, c\rangle \neq 0,$$
$$x \cdot c = -\tfrac{1}{2}c, \quad b(-\tfrac{1}{2})x(-1)\mathbf{vac} = 0, \tag{6.41}$$

(6.36) with $u = c(-\frac{1}{2})\mathbf{vac}$ gives

$$- kY_{1/2}(a(-k)b(-\tfrac{1}{2})\mathbf{vac})c(-\tfrac{1}{2})\mathbf{vac}$$
$$= (k - \tfrac{1}{2})Y_{1/2}(a(-k+1)b(-\tfrac{1}{2})\mathbf{vac})c(-\tfrac{1}{2})\mathbf{vac} \tag{6.42}$$

and, therefore,

$$Y_{1/2}(a(-k)b(-\tfrac{1}{2})\mathbf{vac})c(-\tfrac{1}{2})\mathbf{vac} = \binom{-\frac{1}{2}}{k} \langle a \circ b, c\rangle\mathbf{vac}. \tag{6.43}$$

(c) For each $k \geq 0$, the map sending $a \otimes b \otimes a_1 \in 8_1 \otimes 8_2 \otimes 8_1$ to $Y_0(a(-k)b(-\frac{1}{2})\mathbf{vac})a_1(-\frac{1}{2})\mathbf{vac} \in (\mathbf{U}_2)_{1/2} \cong 8_2$ is an $\mathbf{o}(8)$-module map, so the image must be some linear combination

$$\alpha_k(a \circ (a_1 \circ b))(-\tfrac{1}{2})\mathbf{vac} + \beta_k\langle a, a_1\rangle b(-\tfrac{1}{2})\mathbf{vac}. \tag{6.44}$$

The $k = 0$ case is

$$Y_0(a(0)b(-\tfrac{1}{2})\mathbf{vac})a_1(-\tfrac{1}{2})\mathbf{vac} = Y_0((a \circ b)(-\tfrac{1}{2})\mathbf{vac})a_1(-\tfrac{1}{2})\mathbf{vac}$$
$$= (a \circ b)(0)a_1(-\tfrac{1}{2})\mathbf{vac} = (a_1 \circ (a \circ b))(-\tfrac{1}{2})\mathbf{vac} \tag{6.45}$$
$$= -(a \circ (a_1 \circ b))(-\tfrac{1}{2})\mathbf{vac} + \langle a, a_1\rangle b(-\tfrac{1}{2})\mathbf{vac},$$

so we know that $\alpha_0 = -1$ and $\beta_0 = 1$. Choosing nonzero a, a_1, b and $x = {}^\circ_\circ b\vartheta b{}^\circ_\circ \in \mathbf{o}(8)$ such that

$$x \cdot b = b, \quad x \cdot a = -\tfrac{1}{2}a, \quad x \cdot a_1 = \tfrac{1}{2}a_1, \quad a_1 \circ b = 0,$$
$$\langle a, a_1\rangle \neq 0, \quad b(-\tfrac{1}{2})x(-1)\mathbf{vac} = 0, \tag{6.46}$$

(6.36) with $u = a_1(-\frac{1}{2})\mathbf{vac}$ gives

$$- kY_0(a(-k)b(-\tfrac{1}{2})\mathbf{vac})a_1(-\tfrac{1}{2})\mathbf{vac}$$
$$= kY_0(a(-k+1)b(-\tfrac{1}{2})\mathbf{vac})a_1(-\tfrac{1}{2})\mathbf{vac} \tag{6.47}$$

and, therefore, $-k\beta_k = k\beta_{k-1}$, so $\beta_k = (-1)^k$. Choosing nonzero a, a_1, b and $x = {}^\circ_\circ b\vartheta b^\circ_\circ \in \mathbf{o}(8)$ such that

$$x \cdot b = b, \quad x \cdot a = -\tfrac{1}{2}a, \quad x \cdot a_1 = -\tfrac{1}{2}a_1, \quad a \circ (a_1 \circ b) \neq 0,$$
$$\langle a, a_1 \rangle = 0, \quad b(-\tfrac{1}{2})x(-1)\mathbf{vac} = 0, \tag{6.48}$$

(6.36) with $u = a_1(-\tfrac{1}{2})\mathbf{vac}$ gives

$$-kY_0(a(-k)b(-\tfrac{1}{2})\mathbf{vac})a_1(-\tfrac{1}{2})\mathbf{vac}$$
$$= (k-1)Y_0(a(-k+1)b(-\tfrac{1}{2})\mathbf{vac})a_1(-\tfrac{1}{2})\mathbf{vac} \tag{6.49}$$

and, therefore, $-k\alpha_k = (k-1)\alpha_{k-1}$, so $\alpha_k = 0$ for $k \geq 1$.

(d) For each $k \geq 0$, the map sending $a \otimes b \otimes b_1 \in 8_1 \otimes 8_2 \otimes 8_2$ to $Y_0(a(-k)b(-\tfrac{1}{2})\mathbf{vac})b_1(-\tfrac{1}{2})\mathbf{vac} \in (\mathbf{U}_1)_{1/2} \cong 8_1$ is an $\mathbf{o}(8)$-module map, so the image must be some linear combination

$$\alpha_k((a \circ b_1) \circ b)(-\tfrac{1}{2})\mathbf{vac} + \beta_k\langle b, b_1 \rangle a(-\tfrac{1}{2})\mathbf{vac}. \tag{6.50}$$

The $k = 0$ case is

$$Y_0(a(0)b(-\tfrac{1}{2})\mathbf{vac})b_1(-\tfrac{1}{2})\mathbf{vac} = Y_0((a \circ b)(-\tfrac{1}{2})\mathbf{vac})b_1(-\tfrac{1}{2})\mathbf{vac}$$
$$= (a \circ b)(0)b_1(-\tfrac{1}{2})\mathbf{vac} = ((a \circ b) \circ b_1)(-\tfrac{1}{2})\mathbf{vac} \tag{6.51}$$
$$= -((a \circ b_1) \circ b)(-\tfrac{1}{2})\mathbf{vac} + \langle b, b_1 \rangle a(-\tfrac{1}{2})\mathbf{vac},$$

so we know that $\alpha_0 = -1$ and $\beta_0 = 1$. Choosing nonzero a, b, b_1 and $x = {}^\circ_\circ bb_1{}^\circ_\circ \in \mathbf{o}(8)$ such that

$$x \cdot b = b, \quad x \cdot a = -\tfrac{1}{2}a, \quad x \cdot b_1 = -b_1, \quad a \circ b_1 = 0,$$
$$\langle b, b_1 \rangle = 1, \quad b(-\tfrac{1}{2})x(-1)\mathbf{vac} = 0, \tag{6.52}$$

(6.36) with $u = b_1(-\tfrac{1}{2})\mathbf{vac}$ gives

$$-kY_0(a(-k)b(-\tfrac{1}{2})\mathbf{vac})a_1(-\tfrac{1}{2})\mathbf{vac}$$
$$= (k-\tfrac{3}{2})Y_0(a(-k+1)b(-\tfrac{1}{2})\mathbf{vac})a_1(-\tfrac{1}{2})\mathbf{vac} \tag{6.53}$$

and, therefore, $-k\beta_k = (k-\tfrac{3}{2})\beta_{k-1}$, so $\beta_k = \binom{\frac{1}{2}}{k}$. Choosing nonzero a, b, b_1 and $x = {}^\circ_\circ b\vartheta b^\circ_\circ \in \mathbf{o}(8)$ such that

$$x \cdot b = b, \quad x \cdot a = -\tfrac{1}{2}a, \quad x \cdot b_1 = 0, \quad (a \circ b_1) \circ b \neq 0,$$
$$\langle b, b_1 \rangle = 0, \quad b(-\tfrac{1}{2})x(-1)\mathbf{vac} = 0, \tag{6.54}$$

(6.36) with $u = b_1(-\tfrac{1}{2})\mathbf{vac}$ gives

$$-kY_0(a(-k)b(-\tfrac{1}{2})\mathbf{vac})a_1(-\tfrac{1}{2})\mathbf{vac}$$
$$= (k-\tfrac{1}{2})Y_0(a(-k+1)b(-\tfrac{1}{2})\mathbf{vac})a_1(-\tfrac{1}{2})\mathbf{vac} \tag{6.55}$$

and, therefore, $-k\alpha_k = (k-\tfrac{1}{2})\alpha_{k-1}$, so $\alpha_k = -\binom{-\frac{1}{2}}{k}$. ∎

We can now state and prove the modified versions of the rationality, permutability and associativity theorems as mentioned in the first paragraph of this chapter.

THEOREM 6.6. *(Rationality) Let* $u_i \in \mathbf{U}_{n_i}$, $i = 1, 2$, $n_i \in \{1, 2, 3\}$, $n_1 \neq n_2$, $u \in \mathbf{U}_n$, $u' \in \mathbf{U}$. *For* $0 \leq n \leq 3$, $i = 1, 2$, *let*

$$s_i = \begin{cases} 0 & \text{if } n = 0 \text{ or } n = n_i \\ \\ \frac{1}{2} & \text{if } n \neq 0 \text{ and } n \neq n_i. \end{cases}$$

Then the series

$$\zeta_1^{s_1} \zeta_2^{s_2} \zeta_1^{1/2} (1 - \zeta_2/\zeta_1)^{1/2} (Y(u_1, \zeta_1) Y(u_2, \zeta_2) u, u') \tag{6.56}$$

converges absolutely when $|\zeta_1| > |\zeta_2| > 0$ *to a rational function in the ring*

$$\mathbb{C}[\zeta_1, \zeta_1^{-1}, \zeta_2, \zeta_2^{-1}, (\zeta_1 - \zeta_2)^{-1}]. \tag{6.57}$$

PROOF. It is enough to prove this for u_1, u_2, u, u' homogeneous vectors, that is, each is in some graded piece of $\mathbf{U}$, each is an eigenvector of $D(0)$ with weight equal to the negative of the eigenvalue. Furthermore, the matrix coefficient $(Y(u_1, \zeta_1) Y(u_2, \zeta_2) u, u')$ is zero unless u' is in one particular $\hat{\mathbf{o}}(8)$-module $\mathbf{U}_p$, where n_1, n_2 and n determine p. Let n_3 be such that $\{n_1, n_2, n_3\} = \{1, 2, 3\}$. We prove the theorem by induction on $\mathrm{wt}(u_1) + \mathrm{wt}(u_2) + \mathrm{wt}(u) + \mathrm{wt}(u') \geq \frac{3}{2}$. We begin with the base cases where each vector u_1, u_2, u, u' is of minimal weight in its $\hat{\mathbf{o}}(8)$-module. We use the notation

$$\begin{aligned} \Phi(u_1, \zeta_1; u_2, \zeta_2; u, u') \\ = \zeta_1^{s_1} \zeta_2^{s_2} \zeta_1^{1/2} (1 - \zeta_2/\zeta_1)^{1/2} (Y(u_1, \zeta_1) Y(u_2, \zeta_2) u, u'). \end{aligned} \tag{6.58}$$

Letting $\mathrm{wt}(u_1) = \mathrm{wt}(u_2) = \frac{1}{2}$, so $u_1 = a(-\frac{1}{2})\mathbf{vac} \in (\mathbf{U}_{n_1})_{1/2}$, $u_2 = b(-\frac{1}{2})\mathbf{vac} \in (\mathbf{U}_{n_2})_{1/2}$, there are four base cases, $0 \leq n \leq 3$, which follow from the four parts of Proposition 6.2. When $n = 0$, $u = \mathbf{vac} \in (\mathbf{U}_0)_0$, $p = n_3$, $u' = c(-\frac{1}{2})\mathbf{vac} \in (\mathbf{U}_{n_3})_{1/2}$, and Proposition 6.2 (a) says

$$\begin{aligned} \Phi(u_1, \zeta_1; u_2, \zeta_2; u, u') \\ = \zeta_1^{1/2} (1 - \zeta_2/\zeta_1)^{1/2} \sum_{0 \leq k \in \mathbb{Z}} \zeta_1^{-1/2} (\zeta_2/\zeta_1)^k (a(k) b(-k - \tfrac{1}{2})\mathbf{vac}, c(-\tfrac{1}{2})\mathbf{vac}) \\ = (a \circ b, c) \end{aligned}$$
$$\tag{6.59}$$

which is a constant in the ring (6.57). When $n = n_1$, $u = a_1(-\frac{1}{2})\mathbf{vac} \in (\mathbf{U}_{n_1})_{1/2}$, $p = n_2$, $u' = b_1(-\frac{1}{2})\mathbf{vac} \in (\mathbf{U}_{n_2})_{1/2}$, and Proposition 6.2 (d) says

$$
\begin{aligned}
\Phi(u_1, &\zeta_1; u_2, \zeta_2; u, u') \\
&= (1 - \zeta_2/\zeta_1)^{1/2} \sum_{0 \le k \in \mathbb{Z}} (\zeta_2/\zeta_1)^k (a(k)b(-k)a_1(-\tfrac{1}{2})\mathbf{vac}, b_1(-\tfrac{1}{2})\mathbf{vac}) \\
&= -(\zeta_2/\zeta_1)((a \circ b) \circ a_1, b_1) + (1 - \zeta_2/\zeta_1)(a \circ (b \circ a_1), b_1) \\
&= (a \circ (b \circ a_1), b_1) - (\zeta_2/\zeta_1)\langle a, a_1 \rangle (b, b_1)
\end{aligned}
\tag{6.60}
$$

which is in the ring (6.57). When $n = n_2$, $u = b_1(-\frac{1}{2})\mathbf{vac} \in (\mathbf{U}_{n_2})_{1/2}$, $p = n_1$, $u' = a_1(-\frac{1}{2})\mathbf{vac} \in (\mathbf{U}_{n_1})_{1/2}$, and Proposition 6.2 (c) says

$$
\begin{aligned}
\Phi(u_1, &\zeta_1; u_2, \zeta_2; u, u') \\
&= \zeta_1 (1 - \zeta_2/\zeta_1)^{1/2} \sum_{0 \le k \in \mathbb{Z}} \zeta_2^{-1} (\zeta_2/\zeta_1)^k (a(k - \tfrac{1}{2}) \\
&\qquad\qquad \cdot \; b(-k + \tfrac{1}{2})b_1(-\tfrac{1}{2})\mathbf{vac}, a_1(-\tfrac{1}{2})\mathbf{vac}) \\
&= ((a \circ b) \circ b_1, a_1) + (\zeta_1/\zeta_2 - 1)\langle b, b_1 \rangle (a, a_1) \\
&= -((a \circ b_1) \circ b, a_1) + (\zeta_1/\zeta_2)\langle b, b_1 \rangle (a, a_1)
\end{aligned}
\tag{6.61}
$$

which is in the ring (6.57). When $n = n_3$, $u = c(-\frac{1}{2})\mathbf{vac} \in (\mathbf{U}_{n_3})_{1/2}$, $p = 0$, $u' = \mathbf{vac} \in (\mathbf{U}_0)_0$, and Proposition 6.2 (b) says

$$
\begin{aligned}
\Phi(u_1, &\zeta_1; u_2, \zeta_2; u, u') \\
&= (1 - \zeta_2/\zeta_1)^{1/2} \sum_{0 \le k \in \mathbb{Z}} (\zeta_2/\zeta_1)^k (a(k + \tfrac{1}{2})b(-k)c(-\tfrac{1}{2})\mathbf{vac}, \mathbf{vac}) \\
&= \langle a \circ b, c \rangle
\end{aligned}
\tag{6.62}
$$

which is a constant in the ring (6.57).

To do the inductive step of the proof we need Lemma 6.4. Let v be any one of the vectors u_1, u_2, u, u'. If v is not of minimal weight in its $\hat{\mathrm{o}}(8)$-module then v is a linear combination of vectors of the form $x(-m)v_0$ for $1 \le m \in \mathbb{Z}$, $x \in (\mathbf{U}_0)_1 \cong \mathrm{o}(8)$, $\mathrm{wt}(v_0) < \mathrm{wt}(v)$. Since $\Phi(u_1, \zeta_1; u_2, \zeta_2; u, u')$ is linear in u_1, u_2, u, u', it suffices to do the inductive step with $v = x(-m)v_0$. Using Lemma

6.4 (a) and (b) we get the formula

$$\Phi(x(-m)u_1, \zeta_1; u_2, \zeta_2; u, u') = \zeta_1^{s_1} \zeta_2^{s_2} \zeta_1^{1/2}(1 - \zeta_2/\zeta_1)^{1/2}$$
$$\cdot \left[\sum_{0 \le k \in \mathbb{Z}} \binom{m+k-1}{k} \zeta_1^k (x(-m-k)Y(u_1,\zeta_1)Y(u_2,\zeta_2)u, u') \right.$$
$$\left. -(-1)^m \sum_{0 \le k \in \mathbb{Z}} \binom{m+k-1}{k} \zeta_1^{-m-k}(Y(u_1,\zeta_1)x(k)Y(u_2,\zeta_2)u, u') \right]$$
$$= \sum_{0 \le k \in \mathbb{Z}} \binom{m+k-1}{k} \zeta_1^k \Phi(u_1,\zeta_1; u_2,\zeta_2; u, x^*(m+k)u')$$
$$- (-1)^m \sum_{0 \le k \in \mathbb{Z}} \binom{m+k-1}{k} \zeta_1^{-m-k} \Phi(u_1,\zeta_1; u_2,\zeta_2; x(k)u, u')$$
$$- (-1)^m \sum_{0 \le k \in \mathbb{Z}} \binom{m+k-1}{k} \zeta_1^{-m-k} \zeta_2^k \sum_{0 \le p \in \mathbb{Z}} \zeta_2^{-p} \binom{k}{p} \Phi(u_1,\zeta_1; x(p)u_2, \zeta_2; u, u')$$
$$\tag{6.63}$$

and weight considerations show that the summations over k in the first and second lines are finite, as is the summation over p in the third line. By the induction assumption, the Φ series in all three lines converge absolutely to rational functions in the ring (6.57) when $|\zeta_1| > |\zeta_2| > 0$. Therefore, using the fact that for $0 \le p \in \mathbb{Z}$,

$$\sum_{0 \le k \in \mathbb{Z}} \binom{m+k-1}{k} \binom{k}{p} \zeta^{k-p} = (-1)^p \binom{-m}{p}(1-\zeta)^{-m-p} \tag{6.64}$$

converges absolutely when $|\zeta| < 1$, we may rewrite the last line of (6.63) as

$$-(-1)^m \sum_{0 \le p \in \mathbb{Z}} (-1)^p \binom{-m}{p}(\zeta_1 - \zeta_2)^{-m-p} \Phi(u_1,\zeta_1; x(p)u_2, \zeta_2; u, u') \tag{6.65}$$

giving

$$\Phi(x(-m)u_1, \zeta_1; u_2, \zeta_2; u, u')$$
$$= \sum_{0 \le k \in \mathbb{Z}} \binom{m+k-1}{k} \zeta_1^k \Phi(u_1,\zeta_1; u_2,\zeta_2; u, x^*(m+k)u')$$
$$- (-1)^m \sum_{0 \le k \in \mathbb{Z}} \binom{m+k-1}{k} \zeta_1^{-m-k} \Phi(u_1,\zeta_1; u_2,\zeta_2; x(k)u, u') \tag{6.66}$$
$$- (-1)^m \sum_{0 \le p \in \mathbb{Z}} (-1)^p \binom{-m}{p}(\zeta_1 - \zeta_2)^{-m-p} \Phi(u_1,\zeta_1; x(p)u_2, \zeta_2; u, u').$$

We can similarly derive the additional formulas

$$\Phi(u_1, \zeta_1; x(-m)u_2, \zeta_2; u, u') =$$

$$\sum_{0 \leq k \in \mathbb{Z}} \binom{m+k-1}{k} \zeta_2^k \Phi(u_1, \zeta_1; u_2, \zeta_2; u, x^*(m+k)u')$$

$$- \sum_{0 \leq p \in \mathbb{Z}} \binom{-m}{p} (\zeta_1 - \zeta_2)^{-m-p} \Phi(x(p)u_1, \zeta_1; u_2, \zeta_2; u, u') \qquad (6.67)$$

$$- (-1)^m \sum_{0 \leq k \in \mathbb{Z}} \binom{m+k-1}{k} \zeta_2^{-m-k} \Phi(u_1, \zeta_1; u_2, \zeta_2; x(k)u, u')$$

and

$$\Phi(u_1, \zeta_1; u_2, \zeta_2; x(-m)u, u') = \Phi(u_1, \zeta_1; u_2, \zeta_2; u, x^*(m)u')$$

$$- \sum_{0 \leq p \in \mathbb{Z}} \binom{-m}{p} \zeta_1^{-m-p} \Phi(x(p)u_1, \zeta_1; u_2, \zeta_2; u, u')$$

$$- \sum_{0 \leq p \in \mathbb{Z}} \binom{-m}{p} \zeta_2^{-m-p} \Phi(u_1, \zeta_1; x(p)u_2, \zeta_2; u, u') \qquad (6.68)$$

and

$$\Phi(u_1, \zeta_1; u_2, \zeta_2; u, x(-m)u') = \Phi(u_1, \zeta_1; u_2, \zeta_2; x^*(m)u, u')$$

$$+ \sum_{0 \leq p \in \mathbb{Z}} \binom{m}{p} \zeta_1^{m-p} \Phi(x^*(p)u_1, \zeta_1; u_2, \zeta_2; u, u')$$

$$+ \sum_{0 \leq p \in \mathbb{Z}} \binom{m}{p} \zeta_2^{m-p} \Phi(u_1, \zeta_1; x^*(p)u_2, \zeta_2; u, u'). \qquad (6.69)$$

These four formulas may be viewed either as formal identities with the proviso that powers of $\zeta_1 - \zeta_2$ be expanded in positive powers of ζ_2, or as series which converge absolutely in the domain $0 < |\zeta_2| < |\zeta_1|$. All of the summations in (6.66) - (6.69) are actually finite because of weight considerations, and each term satisfies the inductive hypothesis, giving the result. ∎

THEOREM 6.7. *(Permutability) With the notation as in Theorem 6.6 we have*

$$\zeta_1^{s_1} \zeta_2^{s_2} \zeta_1^{1/2} (1 - \zeta_2/\zeta_1)^{1/2} (Y(u_1, \zeta_1) Y(u_2, \zeta_2) u, u') \sim$$

$$\kappa \zeta_1^{s_1} \zeta_2^{s_2} \zeta_2^{1/2} (1 - \zeta_1/\zeta_2)^{1/2} (Y(u_2, \zeta_2) Y(u_1, \zeta_1) u, u')$$

where $\{n_1, n_2, n_3\} = \{1, 2, 3\}$ *and*

$$\kappa = \begin{cases} 1 & \text{if } n = 0 \text{ or } n = n_3 \\ \\ -1 & \text{if } n = n_1 \text{ or } n = n_2. \end{cases}$$

PROOF. We continue to use the notation set up in the proof of the last theorem, and we proceed by induction on $\mathrm{wt}(u_1) + \mathrm{wt}(u_2) + \mathrm{wt}(u) + \mathrm{wt}(u')$. We begin with the base cases where each vector u_1, u_2, u, u' is of minimal weight in its $\hat{\mathfrak{o}}(8)$-module, so $u_1 = a(-\frac{1}{2})\mathbf{vac} \in (\mathbf{U}_{n_1})_{1/2}$ and $u_2 = b(-\frac{1}{2})\mathbf{vac} \in (\mathbf{U}_{n_2})_{1/2}$. Applying (6.59) with $u = \mathbf{vac} \in (\mathbf{U}_0)_0$, so $\kappa = 1$, and $u' = c(-\frac{1}{2})\mathbf{vac} \in (\mathbf{U}_{n_3})_{1/2}$, we get

$$\Phi(u_2, \zeta_2; u_1, \zeta_1; u, u') = (b \circ a, c) = (a \circ b, c) = \Phi(u_1, \zeta_1; u_2, \zeta_2; u, u'). \tag{6.70}$$

With $u = a_1(-\frac{1}{2})\mathbf{vac} \in (\mathbf{U}_{n_1})_{1/2}$, so $\kappa = -1$, and $u' = b_1(-\frac{1}{2})\mathbf{vac} \in (\mathbf{U}_{n_2})_{1/2}$, we apply (6.61) and then (6.60) to get

$$\begin{aligned}
\Phi(u_2, \zeta_2; u_1, \zeta_1; u, u') &= -((b \circ a_1) \circ a, b_1) + (\zeta_2/\zeta_1)\langle a, a_1 \rangle (b, b_1) \\
&= -[(a \circ (b \circ a_1)), b_1) - (\zeta_2/\zeta_1)\langle a, a_1 \rangle (b, b_1)] \\
&= -\Phi(u_1, \zeta_1; u_2, \zeta_2; u, u').
\end{aligned} \tag{6.71}$$

Finally, with $u = c(-\frac{1}{2})\mathbf{vac} \in (\mathbf{U}_{n_3})_{1/2}$, so $\kappa = 1$, and $u' \in \mathbf{vac} \in (\mathbf{U}_0)_0$, applying (6.62) gives

$$\Phi(u_2, \zeta_2; u_1, \zeta_1; u, u') = \langle b \circ a, c \rangle = \langle a \circ b, c \rangle = \Phi(u_1, \zeta_1; u_2, \zeta_2; u, u'). \tag{6.72}$$

We can use the formulas (6.66) - (6.69) to do the inductive step of the proof as follows. Applying (6.67), we get

$$\begin{aligned}
\Phi(u_2, &\zeta_2; x(-m)u_1, \zeta_1; u, u') = \\
&\sum_{0 \le k \in \mathbb{Z}} \binom{m + k - 1}{k} \zeta_1^k \Phi(u_2, \zeta_2; u_1, \zeta_1; u, x^*(m + k)u') \\
&- \sum_{0 \le p \in \mathbb{Z}} \binom{-m}{p} (\zeta_2 - \zeta_1)^{-m-p} \Phi(x(p)u_2, \zeta_2; u_1, \zeta_1; u, u') \\
&- (-1)^m \sum_{0 \le k \in \mathbb{Z}} \binom{m + k - 1}{k} \zeta_1^{-m-k} \Phi(u_2, \zeta_2; u_1, \zeta_1; x(k)u, u')
\end{aligned} \tag{6.73}$$

for $|\zeta_2| > |\zeta_1| > 0$. Inductively, for $0 \le k, p \in \mathbb{Z}$ we have

$$\begin{aligned}
\kappa \Phi(u_2, \zeta_2; u_1, \zeta_1; u, x^*(m + k)u') &\sim \Phi(u_1, \zeta_1; u_2, \zeta_2; u, x^*(m + k)u'), \\
\kappa \Phi(x(p)u_2, \zeta_2; u_1, \zeta_1; u, u') &\sim \Phi(u_1, \zeta_1; x(p)u_2, \zeta_2; u, u'), \\
\kappa \Phi(u_2, \zeta_2; u_1, \zeta_1; x(k)u, u') &\sim \Phi(u_1, \zeta_1; u_2, \zeta_2; x(k)u, u'),
\end{aligned} \tag{6.74}$$

so comparing with (6.66) we get

$$\kappa \Phi(u_2, \zeta_2; x(-m)u_1, \zeta_1; u, u') \sim \Phi(x(-m)u_1, \zeta_1; u_2\zeta_2; u, u'). \tag{6.75}$$

The cases using (6.68) and (6.69) are each done similarly. ∎

THEOREM 6.8. *(Associativity) With the notation as in Theorem 6.6 we have*

$$\zeta_1^{s_1} \zeta_2^{s_2} \zeta_1^{1/2} (1 - \zeta_2/\zeta_1)^{1/2} (Y(u_1, \zeta_1) Y(u_2, \zeta_2)u, u') \sim$$

$$\kappa' \zeta_2^{s_1+s_2} (\zeta_1 - \zeta_2)^{1/2} \left(1 + \frac{\zeta_1 - \zeta_2}{\zeta_2}\right)^{s_1} (Y(Y(u_1, \zeta_1 - \zeta_2)u_2, \zeta_2)u, u')$$

where

$$
\kappa' =
\begin{cases}
1 & \text{if } n = 0 \text{ or } n = n_2 \text{ or } n = n_3 \\[2mm]
-1 & \text{if } n = n_1
\end{cases}
$$

and the second series converges absolutely when $0 < |\zeta_1 - \zeta_2| < |\zeta_2|$.

PROOF. We continue to use the notation set up in the proofs of the last theorems, and again we proceed by induction on $\mathrm{wt}(u_1)+\mathrm{wt}(u_2)+\mathrm{wt}(u)+\mathrm{wt}(u')$. Letting $\zeta = \zeta_1 - \zeta_2$, define the notation

$$
\Psi(u_1, \zeta; u_2, \zeta_2; u, u') = \zeta_2^{s_1+s_2} \zeta^{1/2}(1 + \zeta/\zeta_2)^{s_1}(Y(Y(u_1, \zeta)u_2, \zeta_2)u, u'). \tag{6.76}
$$

We begin with the base cases where each vector u_1, u_2, u, u' is of minimal weight in its $\hat{\mathrm{o}}(8)$-module, so $u_1 = a(-\tfrac{1}{2})\mathbf{vac} \in (\mathbf{U}_{n_1})_{1/2}$ and $u_2 = b(-\tfrac{1}{2})\mathbf{vac} \in (\mathbf{U}_{n_2})_{1/2}$. When $u = \mathbf{vac} \in (\mathbf{U}_0)_0$, so $s_1 = s_2 = 0$, and $u' = c(-\tfrac{1}{2})\mathbf{vac} \in (\mathbf{U}_{n_3})_{1/2}$, we have

$$
\begin{aligned}
\Psi(u_1, \zeta; u_2, \zeta_2; u, u') &= \zeta^{1/2}(Y(Y(u_1, \zeta)u_2, \zeta_2)u, u') \\
&= \sum_{0 \le k \in \mathbb{Z}} \zeta^k (Y(Y_{-k}(u_1)u_2, \zeta_2)u, u') \\
&= \sum_{0 \le k \in \mathbb{Z}} (\zeta/\zeta_2)^k (Y_{-1/2}(Y_{-k}(u_1)u_2)u, u') \\
&= (a \circ b, c) \\
&= \Phi(u_1, \zeta + \zeta_2; u_2, \zeta_2; u, u')
\end{aligned} \tag{6.77}
$$

by Proposition 6.5 (a) and (6.59). When $u = c(-\tfrac{1}{2})\mathbf{vac} \in (\mathbf{U}_{n_3})_{1/2}$, so $s_1 = s_2 = \tfrac{1}{2}$, and $u' = \mathbf{vac} \in (\mathbf{U}_0)_0$, we have

$$
\begin{aligned}
\Psi(u_1, \zeta; u_2, \zeta_2; u, u') &= \zeta^{1/2}\zeta_2(1 + \zeta/\zeta_2)^{1/2}(Y(Y(u_1, \zeta)u_2, \zeta_2)u, u') \\
&= \zeta_2(1 + \zeta/\zeta_2)^{1/2} \sum_{0 \le k \in \mathbb{Z}} \zeta^k (Y(Y_{-k}(u_1)u_2, \zeta_2)u, u') \\
&= (1 + \zeta/\zeta_2)^{1/2} \sum_{0 \le k \in \mathbb{Z}} (\zeta/\zeta_2)^k (Y_{1/2}(Y_{-k}(u_1)u_2)u, u') \\
&= \langle a \circ b, c \rangle \\
&= \Phi(u_1, \zeta + \zeta_2; u_2, \zeta_2; u, u')
\end{aligned}
$$

$$\tag{6.78}$$

by Proposition 6.5 (b) and (6.62). When $u = a_1(-\tfrac{1}{2})\mathbf{vac} \in (\mathbf{U}_{n_1})_{1/2}$, so $s_1 = 0$,

$s_2 = \frac{1}{2}$, and $u' = b_1(-\frac{1}{2})\mathbf{vac} \in (\mathbf{U}_{n_2})_{1/2}$, we have

$$
\begin{aligned}
\Psi(u_1, \zeta; u_2, \zeta_2; u, u') &= \zeta^{1/2}\zeta_2^{1/2}(Y(Y(u_1, \zeta)u_2, \zeta_2)u, u') \\
&= \zeta_2^{1/2} \sum_{0 \le k \in \mathbb{Z}} \zeta^k (Y(Y_{-k}(u_1)u_2, \zeta_2)u, u') \\
&= \sum_{0 \le k \in \mathbb{Z}} (\zeta/\zeta_2)^k (Y_0(Y_{-k}(u_1)u_2)u, u') \qquad (6.79)\\
&= -(a \circ (b \circ a_1), b_1) + (1 + \zeta/\zeta_2)^{-1}\langle a, a_1\rangle(b, b_1) \\
&= -\Phi(u_1, \zeta + \zeta_2; u_2, \zeta_2; u, u')
\end{aligned}
$$

by Proposition 6.5 (c) and (6.60). When $u = b_1(-\frac{1}{2})\mathbf{vac} \in (\mathbf{U}_{n_2})_{1/2}$, so $s_1 = \frac{1}{2}$, $s_2 = 0$, and $u' = a_1(-\frac{1}{2})\mathbf{vac} \in (\mathbf{U}_{n_1})_{1/2}$, we have

$$
\begin{aligned}
\Psi(u_1, \zeta; u_2, \zeta_2; u, u') &= \zeta^{1/2}\zeta_2^{1/2}(1 + \zeta/\zeta_2)^{1/2}(Y(Y(u_1, \zeta)u_2, \zeta_2)u, u') \\
&= \zeta_2^{1/2}(1 + \zeta/\zeta_2)^{1/2} \sum_{0 \le k \in \mathbb{Z}} \zeta^k (Y(Y_{-k}(u_1)u_2, \zeta_2)u, u') \\
&= (1 + \zeta/\zeta_2)^{1/2} \sum_{0 \le k \in \mathbb{Z}} (\zeta/\zeta_2)^k (Y_0(Y_{-k}(u_1)u_2)u, u') \\
&= (1 + \zeta/\zeta_2)\langle b, b_1\rangle(a, a_1) - ((a \circ b_1) \circ b, a_1) \\
&= \Phi(u_1, \zeta + \zeta_2; u_2, \zeta_2; u, u')
\end{aligned}
$$
$$(6.80)$$

by Proposition 6.5 (d) and (6.61).

To do the inductive step of the proof we use four formulas which are identical to (6.66) - (6.69) with Φ replaced by Ψ. These formulas may be viewed either as formal identities with the proviso that powers of $\zeta_2 + \zeta$ be expanded in positive powers of ζ, or as series which converge absolutely in the domain $0 < |\zeta| < |\zeta_2|$. We use Lemma 6.4 to derive the first formula as follows.

Letting $F = \zeta_2^{s_1 + s_2}\zeta^{1/2}(1 + \zeta/\zeta_2)^{s_1}$ we have

$$
\begin{aligned}
\Psi(x(-m)u_1, \zeta; u_2, \zeta_2; u, u') &= F\ (Y(Y(x(-m)u_1, \zeta)u_2, \zeta_2)u, u') \\
&= F \sum_{0 \le k \in \mathbb{Z}} \binom{m+k-1}{k}[\zeta^k(Y(x(-m-k)Y(u_1, \zeta)u_2, \zeta_2)u, u') \\
&\qquad\qquad - (-1)^m \zeta^{-m-k}(Y(Y(u_1, \zeta)x(k)u_2, \zeta_2)u, u')] \\
&= F \sum_{0 \le k \in \mathbb{Z}} \binom{m+k-1}{k}\zeta^k \sum_{0 \le p \in \mathbb{Z}} \binom{m+k+p-1}{p} \\
&\qquad\qquad \cdot [\zeta_2^p(x(-m-k-p)Y(Y(u_1, \zeta)u_2, \zeta_2)u, u') \\
&\qquad\qquad - (-1)^{m+k}\zeta_2^{-m-k-p}(Y(Y(u_1, \zeta)u_2, \zeta_2)x(p)u, u')] \\
&\quad - (-1)^m \sum_{0 \le k \in \mathbb{Z}} \binom{m+k-1}{k}\zeta^{-m-k}\Psi(u_1, \zeta; x(k)u_2, \zeta_2; u, u')
\end{aligned}
$$

$$= \sum_{0 \le k, p \in \mathbb{Z}} \binom{m+k-1}{k} \binom{m+k+p-1}{p} \zeta^k \zeta_2^p \Psi(u_1, \zeta; u_2, \zeta_2; u, x^*(m+k+p)u')$$

$$- \sum_{0 \le k, p \in \mathbb{Z}} (-1)^{m+k} \binom{m+k-1}{k} \binom{m+k+p-1}{p}$$

$$\cdot \; \zeta^k \zeta_2^{-m-k-p} \Psi(u_1, \zeta; u_2, \zeta_2; x(p)u, u')$$

$$- (-1)^m \sum_{0 \le k \in \mathbb{Z}} \binom{m+k-1}{k} \zeta^{-m-k} \Psi(u_1, \zeta; x(k)u_2, \zeta_2; u, u').$$

$$(6.81)$$

Using some simple rearrangements and the fact that for $0 \le p \in \mathbb{Z}$,

$$\sum_{0 \le k \in \mathbb{Z}} (-1)^k \binom{m+k-1}{k} \binom{m+k+p-1}{p} z^k = \binom{m+p-1}{p} (1+z)^{-m-p}$$

$$(6.82)$$

converges absolutely when $|z| < 1$, we get

$$\Psi(x(-m)u_1, \zeta; u_2, \zeta_2; u, u')$$

$$= \sum_{0 \le k \in \mathbb{Z}} \binom{m+k-1}{k} (\zeta_2 + \zeta)^k \Psi(u_1, \zeta; u_2, \zeta_2; u, x^*(m+k)u')$$

$$- (-1)^m \sum_{0 \le p \in \mathbb{Z}} \binom{m+p-1}{p} (\zeta_2 + \zeta)^{-m-p} \Psi(u_1, \zeta; u_2, \zeta_2; x(p)u, u') \qquad (6.83)$$

$$- (-1)^m \sum_{0 \le k \in \mathbb{Z}} \binom{m+k-1}{k} \zeta^{-m-k} \Psi(u_1, \zeta; x(k)u_2, \zeta_2; u, u').$$

Similarly, one derives the formulas

$$\Psi(u_1, \zeta; x(-m)u_2, \zeta_2; u, u') =$$

$$\sum_{0 \le k \in \mathbb{Z}} \binom{m+k-1}{k} \zeta_2^k \Psi(u_1, \zeta; u_2, \zeta_2; u, x^*(m+k)u')$$

$$- \sum_{0 \le p \in \mathbb{Z}} \binom{-m}{p} \zeta^{-m-p} \Psi(x(p)u_1, \zeta; u_2, \zeta_2; u, u') \qquad (6.84)$$

$$- (-1)^m \sum_{0 \le k \in \mathbb{Z}} \binom{m+k-1}{k} \zeta_2^{-m-k} \Psi(u_1, \zeta; u_2, \zeta_2; x(k)u, u')$$

and

$$\Psi(u_1, \zeta; u_2, \zeta_2; x(-m)u, u') = \Psi(u_1, \zeta; u_2, \zeta_2; u, x^*(m)u')$$

$$- \sum_{0 \le p \in \mathbb{Z}} \binom{-m}{p} (\zeta_2 + \zeta)^{-m-p} \Psi(x(p)u_1, \zeta; u_2, \zeta_2; u, u')$$

$$(6.85)$$

$$- \sum_{0 \le p \in \mathbb{Z}} \binom{-m}{p} \zeta_2^{-m-p} \Psi(u_1, \zeta; x(p)u_2, \zeta_2; u, u')$$

and

$$\Psi(u_1, \zeta; u_2, \zeta_2; u, x(-m)u') = \Psi(u_1, \zeta; u_2, \zeta_2; x^*(m)u, u')$$

$$+ \sum_{0 \leq p \in \mathbb{Z}} \binom{m}{p} (\zeta_2 + \zeta)^{m-p} \Psi(x^*(p)u_1, \zeta; u_2, \zeta_2; u, u')$$

$$+ \sum_{0 \leq p \in \mathbb{Z}} \binom{m}{p} \zeta_2^{m-p} \Psi(u_1, \zeta; x^*(p)u_2, \zeta_2; u, u'). \tag{6.86}$$

All of the summations in (6.83) - (6.86) are finite because of weight considerations. Note that the induction assumption applies to each Ψ in each term of (6.83) - (6.86), and that the form of each reduction formula is identical to the form of the corresponding reduction formula for Φ with ζ_1 replaced by $\zeta_2 + \zeta$ in (6.66) - (6.69). Therefore, taking into account the κ' factor in the base cases, we get the result. $\blacksquare$

Just as the Rationality, Permutability and Associativity Theorems implied the Jacobi identity in Chapter 3, Theorems 6.6 - 6.9 give a modified version of the Jacobi identity in which $f\,sgn$ is replaced by κ, and κ' appears on the right side. This does not give the Jacobi Identity axiom (0.46) for vertex operator para-algebras because κ depends on n_1, n_2 and n, and κ' is not identically 1. To solve this problem we modify these factors by the rescaling

$$Y^\mu(u_1, \zeta)u = \mu(n_1, n)Y(u_1, \zeta)u \tag{6.87}$$

for $u_1 \in \mathbf{U}_{n_1}$, $u \in \mathbf{U}_n$ and $\mu(n_1, n) \in \mathbb{C}^*$. Recall from Chapter 5 that we have given $\Gamma = \{0, 1, 2, 3\}$ the additive group structure $\mathbb{Z}_2 \times \mathbb{Z}_2$. As a 2-cochain, μ has coboundary $\delta\mu(n_1, n_2, n) = \mu(n_2, n)\mu(n_1, n_2 + n)\mu(n_1, n_2)^{-1}\mu(n_1 + n_2, n)^{-1}$. With Y replaced by Y^μ in Theorems 6.7 and 6.8, using (5.45) we see that the factor κ is replaced by

$$\kappa \frac{\mu(n_2, n)\mu(n_1, n_2 + n)}{\mu(n_1, n)\mu(n_2, n_1 + n)} = \kappa \frac{\mu(n_1, n_2)}{\mu(n_2, n_1)} \frac{\delta\mu(n_1, n_2, n)}{\delta\mu(n_2, n_1, n)} \tag{6.88}$$

and the factor κ' is replaced by

$$\kappa' \frac{\mu(n_2, n)\mu(n_1, n_2 + n)}{\mu(n_1, n_2)\mu(n_1 + n_2, n)} = \kappa' \delta\mu(n_1, n_2, n) \tag{6.89}$$

where $0 \neq n_1 \neq n_2 \neq 0$, $0 \leq n \leq 3$ and $n_3 = n_1 + n_2$. We replace Y by Y^μ in the Permutability Theorem 3.29 for two vertex operators, as applied in Chapter 5 to $\mathbf{U}$, with $\{n_1, n_2\} \subseteq \{0, k\}$ for some k, $1 \leq k \leq 3$, $0 \leq n \leq 3$, $u_i \in \mathbf{U}_{n_i}$, $u \in \mathbf{U}_n$, $u' \in \mathbf{U}$ and $s_i = \frac{1}{2}$ if $0 \neq n \neq n_i$, $s_i = 0$ otherwise, $i = 1, 2$. This gives

$$\zeta_1^{s_1} \zeta_2^{s_2} (Y^\mu(u_1, \zeta_1)Y^\mu(u_2, \zeta_2)u, u') \sim$$

$$\frac{\mu(n_2, n)\mu(n_1, n_2 + n)}{\mu(n_1, n)\mu(n_2, n_1 + n)} \zeta_1^{s_1} \zeta_2^{s_2} (Y^\mu(u_2, \zeta_2)Y^\mu(u_1, \zeta_1)u, u'). \tag{6.90}$$

The Associativity Theorem 3.30 gives

$$\zeta_1^{s_1}\zeta_2^{s_2}\left(Y^\mu(u_1,\zeta_1)Y^\mu(u_2,\zeta_2)u,u'\right) \sim$$

$$\delta\mu(n_1,n_2,n)\,\zeta_2^{s_1+s_2}\left(1+\frac{\zeta_1-\zeta_2}{\zeta_2}\right)^{s_1}\left(Y^\mu(Y^\mu(u_1,\zeta_1-\zeta_2)u_2,\zeta_2)u,u'\right). \tag{6.91}$$

In order that Y^μ still gives vertex operator superalgebras and modules, it is necessary that for $\{n_1,n_2\}\subseteq\{0,k\}$ for some k, $1\le k\le 3$, and $0\le n\le 3$, we have

$$\frac{\mu(n_2,n)\mu(n_1,n_2+n)}{\mu(n_1,n)\mu(n_2,n_1+n)}=1=\frac{\mu(n_2,n)\mu(n_1,n_2+n)}{\mu(n_1,n_2)\mu(n_1+n_2,n)}. \tag{6.92}$$

These are equivalent to the conditions $\mu(0,n)=\mu(n,0)=\mu(0,0)$ for $0\le n\le 3$, and we may scale μ so that $\mu(0,0)=1$. We now wish to impose the conditions that, for $0\ne n_1\ne n_2\ne 0$ and $0\le n\le 3$,

$$1=\kappa'\,\frac{\mu(n_2,n)\mu(n_1,n_2+n)}{\mu(n_1,n_2)\mu(n_1+n_2,n)} \tag{6.93}$$

and

$$\beta(n_1,n_2)=\kappa\,\frac{\mu(n_2,n)\mu(n_1,n_2+n)}{\mu(n_1,n)\mu(n_2,n_1+n)} \tag{6.94}$$

is independent of n. It is straightforward to check that these conditions are equivalent to the conditions

$$\mu(k,k)=\mu(k,k')\mu(k,k'')=\mu(k',k)\mu(k'',k),$$
$$\mathbf{i^a}=\mu(k,k')\mu(k',k)^{-1}=-\mu(k',k)\mu(k,k')^{-1} \tag{6.95}$$

for (k,k',k'') any cyclic permutation of $(1,2,3)$ and $\mathbf{a}=\pm1$ independent of k. These imply that a choice of $\mathbf{a}=\pm1$ and any choice of $\mu(1,2)$, $\mu(2,3)$ and $\mu(3,1)$ determines the others. With $n=0$ in (6.94), we see that

$$\beta(k,k')=-\beta(k',k)=\mathbf{i^a}. \tag{6.96}$$

We now extend the definition of $\beta(n_1,n_2)$ to all $0\le n_1,n_2\le 3$ by setting

$$\beta(n_1,n_2)=\begin{cases}1 & \text{if } n_1=0 \text{ or } n_2=0. \\[2ex] -1 & \text{if } n_1=n_2\ne 0.\end{cases} \tag{6.97}$$

The Rationality Theorem is valid for the Y^μ operators, and we now give the Permutability Theorem for two Y^μ operators and the Associativity Theorem which combine the results from Chapters 3 and 6.

THEOREM 6.9. *Let* $u_i\in\mathbf{U}_{n_i}$ *for* $i=1,2$, $u\in\mathbf{U}_n$, $0\le n,n_i\le 3$, $u'\in\mathbf{U}$, $s_i=\frac{1}{2}$ *if* $0\ne n\ne n_i\ne 0$, $s_i=0$ *otherwise, and let* $s_{12}=\frac{1}{2}$ *if* $0\ne n_1\ne n_2\ne 0$, $s_{12}=0$ *otherwise. Then we have*

$$\zeta_1^{s_1}\zeta_2^{s_2}\zeta_1^{s_{12}}(1-\zeta_2/\zeta_1)^{s_{12}}\left(Y^\mu(u_1,\zeta_1)Y^\mu(u_2,\zeta_2)u,u'\right)\sim$$

$$\beta(n_1,n_2)\zeta_1^{s_1}\zeta_2^{s_2}\zeta_2^{s_{12}}(1-\zeta_1/\zeta_2)^{s_{12}}\left(Y^\mu(u_2,\zeta_2)Y^\mu(u_1,\zeta_1)u,u'\right)$$

where Y^μ is given in (6.87), μ satisfies (6.95) and β is defined by (6.96) and (6.97).

THEOREM 6.10. *With the notation as in Theorem 6.9 we have*

$$\zeta_1^{s_1}\zeta_2^{s_2}\zeta_1^{s_{12}}(1-\zeta_2/\zeta_1)^{s_{12}}(Y^\mu(u_1,\zeta_1)Y^\mu(u_2,\zeta_2)u,u') \sim$$

$$\zeta_2^{s_1+s_2}(\zeta_1-\zeta_2)^{s_{12}}\left(1+\frac{\zeta_1-\zeta_2}{\zeta_2}\right)^{s_1}(Y^\mu(Y^\mu(u_1,\zeta_1-\zeta_2)u_2,\zeta_2)u,u')$$

and the series on the right side converges absolutely when $0<|\zeta_1-\zeta_2|<|\zeta_2|$.

We now prove the Jacobi Identity (0.46) for the operators Y^μ. Recall from (3.6) with $l=4$ that $\Delta_0=0$ and $\Delta_k=\frac{1}{2}$ for $1\le k\le 3$, and that $\Delta(n_1,n_2)$ is defined as in (0.41). It is straightforward to check that $\Delta(n_1,n_2)$ is bilinear *mod* $\mathbb{Z}$, that is,

$$\begin{aligned}
\Delta(n_1,n_2,n) &= \Delta(n_1,n)+\Delta(n_2,n)-\Delta(n_1+n_2,n)\\
&= \Delta_{n_1}+\Delta_{n_2}+\Delta_n-\Delta_{n_1+n_2}-\Delta_{n_1+n}-\Delta_{n_2+n}+\Delta_{n_1+n_2+n}
\end{aligned} \tag{6.98}$$

is in $\mathbb{Z}$ for all $n_1,n_2,n\in\Gamma$. Also note that $\Delta(n_1,n_2)=s_{12}$ unless $n_1=n_2\ne 0$, in which case $\Delta(n_1,n_2)=1$ but $s_{12}=0$. For $n_1,n_2\in\Gamma$ define

$$\begin{aligned}
\eta(n_1,n_2) &= \beta(n_1,n_2)\,e^{-\pi i s_{12}}\\
&= \begin{cases}
1 & \text{if } n_1=0 \text{ or } n_2=0\\[1em]
-1 & \text{if } n_1=n_2\ne 0\\[1em]
\mathbf{a} & \text{if } n_1\ne 0 \text{ and } n_2=n_1'\\[1em]
-\mathbf{a} & \text{if } n_1\ne 0 \text{ and } n_2=n_1''
\end{cases}
\end{aligned} \tag{6.99}$$

and check that η is bilinear and that

$$\eta(n_1,n_2)\,e^{\pi i \Delta(n_1,n_2)} \tag{6.100}$$

is skew-symmetric.

THEOREM 6.11. *(The Jacobi Identity) With notations as above, let*

$$f(\zeta_1,\zeta_2)\in\zeta_1^{\Delta(n_1,n)}\zeta_2^{\Delta(n_2,n)}(\zeta_1-\zeta_2)^{\Delta(n_1,n_2)}\mathbb{C}[\zeta_1,\zeta_1^{-1},\zeta_2,\zeta_2^{-1},(\zeta_1-\zeta_2)^{-1}],$$

let $C_i(r)$ be a circle in the ζ_i-plane of radius r centered at the origin, and let $C_1(\zeta_2,\epsilon)$ be a circle in the ζ_1-plane of radius ϵ centered at $\zeta_1=\zeta_2$. Then for

$0 < r < \rho < R$ *and for* $0 < \epsilon < \min(R - \rho, \rho - r)$, *we have*

$$\int_{C_2(\rho)} \int_{C_1(R)} (Y^\mu(u_1, \zeta_1) Y^\mu(u_2, \zeta_2) u, u') f(\zeta_1, \zeta_2) d\zeta_1 d\zeta_2$$

$$- \int_{C_2(\rho)} \int_{C_1(r)} \eta(n_1, n_2)(Y^\mu(u_2, \zeta_2) Y^\mu(u_1, \zeta_1) u, u') f(\zeta_1, \zeta_2) d\zeta_1 d\zeta_2$$

$$= \int_{C_2(\rho)} \int_{C_1(\zeta_2, \epsilon)} (Y^\mu(Y^\mu(u_1, \zeta_1 - \zeta_2) u_2, \zeta_2) u, u') f(\zeta_1, \zeta_2) d\zeta_1 d\zeta_2$$

where the function $f(\zeta_1, \zeta_2)$ *is to be expanded in the three integrands as in (0.47)-(0.49), respectively, with* $z = \zeta_1$ *and* $z_0 = \zeta_2$.

We modify the notation (5.35) to define the components of the vertex operator

$$Y^\mu(u, \zeta) = \sum_{n \in \frac{1}{2}\mathbb{Z}} Y^\mu_{n+1-\text{wt}(u)}(u) \zeta^{-n-1} = \sum_{n \in \frac{1}{2}\mathbb{Z}} \{u; \mu\}_n \zeta^{-n-1}. \tag{6.101}$$

In case $u = a(-\frac{1}{2})\mathbf{vac}(\mathbb{Z} + \frac{1}{2}) \in (\mathbf{U})_{1/2}$ for $a \in \mathbf{C}$, we also modify the notations of (5.37) - (5.39) to define

$$a^\mu(m) = Y^\mu_m(u) = \{u; \mu\}_{m - \frac{1}{2}} \tag{6.102}$$

for $m \in \frac{1}{2}\mathbb{Z}$. Note that if $u \in \mathbf{U}_0$ then $Y^\mu(u, \zeta) = Y(u, \zeta)$ on $\mathbf{U}$, so the components of the vertex operators $Y^\mu(u, \zeta)$ for $wt(u) \in \{0, 1\}$ represent the affine algebra $\hat{\mathfrak{o}}(8)$, and if $\omega = D(-2)\mathbf{vac}$, the components of $Y^\mu(\omega, \zeta)$ represent the Virasoro operators $D(n)$, $n \in \mathbb{Z}$.

COROLLARY 6.12. *With notations as above, for* $p \in \mathbb{Z} + \Delta(n_1, n)$, $q \in \mathbb{Z} + \Delta(n_2, n)$ *and* $r \in \mathbb{Z} + \Delta(n_1, n_2)$, *we have*

$$\sum_{0 \le k \in \mathbb{Z}} \binom{r}{k} (-1)^k [\{u_1; \mu\}_{p+r-k} \{u_2; \mu\}_{q+k} - \eta(n_1, n_2) e^{\pi i r} \{u_2; \mu\}_{q+r-k} \{u_1; \mu\}_{p+k}] u$$

$$= \sum_{0 \le k \in \mathbb{Z}} \binom{p}{k} \{\{u_1; \mu\}_{r+k} u_2; \mu\}_{p+q-k} u. \tag{6.103}$$

PROOF. This follows as in the proof of Corollary 3.34 with $f(\zeta_1, \zeta_2) = \zeta_1^p \zeta_2^q \cdot (\zeta_1 - \zeta_2)^r$. ∎

COROLLARY 6.13. *Let* $\{n_1, n_2, n_3\} = \{1, 2, 3\}$, $a \in \mathbf{C}^{(n_1)}$, $b \in \mathbf{C}^{(n_2)}$, $0 \le n \le 3$, *and let* $\mathbf{a} = \pm 1$ *be as in (6.96). Then for* $s \in \mathbb{Z} + \Delta(n_1, n) + \frac{1}{2}$, $t \in \mathbb{Z} + \Delta(n_2, n)$, *on* $\mathbf{U}_n$ *we have*

$$\mu(n_1, n_2)(a \circ b)^\mu(s + t)$$

$$= \sum_{0 \le k \in \mathbb{Z}} (-1)^k \binom{-\frac{1}{2}}{k} [a^\mu(s - k - \frac{1}{2}) b^\mu(t + k + \frac{1}{2}) + \mathbf{i}^{\pm \mathbf{a}} b^\mu(t - k) a^\mu(s + k)] \tag{6.104}$$

where the $\pm$ is $+$ if (n_1, n_2) is in cyclic order, $-$ otherwise.

PROOF. This follows from Corollary 6.12 where $s = p + \frac{1}{2}$, $t = q$, and $r = -\frac{1}{2}$, using the fact that $\{u; \mu\}_m = u^\mu(m + \frac{1}{2})$ when $\text{wt}(u) = \frac{1}{2}$. $\blacksquare$

For $1 \leq k \leq 3$, with $u_i = a_i(-\frac{1}{2})\mathbf{vac} \in (\mathbf{U}_k)_{1/2}$ and $r = 0$, Corollary 6.12 says that

$$a_1^\mu(p)a_2^\mu(q) + a_2^\mu(q)a_1^\mu(p) = \mu(k, k)\langle a_1, a_2 \rangle \delta_{p, -q} \tag{6.105}$$

for $p, q \in \mathbb{Z} + \frac{1}{2}$ on $\mathbf{U}_0 \oplus \mathbf{U}_k$ and for $p, q \in \mathbb{Z}$ on $\mathbf{U}_{k'} \oplus \mathbf{U}_{k''}$. This means that the rescaled operators (6.102) for $p \in Z$ represent on $\mathbf{CM}^{(k)}(Z)$ the Clifford algebra $\mathbf{Cliff}^{(k)}(Z; \mu)$ generated by $\mathbf{C}^{(k)}(Z)$ and the rescaled form

$$\langle a_1(p), a_2(q) \rangle_\mu = \mu(k, k)\langle a_1, a_2 \rangle \delta_{p, -q}. \tag{6.106}$$

Let us see how the rescaling of Y to Y^μ affects the intertwining of its action with the action of the triality group G. For $g \in G$, Theorem 5.7 gives

$$\hat{g}Y^\mu(u_1, \zeta)u = \frac{\mu(n_1, n)}{\mu(gn_1, gn)} Y^\mu(\hat{g}u_1, \zeta)\hat{g}u \tag{6.107}$$

for $u_1 \in \mathbf{U}_{n_1}$, $u \in \mathbf{U}_n$. If $n_1 = 0$ or $n = 0$ then the factor $\mu(n_1, n)\mu(gn_1, gn)^{-1}$ equals 1 for all $g \in G$. For all $n_1 = n \neq 0$, this factor will be 1 when $\mu(1, 2) = \mu(2, 3) = \mu(3, 1)$. With this assumption on μ, we see from (6.95) that the factor is 1 when $g = \sigma$ for all $0 \leq n_1, n \leq 3$, but no choice of μ can make the factor 1 for all $0 \leq n_1, n \leq 3$ when $g = \tau$.

THEOREM 6.14. *($D_4^{(1)}$ Vertex Operator Para-algebra Theorem) Let $\Gamma = \mathbb{Z}_2 \times \mathbb{Z}_2 = \{0, 1, 2, 3\}$ and let $\Delta_0 = 0$, $\Delta_k = \frac{1}{2}$ for $1 \leq k \leq 3$. With μ as given in (6.95), η as given in (6.99), $\mathbf{1} = \mathbf{vac}(\mathbb{Z} + \frac{1}{2})$ and $\omega = D(-2)\mathbf{1}$, $(\mathbf{U}, Y^\mu(\ , z), \mathbf{1}, \omega, \Gamma, \Delta, \eta)$ is a vertex operator para-algebra as defined in the Introduction.*

CHAPTER 7

Spinor Construction of E_8

We now give a spinor construction of E_8 from two copies of D_4 and their three 8-dimensional representations. For $i = 1, 2$, let $\mathbf{A}_i \cong \mathbb{C}^8$ be a vector space with a nondegenerate symmetric bilinear form $\langle \, , \, \rangle$ and a polarization $\mathbf{A}_i = \mathbf{A}_i^+ \oplus \mathbf{A}_i^-$ into maximally isotropic subspaces. Let $\vartheta_i : \mathbf{A}_i \to \mathbf{A}_i$ be a $\mathbb{C}$-antilinear involution such that $\vartheta_i(\mathbf{A}_i^+) = \mathbf{A}_i^-$, $\langle \vartheta_i a, \vartheta_i b \rangle = \overline{\langle a, b \rangle}$ for $a, b \in \mathbf{A}_i$ and $0 < \langle a, \vartheta_i a \rangle \in \mathbb{R}$ for $0 \neq a \in \mathbf{A}_i$. Let $\mathbf{A}_i$ and the form $\langle \, , \, \rangle$ generate a Clifford algebra $\mathbf{Cliff}_i$. Construct the $\mathbf{Cliff}_i$-module $\mathbf{CM}_i = (\wedge \mathbf{A}_i^-) \cdot \mathbf{vac}_i$, $0 \neq \mathbf{vac}_i \in \wedge^4 \mathbf{A}_i^+$, from the polarization of $\mathbf{A}_i$ and let $\mathbf{C}_i = \mathbf{A}_i \oplus \mathbf{CM}_i$. In Chapter 4 we showed how Chevalley extends ϑ_i and the form $\langle \, , \, \rangle$ to $\mathbf{C}_i$, puts a commutative nonassociative product $\circ$ on $\mathbf{C}_i$, and constructs the triality group $G_i \cong S_3$ on $\mathbf{C}_i$ as automorphisms of $(\mathbf{C}_i, \circ)$ commuting with ϑ_i. Let $\mathbf{C}_i^{(1)} = \mathbf{A}_i$, $\mathbf{C}_i^{(2)} = \mathbf{CM}_i^0$, $\mathbf{C}_i^{(3)} = \mathbf{CM}_i^1$. Let $\sigma_i \in G_i$ be the element which cyclically permutes $\mathbf{C}_i^{(1)}$, $\mathbf{C}_i^{(2)}$, $\mathbf{C}_i^{(3)}$, and let $\tau_i \in G_i$ be the element which preserves $\mathbf{C}_i^{(1)}$ and switches $\mathbf{C}_i^{(2)}$ and $\mathbf{C}_i^{(3)}$. For $1 \leq k \leq 3$, the Clifford algebra $\mathbf{Cliff}_i^{(k)}$ generated by $\mathbf{C}_i^{(k)}$ and $\langle \, , \, \rangle$ has irreducible module $\mathbf{CM}_i^{(k)} = \mathbf{C}_i^{(k')} \oplus \mathbf{C}_i^{(k'')}$ with vacuum vector $\mathbf{vac}_i^{(k)} = \sigma_i^{k-1} \mathbf{vac}_i$ with respect to the polarization of $\mathbf{C}_i^{(k)}$ transported by σ_i from $\mathbf{C}_i^{(1)}$. We have seen how the Lie algebra $\mathbf{o}_i(8)$ is represented on $\mathbf{C}_i$ by the span of normally ordered quadratic elements ${}^\circ_\circ a_i b_i {}^\circ_\circ$, $a_i, b_i \in \mathbf{C}_i^{(k)}$ and how G_i acts as Lie algebra automorphisms on $\mathbf{o}_i(8)$ by $g_i \cdot {}^\circ_\circ a_i b_i {}^\circ_\circ = {}^\circ_\circ (g_i a_i)(g_i b_i) {}^\circ_\circ$. We have also seen in (4.65) - (4.67) how G_i acts as isomorphisms permuting the Clifford algebras, $g_i^{(k)} : \mathbf{Cliff}_i^{(k)} \to \mathbf{Cliff}_i^{(g(k))}$, induced by $g_i^{(k)}(a_i) = g_i a_i$ for $a_i \in \mathbf{C}_i^{(k)}$, intertwining the $\circ$ action of $\mathbf{Cliff}_i^{(k)}$ on $\mathbf{CM}_i^{(k)}$.

Let

$$\mathbf{A}^{(k)} = \mathbf{C}_1^{(k)} \oplus \mathbf{C}_2^{(k)} \tag{7.1}$$

with the form $\langle \, , \, \rangle$ extended so that $\langle \mathbf{C}_1^{(k)}, \mathbf{C}_2^{(k)} \rangle = 0$, with the polarization $\mathbf{A}^{(k)} = \mathbf{A}^{(k)+} \oplus \mathbf{A}^{(k)-}$ where $\mathbf{A}^{(k)\pm} = \mathbf{C}_1^{(k)\pm} \oplus \mathbf{C}_2^{(k)\pm}$, and with the $\mathbb{C}$-antilinear involution $\vartheta = \vartheta_1 \oplus \vartheta_2$. The Clifford algebra generated by $\mathbf{A}^{(k)}$ and $\langle \, , \, \rangle$ is the $\mathbb{Z}_2$-graded tensor product $\mathbf{Cliff}_1^{(k)} \otimes \mathbf{Cliff}_2^{(k)}$ having the $\mathbb{Z}_2$-graded tensor product $\mathbf{CM}_1^{(k)} \otimes \mathbf{CM}_2^{(k)}$ as irreducible module. This means that for $a_1 \in \mathbf{C}_1^{(k)}$, $a_2 \in \mathbf{C}_2^{(k)}$, $u \in \mathbf{Cliff}_1^{(k)}$, $v \in \mathbf{Cliff}_2^{(k)}$, $x \in \mathbf{CM}_1^{(k)}$, $y \in \mathbf{CM}_2^{(k)}$, we have

$$a_1(u \otimes v) = a_1 u \otimes v, \quad a_2(u \otimes v) = \pm u \otimes a_2 v, \tag{7.2}$$

128

where the sign is $+$ (or $-$) if u is in the even (or odd) part of $\mathbf{Cliff}_1^{(k)}$, and

$$a_1(x \otimes y) = a_1 x \otimes y, \quad a_2(x \otimes y) = \pm x \otimes a_2 y, \tag{7.3}$$

where the sign is $+$ if $x \in \mathbf{CM}_1^{(k)0}$, $-$ if $x \in \mathbf{CM}_1^{(k)1}$. The span of normally ordered quadratic elements ${}^\circ_\circ ab {}^\circ_\circ$ for $a, b \in \mathbf{A}^{(k)}$ represents a type D_8 Lie algebra $\mathbf{o}_k(16)$ on $\mathbf{A}^{(k)}$ and on the even and odd subspaces of $\mathbf{CM}_1^{(k)} \otimes \mathbf{CM}_2^{(k)}$, which are $(\mathbf{C}_1^{(k')} \otimes \mathbf{C}_2^{(k')}) \oplus (\mathbf{C}_1^{(k'')} \otimes \mathbf{C}_2^{(k'')})$ and $(\mathbf{C}_1^{(k')} \otimes \mathbf{C}_2^{(k'')}) \oplus (\mathbf{C}_1^{(k'')} \otimes \mathbf{C}_2^{(k')})$, respectively.

THEOREM 7.1. *Let*

$$\mathbf{k}_0 = \mathbf{o}_1(8) \oplus \mathbf{o}_2(8), \tag{7.4}$$

$$\mathbf{p}_k = \mathbf{C}_1^{(k)} \otimes \mathbf{C}_2^{(k)}, \quad 1 \le k \le 3, \tag{7.5}$$

$$\mathbf{g} = \mathbf{k}_0 \oplus \mathbf{p}_1 \oplus \mathbf{p}_2 \oplus \mathbf{p}_3. \tag{7.6}$$

Then for $1 \le k \le 3$,

$$\mathbf{o}_k(16) = \mathbf{k}_0 \oplus \mathbf{p}_k \tag{7.7}$$

is a type D_8 Lie algebra under the brackets

$$[x_1, a \otimes b] = (x_1 \cdot a) \otimes b, \quad [x_2, a \otimes b] = a \otimes (x_2 \cdot b),$$
$$[a \otimes b, c \otimes d] = -\langle a, c \rangle {}^\circ_\circ bd {}^\circ_\circ - \langle b, d \rangle {}^\circ_\circ ac {}^\circ_\circ, \tag{7.8}$$

where $x_1 \in \mathbf{o}_1(8)$, $x_2 \in \mathbf{o}_2(8)$, $a \otimes b, c \otimes d \in \mathbf{p}_k$. Furthermore, $\mathbf{g}$ is a type E_8 Lie algebra under the additional brackets

$$[a \otimes b, c \otimes d] = \kappa(a \circ c) \otimes (b \circ d) \tag{7.9}$$

where

$$\kappa = \begin{cases} 1 & \text{if } a \otimes b \in \mathbf{p}_k, \ c \otimes d \in \mathbf{p}_{k'} \\ \\ -1 & \text{if } a \otimes b \in \mathbf{p}_k, \ c \otimes d \in \mathbf{p}_{k''}. \end{cases} \tag{7.10}$$

PROOF. The brackets (7.8) correspond to those of the spinor construction of $\mathbf{o}_k(16)$ from $\mathbf{A}^{(k)}$ and the form $\langle \ , \ \rangle$, and (7.9) makes $\mathbf{p} = \mathbf{p}_{k'} \oplus \mathbf{p}_{k''}$ the even semispinor representation of $\mathbf{o}_k(16)$. Therefore, if we verify the Jacobi identity for $\mathbf{g}$, then it will be a Lie algebra of type E_8. One only needs to check brackets of four types; $[\mathbf{k}_0, [\mathbf{p}_1, \mathbf{p}_2]]$, $[\mathbf{p}_3, [\mathbf{p}_1, \mathbf{p}_2]]$, $[\mathbf{p}_1, [\mathbf{p}_1, \mathbf{p}_2]]$ and $[\mathbf{p}_2, [\mathbf{p}_1, \mathbf{p}_2]]$.

Let $x \in \mathbf{o}_1(8)$, $a \otimes b \in \mathbf{p}_1$, $c \otimes d \in \mathbf{p}_2$, so that $[a \otimes b, c \otimes d] = (a \circ c) \otimes (b \circ d) \in \mathbf{p}_3$. Then, using the fact that $\mathbf{o}_1(8)$ acts as derivations of $\circ$ on $\mathbf{C}_1$ we have

$$[x, [a \otimes b, c \otimes d]] = [x, (a \circ c) \otimes (b \circ d)]$$
$$= (x \cdot (a \circ c)) \otimes (b \circ d)$$
$$= (x \cdot a) \circ c \otimes (b \circ d) + a \circ (x \cdot c) \otimes (b \circ d)$$
$$= [[x, a \otimes b], c \otimes d] + [a \otimes b, [x, c \otimes d]].$$

A similar calculation takes care of the case when $x \in \mathbf{o}_2(8)$.

Let $a \otimes b \in \mathbf{p}_1$, $c \otimes d \in \mathbf{p}_2$, $e \otimes f \in \mathbf{p}_3$. Then we have

$$[e \otimes f, [a \otimes b, c \otimes d]] = [e \otimes f, (a \circ c) \otimes (b \circ d)]$$
$$= -\langle e, a \circ c \rangle_\circ^\circ f(b \circ d)_\circ^\circ - \langle f, b \circ d \rangle_\circ^\circ e(a \circ c)_\circ^\circ,$$
$$[a \otimes b, [c \otimes d, e \otimes f]] = [a \otimes b, (c \circ e) \otimes (d \circ f)]$$
$$= -\langle a, c \circ e \rangle_\circ^\circ b(d \circ f)_\circ^\circ - \langle b, d \circ f \rangle_\circ^\circ a(c \circ e)_\circ^\circ,$$
$$[c \otimes d, [e \otimes f, a \otimes b]] = [c \otimes d, (e \circ a) \otimes (f \circ b)]$$
$$= -\langle c, e \circ a \rangle_\circ^\circ d(f \circ b)_\circ^\circ - \langle d, f \circ b \rangle_\circ^\circ c(e \circ a)_\circ^\circ.$$

The sum of these three expressions is zero because

$$\langle a, c \circ e \rangle = \langle a \circ c, e \rangle = \langle c, e \circ a \rangle,$$
$$\langle b, d \circ f \rangle = \langle f, b \circ d \rangle = \langle d, f \circ b \rangle,$$

and from Proposition 4.14 we have

$$_\circ^\circ a(c \circ e)_\circ^\circ + {}_\circ^\circ c(e \circ a)_\circ^\circ + {}_\circ^\circ e(a \circ c)_\circ^\circ = 0$$

and

$$_\circ^\circ b(d \circ f)_\circ^\circ + {}_\circ^\circ d(f \circ b)_\circ^\circ + {}_\circ^\circ f(b \circ d)_\circ^\circ = 0.$$

Let $a \otimes b$, $c \otimes d \in \mathbf{p}_1$, $e \otimes f \in \mathbf{p}_2$. Then we have

$$[a \otimes b, [c \otimes d, e \otimes f]] = [a \otimes b, (c \circ e) \otimes (d \circ f)] = -a \circ (c \circ e) \otimes b \circ (d \circ f),$$
$$[c \otimes d, [e \otimes f, a \otimes b]] = [c \otimes d, -(e \circ a) \otimes (f \circ b)] = c \circ (e \circ a) \otimes d \circ (f \circ b),$$
$$[e \otimes f, [a \otimes b, c \otimes d]] = [e \otimes f, -\langle a, c \rangle_\circ^\circ bd_\circ^\circ - \langle b, d \rangle_\circ^\circ ac_\circ^\circ]$$
$$= \langle a, c \rangle e \otimes ({}_\circ^\circ bd_\circ^\circ \cdot f) + \langle b, d \rangle ({}_\circ^\circ ac_\circ^\circ \cdot e) \otimes f$$
$$= \langle a, c \rangle e \otimes (b \circ (d \circ f) - \tfrac{1}{2} \langle b, d \rangle f)$$
$$+ \langle b, d \rangle (a \circ (c \circ e) - \tfrac{1}{2} \langle a, c \rangle e) \otimes f$$
$$= \langle a, c \rangle e \otimes b \circ (d \circ f) + \langle b, d \rangle a \circ (c \circ e) \otimes f - \langle a, c \rangle \langle b, d \rangle e \otimes f.$$

We wish to show that

$$- a \circ (c \circ e) \otimes b \circ (d \circ f) + c \circ (e \circ a) \otimes d \circ (f \circ b) + \langle a, c \rangle e \otimes b \circ (d \circ f)$$
$$+ \langle b, d \rangle a \circ (c \circ e) \otimes f - \langle a, c \rangle \langle b, d \rangle e \otimes f$$

equals zero. Using $a \circ (c \circ e) + c \circ (a \circ e) = \langle a, c \rangle e$, the first and third terms combine to give $c \circ (a \circ e) \otimes b \circ (d \circ f)$. Also, the second term equals $c \circ (a \circ e) \otimes d \circ (b \circ f)$, so we get

$$c \circ (a \circ e) \otimes b \circ (d \circ f) + c \circ (a \circ e) \otimes d \circ (b \circ f)$$
$$+ \langle b, d \rangle a \circ (c \circ e) \otimes f - \langle a, c \rangle \langle b, d \rangle e \otimes f$$
$$= c \circ (a \circ e) \otimes \langle b, d \rangle f + \langle b, d \rangle a \circ (c \circ e) \otimes f - \langle a, c \rangle \langle b, d \rangle e \otimes f = 0.$$

Let $a \otimes b \in \mathbf{p}_1$, $c \otimes d$, $e \otimes f \in \mathbf{p}_2$. Then we have

$$[e \otimes f, [a \otimes b, c \otimes d]] = [e \otimes f, (a \circ c) \otimes (b \circ d)] = e \circ (a \circ c) \otimes f \circ (b \circ d),$$

$$[a \otimes b, [c \otimes d, e \otimes f]] = [a \otimes b, -\langle c, e\rangle{}_\circ^\circ df{}_\circ^\circ - \langle d, f\rangle{}_\circ^\circ ce{}_\circ^\circ]$$

$$= \langle c, e\rangle a \otimes ({}_\circ^\circ df{}_\circ^\circ \cdot b) + \langle d, f\rangle ({}_\circ^\circ ce{}_\circ^\circ \cdot a) \otimes b$$

$$= \langle c, e\rangle a \otimes (d \circ (f \circ b) - \tfrac{1}{2}\langle d, f\rangle b)$$

$$+ \langle d, f\rangle (c \circ (e \circ a) - \tfrac{1}{2}\langle c, e\rangle a) \otimes b$$

$$= \langle c, e\rangle a \otimes d \circ (f \circ b) + \langle d, f\rangle c \circ (e \circ a) \otimes b - \langle c, e\rangle\langle d, f\rangle a \otimes b,$$

$$[c \otimes d, [e \otimes f, a \otimes b]] = [c \otimes d, -(e \circ a) \otimes (f \circ b)] = -c \circ (e \circ a) \otimes d \circ (f \circ b).$$

We wish to show that

$$e \circ (a \circ c) \otimes f \circ (b \circ d) - c \circ (e \circ a) \otimes d \circ (f \circ b) + \langle c, e\rangle a \otimes d \circ (f \circ b)$$

$$+ \langle d, f\rangle c \circ (e \circ a) \otimes b - \langle c, e\rangle\langle d, f\rangle a \otimes b$$

equals zero. Using $c \circ (e \circ a) + e \circ (c \circ a) = \langle c, e\rangle a$, the second and third terms combine to give $e \circ (c \circ a) \otimes d \circ (f \circ b)$. Also, the first term equals $e \circ (c \circ a) \otimes f \circ (d \circ b)$, so we get

$$e \circ (c \circ a) \otimes f \circ (d \circ b) + e \circ (c \circ a) \otimes d \circ (f \circ b)$$

$$+ \langle d, f\rangle c \circ (e \circ a) \otimes b - \langle c, e\rangle\langle d, f\rangle a \otimes b$$

$$= e \circ (c \circ a) \otimes \langle d, f\rangle b + \langle d, f\rangle c \circ (e \circ a) \otimes b - \langle c, e\rangle\langle d, f\rangle a \otimes b = 0. \qquad \blacksquare$$

Remark. Note that $-\kappa$ could have been used in place of κ.

Defining $\vartheta = \vartheta_1 \oplus \vartheta_2$ on $\mathbf{C}_1 \oplus \mathbf{C}_2$ allows us to define the $\mathbb{C}$-antilinear antiautomorphism ϑ on $\mathbf{g}$ by

$$\vartheta({}_\circ^\circ ab{}_\circ^\circ) = {}_\circ^\circ \vartheta b \vartheta a{}_\circ^\circ \quad \text{for} \quad {}_\circ^\circ ab{}_\circ^\circ \in \mathbf{k}_0, \tag{7.11}$$

$$\vartheta(a \otimes b) = -\vartheta a \otimes \vartheta b \quad \text{for} \quad a \otimes b \in \mathbf{p}_k, \quad 1 \le k \le 3. \tag{7.12}$$

We have a nondegenerate symmetric invariant (see (2.12)) bilinear form $\langle x, y\rangle$ defined in (2.36) for $x, y \in \mathbf{o}_i(8)$, $i = 1, 2$. We extend this to all of $\mathbf{g}$ by defining

$$\langle a_1 \otimes a_2, b_1 \otimes b_2\rangle = -\langle a_1, b_1\rangle\langle a_2, b_2\rangle \tag{7.13}$$

for $a_1 \otimes a_2$, $b_1 \otimes b_2 \in \mathbf{p}_k$, $1 \le k \le 3$, and letting all other pairings be zero. It is straightforward to check that this defines a nondegenerate symmetric invariant bilinear form on all of $\mathbf{g}$. We also define the positive Hermitian form

$$(x, y) = \langle x, \vartheta y\rangle \quad \text{for } x, y \in \mathbf{g}. \tag{7.14}$$

We have an action of G on $\mathbf{g}$ as Lie algebra automorphisms by

$$g \cdot {}_\circ^\circ ab{}_\circ^\circ = {}_\circ^\circ (g_i a)(g_i b){}_\circ^\circ \quad \text{for} \quad {}_\circ^\circ ab{}_\circ^\circ \in \mathbf{o}_i(8), \tag{7.15}$$

$$g \cdot a \otimes b = g_1 a \otimes g_2 b \quad \text{for} \quad a \otimes b \in \mathbf{p}_k, \quad 1 \le k \le 3, \quad g \in G. \tag{7.16}$$

CHAPTER 8

Spinor Construction of Vertex Operator Algebras for $E_8^{(1)}$

In this chapter we will use the results of Chapters 3, 5 and 6 to obtain a spinor construction of a vertex operator algebra of type $E_8^{(1)}$, and a vertex operator para-algebra containing that algebra and various superalgebras of type $D_8^{(1)}$. The Rationality, Permutability and Associativity Theorems for the type $E_8^{(1)}$ vertex operator algebra are proved easily from those theorems for type $D_4^{(1)}$.

We first set up the notations for two copies of the four level 1 representations of type $D_4^{(1)}$ and associated structures as given in Chapters 4, 5 and 6. For $i = 1, 2$, $1 \leq k \leq 3$, $Z = \mathbb{Z}$ or $\mathbb{Z} + \frac{1}{2}$, let $(\mathbf{C}_i, \circ)$, $G_i \cong S_3$ and ϑ_i be as described at the beginning of Chapter 7. Let $\mathbf{Cliff}_i^{(k)}(Z)$ be the Clifford algebra generated by $\mathbf{C}_i^{(k)}(Z)$ (see (5.40)) and the form (2.39), with irreducible module $\mathbf{CM}_i^{(k)}(Z)$ having vacuum vector $\mathbf{vac}_i^{(k)}(Z)$. We use the notation $\mathbf{vac}_i = \mathbf{vac}_i^{(k)}(\mathbb{Z} + \frac{1}{2})$ which is independent of k. In Chapter 5 we have seen how to identify as one space

$$\mathbf{U} = \mathbf{CM}_1^{(k)}(\mathbb{Z} + \tfrac{1}{2}) \oplus \mathbf{CM}_1^{(k)}(\mathbb{Z}), \tag{8.1}$$

independent of k, $1 \leq k \leq 3$, such that $\mathbf{U}$ decomposes into 4 irreducible $\hat{\mathbf{o}}_1(8)$-modules

$$\mathbf{U} = \mathbf{U}_0 \oplus \mathbf{U}_1 \oplus \mathbf{U}_2 \oplus \mathbf{U}_3 \tag{8.2}$$

where

$$\begin{aligned}
\mathbf{U}_0 &= \mathbf{CM}_1^{(k)}(\mathbb{Z} + \tfrac{1}{2})^0, & \mathbf{U}_k &= \mathbf{CM}_1^{(k)}(\mathbb{Z} + \tfrac{1}{2})^1, \\
\mathbf{U}_{k'} &= \mathbf{CM}_1^{(k)}(\mathbb{Z})^0, & \mathbf{U}_{k''} &= \mathbf{CM}_1^{(k)}(\mathbb{Z})^1,
\end{aligned} \tag{8.3}$$

are the even $(^0)$ and odd $(^1)$ parity subspaces of these Clifford modules, and (k, k', k'') is a cyclic permutation of $(1, 2, 3)$. Let $\mathbf{Vir}_1$ be the Virasoro algebra represented on $\mathbf{U}$ by the operators $D_1(n)$, $n \in \mathbb{Z}$. We similarly have

$$\mathbf{W} = \mathbf{CM}_2^{(k)}(\mathbb{Z} + \tfrac{1}{2}) \oplus \mathbf{CM}_2^{(k)}(\mathbb{Z}), \tag{8.4}$$

independent of k, $1 \leq k \leq 3$, such that $\mathbf{W}$ decomposes into 4 irreducible $\hat{\mathbf{o}}_2(8)$-modules

$$\mathbf{W} = \mathbf{W}_0 \oplus \mathbf{W}_1 \oplus \mathbf{W}_2 \oplus \mathbf{W}_3 \tag{8.5}$$

132

where

$$
\mathbf{W}_0 = \mathbf{CM}_2^{(k)}(\mathbb{Z} + \tfrac{1}{2})^0, \qquad \mathbf{W}_k = \mathbf{CM}_2^{(k)}(\mathbb{Z} + \tfrac{1}{2})^1,
$$
$$
\mathbf{W}_{k'} = \mathbf{CM}_2^{(k)}(\mathbb{Z})^0, \qquad \mathbf{W}_{k''} = \mathbf{CM}_2^{(k)}(\mathbb{Z})^1,
\tag{8.6}
$$

Let $\mathbf{Vir}_2$ be the Virasoro algebra represented on $\mathbf{W}$ by the operators $D_2(n)$, $n \in \mathbb{Z}$. As defined in Chapter 2, $\mathbf{U}$ and $\mathbf{W}$ are each equipped with a Hermitian form $(\ ,\ )$. Let

$$
\Upsilon_1 : \mathbf{C}_1 \to (\mathbf{U})_{1/2}, \qquad \Upsilon_2 : \mathbf{C}_2 \to (\mathbf{W})_{1/2},
$$
$$
\Upsilon_1' : \mathbf{o}_1(8) \to (\mathbf{U})_1, \qquad \Upsilon_2' : \mathbf{o}_2(8) \to (\mathbf{W})_1
\tag{8.7}
$$

be unitary isomorphisms as defined in Chapter 5. For $g_i \in G_i$, let

$$
\hat{g}_1 : \mathbf{U} \to \mathbf{U}, \qquad \hat{g}_2 : \mathbf{W} \to \mathbf{W}
\tag{8.8}
$$

be the automorphisms defined in Theorem 5.5 lifting the actions of G_1 and G_2 on $\mathbf{C}_1$ and $\mathbf{C}_2$. For $u \in \mathbf{U}$, $w \in \mathbf{W}$, we have constructed vertex operators $Y(u, \zeta)$ on $\mathbf{U}$, and $Y(w, \zeta)$ on $\mathbf{W}$, satisfying the theorems in Chapters 3, 5 and 6. Let $\Gamma_1 = \Gamma_2 = \{0, 1, 2, 3\}$ be given the additive group structure $\mathbb{Z}_2 \times \mathbb{Z}_2$, and recall from (5.45) that G_i acts on Γ_i. Define Δ on $\Gamma_1 = \Gamma_2$ as in Theorem 6.14, and let μ_1 and μ_2 be functions from Γ_1 and Γ_2 to $\mathbb{C}^*$ satisfying the conditions (6.95) with $\mathbf{a}_1$ and $\mathbf{a}_2$, respectively, in place of $\mathbf{a}$. Let β_1 and β_2 be the corresponding functions obtained from (6.96) and (6.97), and let η_1 and η_2 be similarly obtained from (6.99). Let $Y^1(u, \zeta) = Y^{\mu_1}(u, \zeta)$ and $Y^2(w, \zeta) = Y^{\mu_2}(w, \zeta)$ be the rescaled vertex operators as defined in (6.87). Then these operators satisfy Theorems 6.9 - 6.11 and Corollaries 6.12 - 6.13 with the notations

$$
Y^1(u, \zeta) = \sum_{n \in \frac{1}{2}\mathbb{Z}} Y_{n+1-\mathrm{wt}(u)}^1(u)\zeta^{-n-1} = \sum_{n \in \frac{1}{2}\mathbb{Z}} \{u; \mu_1\}_n \zeta^{-n-1}
\tag{8.9}
$$

and

$$
Y^2(w, \zeta) = \sum_{n \in \frac{1}{2}\mathbb{Z}} Y_{n+1-\mathrm{wt}(w)}^2(w)\zeta^{-n-1} = \sum_{n \in \frac{1}{2}\mathbb{Z}} \{w; \mu_2\}_n \zeta^{-n-1}.
\tag{8.10}
$$

Recall from (3.37) and (3.64) that for $\omega_i = D_i(-2)\mathbf{vac}_i$, we have

$$
Y^i(\omega_i, \zeta) = \zeta^{-2}D_i(\zeta) = \sum_{n \in \mathbb{Z}} D_i(n)\zeta^{-n-2}.
\tag{8.11}
$$

Define $\Gamma = \Gamma_1 \times \Gamma_2$ and Δ on Γ as follows. For $\gamma = (m, n) \in \Gamma$, let

$$
\Delta_\gamma = \Delta_m + \Delta_n.
\tag{8.12}
$$

and for $\gamma_1 = (m_1, n_1), \gamma_2 = (m_2, n_2) \in \Gamma$, as in (0.41), let

$$
\Delta(\gamma_1, \gamma_2) = \Delta_{\gamma_1} + \Delta_{\gamma_2} - \Delta_{\gamma_1 + \gamma_2}
$$
$$
= \Delta(m_1, m_2) + \Delta(n_1, n_2)
\tag{8.13}
$$

which is bilinear *mod* $\mathbb{Z}$. Define

$$\mathbf{V} = \mathbf{U} \otimes \mathbf{W} = \bigoplus_{0 \le m,n \le 3} \mathbf{U}_m \otimes \mathbf{W}_n \tag{8.14}$$

and give $\mathbf{V}$ the product Hermitian form

$$(u_1 \otimes w_1, u_2 \otimes w_2) = (u_1, u_2)(w_1, w_2). \tag{8.15}$$

For homogeneous vectors $u \in \mathbf{U}$, $w \in \mathbf{W}$, define

$$\mathrm{wt}(u \otimes w) = \mathrm{wt}(u) + \mathrm{wt}(w) \tag{8.16}$$

and if $\mathbf{S}$ is any subspace of $\mathbf{V}$, for $r \in \frac{1}{2}\mathbb{Z}$ let

$$(\mathbf{S})_r = \{v \in \mathbf{S} \mid \mathrm{wt}(v) = r.\} \tag{8.17}$$

For $\gamma = (m,n) \in \Gamma$ let

$$\mathbf{V}_\gamma = \mathbf{U}_m \otimes \mathbf{W}_n, \tag{8.18}$$

and note that Δ_γ is the minimal weight of $\mathbf{V}_\gamma$. Define four subspaces of $\mathbf{V}$ by

$$\mathbf{V}_0 = (\mathbf{U}_0 \otimes \mathbf{W}_0) \oplus (\mathbf{U}_1 \otimes \mathbf{W}_1) \oplus (\mathbf{U}_2 \otimes \mathbf{W}_2) \oplus (\mathbf{U}_3 \otimes \mathbf{W}_3), \tag{8.19}$$

$$\mathbf{V}_k = (\mathbf{U}_0 \otimes \mathbf{W}_k) \oplus (\mathbf{U}_k \otimes \mathbf{W}_0) \oplus (\mathbf{U}_{k'} \otimes \mathbf{W}_{k''}) \oplus (\mathbf{U}_{k''} \otimes \mathbf{W}_{k'}), \tag{8.20}$$

for $1 \le k \le 3$, so that

$$\mathbf{V} = \mathbf{V}_0 \oplus \mathbf{V}_1 \oplus \mathbf{V}_2 \oplus \mathbf{V}_3. \tag{8.21}$$

From the $\frac{1}{2}\mathbb{Z}$-gradings of $\mathbf{U}$ and $\mathbf{W}$ given in (5.9) - (5.10), we have the gradings

$$\mathbf{V}_0 = (\mathbf{V}_0)_0 \oplus (\mathbf{V}_0)_1 \oplus (\mathbf{V}_0)_2 \oplus \cdots , \tag{8.22}$$

$$\mathbf{V}_k = (\mathbf{V}_k)_{1/2} \oplus (\mathbf{V}_k)_1 \oplus (\mathbf{V}_k)_{3/2} \oplus (\mathbf{V}_k)_2 \oplus (\mathbf{V}_k)_{5/2} \oplus \cdots \tag{8.23}$$

for $1 \le k \le 3$.

We now define vertex operators $Y(v,\zeta)$ on $\mathbf{V}$ for $v \in \mathbf{V}$. For $u_1, u \in \mathbf{U}$, $w_1, w \in \mathbf{W}$, define

$$Y^1(u_1,\zeta)(u \otimes w) = (Y^1(u_1,\zeta)u) \otimes w, \tag{8.24}$$

and, for $u \in \mathbf{U}_m$, $w_1 \in \mathbf{W}_{n_1}$, define

$$Y^2(w_1,\zeta)(u \otimes w) = \epsilon \, u \otimes Y^2(w_1,\zeta)w \tag{8.25}$$

where

$$\epsilon = \epsilon(m,n_1) = \begin{cases} -1 & \text{if } m \ne 0,\ n_1 \ne 0 \text{ and } m \ne n_1' \\[2ex] 1 & \text{otherwise.} \end{cases} \tag{8.26}$$

Note that ϵ is bilinear, that is, for $m, m_1, m_2, n, n_1, n_2 \in \{0,1,2,3\}$ we have

$$\epsilon(m_1 + m_2, n) = \epsilon(m_1, n)\epsilon(m_2, n), \quad \epsilon(m, n_1 + n_2) = \epsilon(m, n_1)\epsilon(m, n_2). \tag{8.27}$$

For $v = u \otimes w \in \mathbf{U}_m \otimes \mathbf{W}_n$, define

$$Y(v,\zeta) = Y^1(u,\zeta)Y^2(w,\zeta) = \epsilon(m,n) \, Y^2(w,\zeta)Y^1(u,\zeta) \tag{8.28}$$

on $\mathbf{V}$, and extend this definition to all $v \in \mathbf{V}$ by linearity. Then with the notation

$$Y(v, \zeta) = \sum_{n \in \frac{1}{2}\mathbb{Z}} \{v\}_n \zeta^{-n-1} \tag{8.29}$$

we have

$$\{u \otimes w\}_n = \sum_{k \in \frac{1}{2}\mathbb{Z}} \{u; \mu_1\}_k \{w; \mu_2\}_{n-k-1}. \tag{8.30}$$

Let $\omega = \omega_1 \otimes \mathbf{vac}_2 + \mathbf{vac}_1 \otimes \omega_2$ and define the operators $D(n) = \{\omega\}_{n+1}$ by

$$Y(\omega, \zeta) = \sum_{n \in \mathbb{Z}} D(n) \zeta^{-n-2}. \tag{8.31}$$

It follows from (8.28) that

$$Y(\omega, \zeta) = Y^1(\omega_1, \zeta) + Y^2(\omega_2, \zeta) \tag{8.32}$$

so

$$D(n) = D_1(n) + D_2(n) \tag{8.33}$$

for $n \in \mathbb{Z}$ are Virasoro operators on $\mathbf{V}$ satisfying (2.62) with $l = 8$.

We wish to derive the Rationality, Permutability, Associativity and Jacobi Identity theorems for the operators (8.29) on $\mathbf{V}$ from those theorems on $\mathbf{U}$ and $\mathbf{W}$. For $0 \leq m, n \leq 3$ define

$$s(m, n) = \begin{cases} \frac{1}{2} & \text{if } 0 \neq m \neq n \neq 0 \\[2ex] 0 & \text{otherwise,} \end{cases} \tag{8.34}$$

and note that $s(m, n) = \Delta(m, n)$ unless $m = n \neq 0$, in which case $s(m, n) = 0$ but $\Delta(m, n) = 1$. For $u_i \in \mathbf{U}_{m_i}$, $u \in \mathbf{U}_m$, $w_i \in \mathbf{W}_{n_i}$, $w \in \mathbf{W}_n$, $1 \leq i \leq r$, let

$$\begin{aligned} \mathbf{s} &= \mathbf{s}(u_1, \cdots, u_r; u) = (s_1, \cdots, s_r), \\ \mathbf{t} &= \mathbf{t}(w_1, \cdots, w_r; w) = (t_1, \cdots, t_r) \end{aligned} \tag{8.35}$$

be defined by

$$s_i = s(m_i, m), \qquad t_i = s(n_i, n). \tag{8.36}$$

(Note that this generalizes (3.44) - (3.45).) For $1 \leq i, j \leq r$ define

$$s_{ij} = s(m_i, m_j), \qquad t_{ij} = s(n_i, n_j) \tag{8.37}$$

and let

$$\zeta^{\mathbf{s}+\mathbf{t}} = \zeta_1^{s_1+t_1} \cdots \zeta_r^{s_r+t_r}. \tag{8.38}$$

THEOREM 8.1. *(Rationality) For $v_i \in \mathbf{U}_{m_i} \otimes \mathbf{W}_{n_i}$, $v \in \mathbf{U}_m \otimes \mathbf{W}_n$, $v' \in \mathbf{V}$, $1 \leq i \leq 2$, let $\mathbf{s} = (s_1, s_2)$ and $\mathbf{t} = (t_1, t_2)$ be defined by $s_i = s(m_i, m)$, $t_i = s(n_i, n)$, and let $s_{12} = s(m_1, m_2)$, $t_{12} = s(n_1, n_2)$. Then the series*

$$\zeta^{\mathbf{s}+\mathbf{t}} \zeta_1^{s_{12}+t_{12}} (1 - \zeta_2/\zeta_1)^{s_{12}+t_{12}} (Y(v_1, \zeta_1) Y(v_2, \zeta_2) v, v') \tag{8.39}$$

converges absolutely when $|\zeta_1| > |\zeta_2| > 0$ *to a rational function in the ring*

$$\mathbb{C}[\zeta_1, \zeta_1^{-1}, \zeta_2, \zeta_2^{-1}, (\zeta_1 - \zeta_2)^{-1}]. \tag{8.40}$$

PROOF. By linearity of the vertex operators it suffices to prove the theorem for $v_i = u_i \otimes w_i \in \mathbf{U}_{m_i} \otimes \mathbf{W}_{n_i}$, $v = u \otimes w \in \mathbf{U}_m \otimes \mathbf{W}_n$ and $v' = u' \otimes w' \in \mathbf{V}$. From the definitions, we have

$$(Y(u_1 \otimes w_1, \zeta_1)Y(u_2 \otimes w_2, \zeta_2)u \otimes w, u' \otimes w')$$
$$= (Y^1(u_1, \zeta_1)Y^2(w_1, \zeta_1)Y^1(u_2, \zeta_2)Y^2(w_2, \zeta_2)u \otimes w, u' \otimes w')$$
$$= \epsilon(m, n_2)(Y^1(u_1, \zeta_1)Y^2(w_1, \zeta_1)Y^1(u_2, \zeta_2)u \otimes Y^2(w_2, \zeta_2)w, u' \otimes w')$$
$$= \epsilon(m, n_2)\epsilon(m + m_2, n_1)(Y^1(u_1, \zeta_1)Y^1(u_2, \zeta_2)u \tag{8.41}$$
$$\otimes\ Y^2(w_1, \zeta_1)Y^2(w_2, \zeta_2)w, u' \otimes w')$$
$$= \epsilon(m, n_2)\epsilon(m + m_2, n_1)(Y^1(u_1, \zeta_1)Y^1(u_2, \zeta_2)u, u')$$
$$\cdot\ (Y^2(w_1, \zeta_1)Y^2(w_2, \zeta_2)w, w').$$

If $\{m_1, m_2\} \subseteq \{0, k\}$ for some k, $1 \le k \le 3$, then $s_{12} = 0$ and we can apply Theorem 3.28 to get that the series

$$\zeta^{\mathbf{s}}(Y^1(u_1, \zeta_1)Y^1(u_2, \zeta_2)u, u') \tag{8.42}$$

converges absolutely to a rational function in the ring (8.40) when $|\zeta_1| > |\zeta_2| > 0$. Otherwise, $\{m_1, m_2\} \subseteq \{1, 2, 3\}$, $m_1 \ne m_2$, $s_{12} = \frac{1}{2}$ and we can apply Theorem 6.6 to get that the series

$$\zeta^{\mathbf{s}}\zeta_1^{s_{12}}(1 - \zeta_2/\zeta_1)^{s_{12}}(Y^1(u_1, \zeta_1)Y^1(u_2, \zeta_2)u, u') \tag{8.43}$$

converges absolutely to a rational function in the ring (8.40) when $|\zeta_1| > |\zeta_2| > 0$. The same statements are true with m_1, m_2, m, s_{12}, $\mathbf{s}$, u_1, u_2, u, and u' replaced by n_1, n_2, n, t_{12}, $\mathbf{t}$, w_1, w_2, w and w', respectively, giving the result. $\blacksquare$

Remark. It is obvious how to modify the statement of this theorem to assert rationality of matrix coefficients for a product of more than two vertex operators, and we proved Theorem 3.28 in that generality. But since we only proved Theorem 6.6 for two vertex operators, we can only prove this theorem for two operators. While the inductive steps of the proof of Theorem 6.6 seem to work just as well for more than two operators, the base cases seem to become more and more complex as the number of operators increases. We believe that in principle our techniques could establish these base cases, but in order to prove the Jacobi Identity we only need the above theorem, so we will not pursue the matter here.

THEOREM 8.2. *(Permutability) With notation as in Theorem 8.1, we have*

$$\zeta^{\mathbf{s}+\mathbf{t}}\zeta_1^{s_{12}+t_{12}}(1 - \zeta_2/\zeta_1)^{s_{12}+t_{12}}(Y(v_1, \zeta_1)Y(v_2, \zeta_2)v, v') \sim$$
$$\kappa\ \zeta^{\mathbf{s}+\mathbf{t}}\zeta_2^{s_{12}+t_{12}}(1 - \zeta_1/\zeta_2)^{s_{12}+t_{12}}(Y(v_2, \zeta_2)Y(v_1, \zeta_1)v, v')$$

where

$$\kappa = \epsilon(m_1, n_2)\epsilon(m_2, n_1)^{-1}\beta_1(m_1, m_2)\beta_2(n_1, n_2).$$

PROOF. As in the proof of Theorem 8.1, it is enough to take $v_i = u_i \otimes w_i \in \mathbf{U}_{m_i} \otimes \mathbf{W}_{n_i}$, $v = u \otimes w \in \mathbf{U}_m \otimes \mathbf{W}_n$ and $v' = u' \otimes w' \in \mathbf{V}$. As in (8.41), we have

$$(Y(u_1 \otimes w_1, \zeta_1)Y(u_2 \otimes w_2, \zeta_2)u \otimes w, u' \otimes w')$$
$$= \epsilon(m, n_2)\epsilon(m + m_2, n_1)(Y^1(u_1, \zeta_1)Y^1(u_2, \zeta_2)u, u')(Y^2(w_1, \zeta_1)Y^2(w_2, \zeta_2)w, w') \tag{8.44}$$

and

$$(Y(u_2 \otimes w_2, \zeta_2)Y(u_1 \otimes w_1, \zeta_1)u \otimes w, u' \otimes w')$$
$$= \epsilon(m, n_1)\epsilon(m + m_1, n_2)(Y^1(u_2, \zeta_2)Y^1(u_1, \zeta_1)u, u')(Y^2(w_2, \zeta_2)Y^2(w_1, \zeta_1)w, w'). \tag{8.45}$$

Applying Theorem 6.9, we get

$$\zeta^{\mathbf{s}}\zeta_1^{s_{12}}(1 - \zeta_2/\zeta_1)^{s_{12}}(Y^1(u_1, \zeta_1)Y^1(u_2, \zeta_2)u, u') \sim$$
$$\beta_1(m_1, m_2)\zeta^{\mathbf{s}}\zeta_2^{s_{12}}(1 - \zeta_1/\zeta_2)^{s_{12}}(Y^1(u_2, \zeta_2)Y^1(u_1, \zeta_1)u, u'). \tag{8.46}$$

The same statements are true with Y^1, m_1, m_2, m, s_{12}, $\mathbf{s}$, β_1, u_1, u_2, u and u' replaced by Y^2, n_1, n_2, n, t_{12}, $\mathbf{t}$, β_2, w_1, w_2, w and w', respectively, giving the result. $\blacksquare$

THEOREM 8.3. *(Associativity) With notation as in Theorem 8.1, we have*

$$\zeta^{\mathbf{s}+\mathbf{t}}\zeta_1^{s_{12}+t_{12}}(1 - \zeta_2/\zeta_1)^{s_{12}+t_{12}}(Y(v_1, \zeta_1)Y(v_2, \zeta_2)v, v') \sim$$
$$\zeta_2^{s_1+s_2+t_1+t_2}(\zeta_1 - \zeta_2)^{s_{12}+t_{12}}(1 + (\zeta_1 - \zeta_2)/\zeta_2)^{s_1+t_1}(Y(Y(v_1, \zeta_1 - \zeta_2)v_2, \zeta_2)v, v')$$

and the second series converges absolutely when $0 < |\zeta_1 - \zeta_2| < |\zeta_2|$.

PROOF. Using the notation of the last two proofs, in addition to (8.44), we have

$$(Y(Y(u_1 \otimes w_1, \zeta_1 - \zeta_2)u_2 \otimes w_2, \zeta_2)u \otimes w, u' \otimes w')$$
$$= \epsilon(m_2, n_1)(Y(Y^1(u_1, \zeta_1 - \zeta_2)u_2 \otimes Y^2(w_1, \zeta_1 - \zeta_2)w_2, \zeta_2)u \otimes w, u' \otimes w')$$
$$= \epsilon(m_2, n_1)\epsilon(m, n_1 + n_2)(Y^1(Y^1(u_1, \zeta_1 - \zeta_2)u_2, \zeta_2)u, u')$$
$$\cdot (Y^2(Y^2(w_1, \zeta_1 - \zeta_2)w_2, \zeta_2)w, w'). \tag{8.47}$$

Applying Theorem 6.10, we get

$$\zeta^{\mathbf{s}}\zeta_1^{s_{12}}(1 - \zeta_2/\zeta_1)^{s_{12}}(Y^1(u_1, \zeta_1)Y^1(u_2, \zeta_2)u, u') \sim$$
$$\zeta_2^{s_1+s_2}(\zeta_1 - \zeta_2)^{s_{12}}(1 + (\zeta_1 - \zeta_2)/\zeta_2)^{s_1}(Y^1(Y^1(u_1, \zeta_1 - \zeta_2)u_2, \zeta_2)u, u') \tag{8.48}$$

and the series on the right side converges absolutely when $0 < |\zeta_1 - \zeta_2| < |\zeta_2|$. The same statements are true with Y^1, m_1, m_2, m, s_{12}, $\mathbf{s}$, u_1, u_2, u and u' replaced by Y^2, n_1, n_2, n, t_{12}, $\mathbf{t}$, w_1, w_2, w, and w', respectively, giving the result. $\blacksquare$

For $i = 1, 2$ let $\gamma_i = (m_i, n_i) \in \Gamma$ and let $\kappa(\gamma_1, \gamma_2)$ be the permutability factor defined in Theorem 8.2. Define

$$\eta(\gamma_1, \gamma_2) = \kappa(\gamma_1, \gamma_2)\, e^{\pi i (s_{12} + t_{12})} = \frac{\epsilon(m_1, n_2)}{\epsilon(m_2, n_1)}\, \eta_1(m_1, m_2)\, \eta_2(n_1, n_2) \qquad (8.49)$$

and note that η is bilinear since ϵ, η_1 and η_2 are bilinear. Furthermore, from (8.13) we see that $\eta(\gamma_1, \gamma_2)\, e^{\pi i \Delta(\gamma_1, \gamma_2)}$ is skew-symmetric.

THEOREM 8.4. *(Jacobi Identity) Let $v_i \in \mathbf{V}_{\gamma_i}$ for $i = 1, 2$, $v \in \mathbf{V}_\gamma$, $v' \in \mathbf{V}$. Let*

$$f(\zeta_1, \zeta_2) \in \zeta_1^{\Delta(\gamma_1, \gamma)}\, \zeta_2^{\Delta(\gamma_2, \gamma)}\, (\zeta_1 - \zeta_2)^{\Delta(\gamma_1, \gamma_2)}\, \mathbb{C}[\zeta_1, \zeta_1^{-1}, \zeta_2, \zeta_2^{-1}(\zeta_1 - \zeta_2)^{-1}],$$

let $C_i(r)$ be a circle in the ζ_i-plane of radius r centered at the origin, and let $C_1(\zeta_2, \epsilon)$ be a circle in the ζ_1-plane of radius ϵ centered at $\zeta_1 = \zeta_2$. Then for $0 < r < \rho < R$ and for $0 < \epsilon < \min(R - \rho, \rho - r)$ we have

$$\int_{C_2(\rho)} \int_{C_1(R)} (Y(v_1, \zeta_1) Y(v_2, \zeta_2) v, v') f(\zeta_1, \zeta_2) d\zeta_1 d\zeta_2$$

$$- \int_{C_2(\rho)} \int_{C_1(r)} \eta(\gamma_1, \gamma_2)\, (Y(v_2, \zeta_2) Y(v_1, \zeta_1) v, v') f(\zeta_1, \zeta_2) d\zeta_1 d\zeta_2$$

$$= \int_{C_2(\rho)} \int_{C_1(\zeta_2, \epsilon)} (Y(Y(v_1, \zeta_1 - \zeta_2) v_2, \zeta_2) v, v') f(\zeta_1, \zeta_2) d\zeta_1 d\zeta_2$$

where the function $f(\zeta_1, \zeta_2)$ is to be expanded in the three integrands as in (0.48)-(0.50), respectively, with $z = \zeta_1$ and $z_0 = \zeta_2$.

COROLLARY 8.5. *With notation as above, for $p \in \mathbb{Z} + \Delta(\gamma_1, \gamma)$, $q \in \mathbb{Z} + \Delta(\gamma_2, \gamma)$, $r \in \mathbb{Z} + \Delta(\gamma_1, \gamma_2)$, we have*

$$\sum_{0 \leq k \in \mathbb{Z}} \binom{r}{k} (-1)^k [\{v_1\}_{p+r-k} \{v_2\}_{q+k} - \eta(\gamma_1, \gamma_2)\, e^{\pi i r}\, \{v_2\}_{q+r-k} \{v_1\}_{p+k}] v$$

$$= \sum_{0 \leq k \in \mathbb{Z}} \binom{p}{k} \{\{v_1\}_{r+k} v_2\}_{p+q-k} v.$$

PROOF. With $f(\zeta_1, \zeta_2) = \zeta_1^p \zeta_2^q (\zeta_1 - \zeta_2)^r$ this follows as in the proof of Corollary 3.34. ∎

THEOREM 8.6. *With notation as above, let $m_1 = n_1$ and $m_2 = n_2$ so that $v_1, v_2 \in \mathbf{V}_0$. If $\mathbf{a}_1 \mathbf{a}_2 = -1$ then we have*

$$[\{v_1\}_p, \{v_2\}_q] v = \sum_{0 \leq k \in \mathbb{Z}} \binom{p}{k} \{\{v_1\}_k v_2\}_{p+q-k} v$$

where $p \in \mathbb{Z} + \Delta(\gamma_1, \gamma)$, $q \in \mathbb{Z} + \Delta(\gamma_2, \gamma)$.

PROOF. With $m_1 = n_1$ and $m_2 = n_2$ we have $\Delta(\gamma_1, \gamma_2) = 2\, \Delta(m_1, m_2) \in \mathbb{Z}$, so we may take $r = 0$ in Corollary 8.5. From (8.25) and (8.33) we see that

$$\epsilon(m_1, m_2) \epsilon(m_2, m_1)^{-1} = (-1)^{2\Delta(\gamma_1, \gamma_2)}. \qquad (8.50)$$

So from (6.99) with $\mathbf{a}_1$ and $\mathbf{a}_2$ in place of $\mathbf{a}$ we have

$$\eta(\gamma_1, \gamma_2) = (-1)^{2\Delta(m_1, m_2)}\, \eta_1(m_1, m_2)\, \eta_2(m_1, m_2)$$

$$= \begin{cases} 1 & \text{if } m_1 = 0 \text{ or } m_2 = 0 \text{ or } m_1 = m_2 \\[2ex] -\mathbf{a}_1\mathbf{a}_2 & \text{if } 0 \neq m_1 \neq m_2 \neq 0 \end{cases} \tag{8.51}$$

which equals 1 in all cases if and only if $\mathbf{a}_1\mathbf{a}_2 = -1$. $\blacksquare$

Let

$$\hat{\mathbf{V}}_0 = \mathbf{V}_0 \otimes \mathbb{C}[t, t^{-1}], \tag{8.52}$$

for $1 \leq k \leq 3$ let

$$\hat{\mathbf{V}}_k = ((\mathbf{U}_0 \otimes \mathbf{W}_0) \oplus (\mathbf{U}_k \otimes \mathbf{W}_k)) \otimes \mathbb{C}[t, t^{-1}] \oplus$$
$$((\mathbf{U}_{k'} \otimes \mathbf{W}_{k'}) \oplus (\mathbf{U}_{k''} \otimes \mathbf{W}_{k''})) \otimes t^{1/2}\mathbb{C}[t, t^{-1}], \tag{8.53}$$

and let $D(-1)$ act on these spaces as in (0.21). It can be shown that for each k, $0 \leq k \leq 3$, $\hat{\mathbf{V}}_k/D(-1)\hat{\mathbf{V}}_k$ has the structure of a Lie algebra and is represented on $\mathbf{V}_k$ so that the coset of $v \otimes t^n$ is represented by the operator $\{v\}_n$. We will now identify $\mathbf{V}_0$ as the homogeneously graded basic representation of the affine algebra $E_8^{(1)}$, and $\mathbf{V}_1$, $\mathbf{V}_2$ and $\mathbf{V}_3$ as "k-p" graded basic representations of $E_8^{(1)}$.

In (2.17)-(2.18) we defined the homogeneously graded affine algebra $\hat{\mathbf{g}}$ associated with a finite-dimensional Lie algebra $\mathbf{g}$. Recall from (2.29)-(2.31) that when $\mathbf{g}$ has a decomposition

$$\mathbf{g} = \mathbf{k} \oplus \mathbf{p} \tag{8.54}$$

such that

$$[\mathbf{k}, \mathbf{k}] \subseteq \mathbf{k}, \quad [\mathbf{k}, \mathbf{p}] \subseteq \mathbf{p}, \quad [\mathbf{p}, \mathbf{p}] \subseteq \mathbf{k}, \tag{8.55}$$

then we can define a "k-p" graded affine algebra associated with (8.54),

$$\hat{\mathbf{g}}' = \mathbf{k} \otimes \mathbb{C}[t, t^{-1}] \oplus \mathbf{p} \otimes t^{1/2}\mathbb{C}[t, t^{-1}] \oplus \mathbb{C}c \oplus \mathbb{C}d \tag{8.56}$$

where the brackets are as in (2.18). From Theorem 7.1, $\mathbf{g}$ of type E_8 has three such decompositions,

$$\mathbf{g} = \mathbf{k}^{(k)} \oplus \mathbf{p}^{(k)} \tag{8.57}$$

where

$$\mathbf{k}^{(k)} = \mathbf{k}_0 \oplus \mathbf{p}_k = \mathbf{o}_k(16), \tag{8.58}$$
$$\mathbf{p}^{(k)} = \mathbf{p}_{k'} \oplus \mathbf{p}_{k''}. \tag{8.59}$$

Let the three corresponding "k-p" graded affine algebras be denoted $\hat{\mathbf{g}}^{(k)}$, $1 \leq k \leq 3$, and write $\hat{\mathbf{g}}^{(0)}$ for the homogeneously graded affine algebra $\hat{\mathbf{g}}$. Using the following notations,

$$\hat{\mathbf{k}}_0 = \mathbf{k}_0 \otimes \mathbb{C}[t, t^{-1}] \oplus \mathbb{C}c \oplus \mathbb{C}d, \tag{8.60}$$

$$\hat{\mathbf{p}}_k = \mathbf{p}_k \otimes \mathbb{C}[t, t^{-1}], \quad \hat{\mathbf{p}}_k' = \mathbf{p}_k \otimes t^{1/2}\mathbb{C}[t, t^{-1}], \tag{8.61}$$
$$\hat{\mathbf{k}}^{(k)} = \hat{\mathbf{k}}_0 \oplus \hat{\mathbf{p}}_k = \hat{\mathbf{o}}_k(16), \quad \hat{\mathbf{k}}^{(k)'} = \hat{\mathbf{k}}_0 \oplus \hat{\mathbf{p}}_k' = \hat{\mathbf{o}}_k'(16), \tag{8.62}$$
$$\hat{\mathbf{p}}^{(k)} = \hat{\mathbf{p}}_{k'} \oplus \hat{\mathbf{p}}_{k''}, \quad \hat{\mathbf{p}}^{(k)'} = \hat{\mathbf{p}}_{k'}' \oplus \hat{\mathbf{p}}_{k''}', \tag{8.63}$$

for $1 \le k \le 3$, we may write

$$\hat{\mathbf{g}}^{(0)} = \hat{\mathbf{k}}_0 \oplus \hat{\mathbf{p}}_1 \oplus \hat{\mathbf{p}}_2 \oplus \hat{\mathbf{p}}_3 = \hat{\mathbf{k}}^{(k)} \oplus \hat{\mathbf{p}}^{(k)} \tag{8.64}$$

and

$$\hat{\mathbf{g}}^{(k)} = \hat{\mathbf{k}}_0 \oplus \hat{\mathbf{p}}_k \oplus \hat{\mathbf{p}}'_{k'} \oplus \hat{\mathbf{p}}'_{k''} = \hat{\mathbf{k}}^{(k)} \oplus \hat{\mathbf{p}}^{(k)'}. \tag{8.65}$$

Recall from (5.12)-(5.15) that $(\mathbf{U}_0)_0$ is one dimensional with basis $\{\mathbf{vac}_1\}$, that $(\mathbf{U}_0)_1$ is 28-dimensional, spanned by

$$\{a(-\tfrac{1}{2})b(-\tfrac{1}{2})\mathbf{vac}_1 \mid a, b \in \mathbf{C}_1^{(k)}\}, \tag{8.66}$$

and that $(\mathbf{U}_k)_{1/2}$ is 8-dimensional, spanned by

$$\{a(-\tfrac{1}{2})\mathbf{vac}_1 \mid a \in \mathbf{C}_1^{(k)}\}. \tag{8.67}$$

Furthermore, $(\mathbf{U}_{k'})_{1/2}$ is 8-dimensional, spanned by

$$\{a_1(0)\cdots a_r(0)\mathbf{vac}_1(\mathbb{Z}) \mid a_i \in \mathbf{C}_1^{(k)-}, \quad 1 \le i \le r, \quad r = 0, 2, 4\}, \tag{8.68}$$

and $(\mathbf{U}_{k''})_{1/2}$ is 8-dimensional, spanned by

$$\{a_1(0)\cdots a_r(0)\mathbf{vac}_1(\mathbb{Z}) \mid a_i \in \mathbf{C}_1^{(k)-}, \quad 1 \le i \le r, \quad r = 1, 3\}. \tag{8.69}$$

Similar statements hold for $\mathbf{W}$.

Combining these results for $\mathbf{U}$ and $\mathbf{W}$ we see that $(\mathbf{V}_0)_0$ is one-dimensional with basis $\{\mathbf{vac}_1 \otimes \mathbf{vac}_2\}$,

$$\begin{aligned}(\mathbf{V}_0)_1 = (\mathbf{U}_0)_0 \otimes (\mathbf{W}_0)_1 &\oplus (\mathbf{U}_0)_1 \otimes (\mathbf{W}_0)_0 \oplus (\mathbf{U}_1)_{1/2} \otimes (\mathbf{W}_1)_{1/2} \\ \oplus (\mathbf{U}_2)_{1/2} \otimes (\mathbf{W}_2)_{1/2} &\oplus (\mathbf{U}_3)_{1/2} \otimes (\mathbf{W}_3)_{1/2}\end{aligned} \tag{8.70}$$

is 248-dimensional, and for any i, $1 \le i \le 3$, it is spanned by

$$\begin{aligned}&\{\mathbf{vac}_1 \otimes a_2(-\tfrac{1}{2})b_2(-\tfrac{1}{2})\mathbf{vac}_2 \mid a_2, b_2 \in \mathbf{C}_2^{(i)}\} \\ &\cup \{a_1(-\tfrac{1}{2})b_1(-\tfrac{1}{2})\mathbf{vac}_1 \otimes \mathbf{vac}_2 \mid a_1, b_1 \in \mathbf{C}_1^{(i)}\} \\ &\bigcup_{1 \le k \le 3} \{a_1(-\tfrac{1}{2})\mathbf{vac}_1 \otimes a_2(-\tfrac{1}{2})\mathbf{vac}_2 \mid a_i \in \mathbf{C}_i^{(k)}\}.\end{aligned} \tag{8.71}$$

For $1 \le k \le 3$ we also see that

$$(\mathbf{V}_k)_{1/2} = (\mathbf{U}_0)_0 \otimes (\mathbf{W}_k)_{1/2} \oplus (\mathbf{U}_k)_{1/2} \otimes (\mathbf{W}_0)_0 \tag{8.72}$$

is 16-dimensional, spanned by

$$\{\mathbf{vac}_1 \otimes a_2(-\tfrac{1}{2})\mathbf{vac}_2 \mid a_2 \in \mathbf{C}_2^{(k)}\} \cup \{a_1(-\tfrac{1}{2})\mathbf{vac}_1 \otimes \mathbf{vac}_2 \mid a_1 \in \mathbf{C}_1^{(k)}\}. \tag{8.73}$$

We can use the isomorphisms in (8.7) to define an isomorphism

$$\Upsilon : \mathbf{g} \to (\mathbf{V}_0)_1 \tag{8.74}$$

by

$$\Upsilon({}^{\circ}_{\circ}a_1b_1{}^{\circ}_{\circ}) = \Upsilon'_1({}^{\circ}_{\circ}a_1b_1{}^{\circ}_{\circ}) \otimes \mathbf{vac}_2,$$

$$\Upsilon({}^{\circ}_{\circ}a_2b_2{}^{\circ}_{\circ}) = \mathbf{vac}_1 \otimes \Upsilon'_2({}^{\circ}_{\circ}a_2b_2{}^{\circ}_{\circ}), \tag{8.75}$$

$$\Upsilon(a_1 \otimes a_2) = \Upsilon_1(a_1) \otimes \Upsilon_2(a_2)$$

for $a_1, b_1 \in \mathbf{C}_1^{(k)}$, $a_2, b_2 \in \mathbf{C}_2^{(k)}$, $1 \leq k \leq 3$. For $x \in \mathbf{g}$, $n \in \frac{1}{2}\mathbb{Z}$, define the notation

$$x(n) = Y_n(\Upsilon(x)) = \{\Upsilon(x)\}_n. \tag{8.76}$$

The following theorem gives spinor constructions of four basic $E_8^{(1)}$ representations when $\mu_1 = \mu_2^{-1}$, an assumption we make for the rest of Chapter 8.

THEOREM 8.7. *Let* $\mu_1(i,j) = \mu_2(i,j)^{-1}$ *for* $0 \leq i, j \leq 3$. *The homogeneous affine algebra* $\hat{\mathbf{g}}^{(0)}$ *(8.64) of type* $E_8^{(1)}$ *is represented on* $\mathbf{V}_0$ *by the identity operator,* $D(0)$ *and the operators* $x(q)$, $x \in \mathbf{g}$, $q \in \mathbb{Z}$. *For* $1 \leq k \leq 3$, *the "k-p" graded affine algebra* $\hat{\mathbf{g}}^{(k)}$ *(8.65) of type* $E_8^{(1)}$ *is represented on* $\mathbf{V}_k$ *by the identity operator,* $D(0)$ *and the operators* $x(q)$, *where* $q \in \mathbb{Z}$ *if* $x \in \mathbf{k}^{(k)} = \mathbf{o}_k(16)$, *and* $q \in \mathbb{Z} + \frac{1}{2}$ *if* $x \in \mathbf{p}^{(k)}$.

PROOF. For $x, y \in \mathbf{g}$, $v \in \mathbf{U}_m \otimes \mathbf{W}_n$, Theorem 8.6 says that

$$[x(p), y(q)]v = \{x(0)\Upsilon(y)\}_{p+q}v + p\{x(1)\Upsilon(y)\}_{p+q-1}v \tag{8.77}$$

for $p \in \mathbb{Z} + \Delta(\gamma_1, \gamma)$, $q \in \mathbb{Z} + \Delta(\gamma_2, \gamma)$. Using (5.21), (5.26), (7.8) and (7.9) it is straightforward to check that when

$$\mu_1(i,j) = \mu_2(i,j)^{-1} \quad \text{for} \ \ 0 \leq i, j \leq 3 \tag{8.78}$$

we have

$$x(0)\Upsilon(y) = \Upsilon([x,y]) \quad \text{and} \quad x(1)\Upsilon(y) = \langle x, y \rangle \mathbf{vac}_1 \otimes \mathbf{vac}_2 \tag{8.79}$$

where $\langle x, y \rangle$ is given in Chapter 7 (see (2.36) and (7.13)). Therefore, from (8.77) we get

$$[x(p), y(q)]v = [x,y](p+q)v + p\langle x, y \rangle \delta_{p,-q}v. \tag{8.80}$$

For $v \in \mathbf{V}_0$, $\Delta(\gamma_1, \gamma)$, $\Delta(\gamma_2, \gamma) \in \mathbb{Z}$, so (8.80) is valid on $\mathbf{V}_0$ for $p, q \in \mathbb{Z}$ and all $x, y \in \mathbf{g}$, showing that $\hat{\mathbf{g}}^{(0)}$ is represented on $\mathbf{V}_0$. For fixed k, $1 \leq k \leq 3$, if $x \in \mathbf{p}^{(k)}$ then $\Delta(\gamma_1, \gamma) \in \mathbb{Z} + \frac{1}{2}$ for any $v \in \mathbf{V}_k$, so (8.80) holds for $p \in \mathbb{Z} + \frac{1}{2}$. If $x \in \mathbf{k}^{(k)} = \mathbf{o}_k(16)$ then $\Delta(\gamma_1, \gamma) \in \mathbb{Z}$ for any $v \in \mathbf{V}_k$, so (8.80) holds for $p \in \mathbb{Z}$. Similar statements hold for y and q, so $\hat{\mathbf{g}}^{(k)}$ is represented on $\mathbf{V}_k$. From Theorem 7.1, $\hat{\mathbf{g}}^{(0)}$ and each $\hat{\mathbf{g}}^{(k)}$ is of type $E_8^{(1)}$. $\blacksquare$

Remark. It is possible to prove Theorem 8.7 using only Proposition 6.2, Corollary 6.3 and an induction argument. Using the Hermitian form, the induction reduces the verification of (8.80) to those v for which $wt(v)$ is minimal in each $\hat{\mathbf{k}}_0$-module $\mathbf{U}_m \otimes \mathbf{W}_n$, $0 \leq m, n \leq 3$, and to those p, q such that $p+q$ ranges over a small finite set. This method of proof is more elementary in that it uses the theory of the affine algebra $D_4^{(1)}$ and triality (as in Chapters 2 and 5), and avoids the vertex operator superalgebra constructions in Chapter 3. It's disadvantage is its length, there being many cases to check.

For $g = (g_1, g_2) \in G_1 \times G_2$ we define $\hat{g} : \mathbf{V} \to \mathbf{V}$ by

$$\hat{g}(u \otimes w) = \hat{g}_1 u \otimes \hat{g}_2 w \quad \text{for } u \in \mathbf{U}, \ w \in \mathbf{W}. \tag{8.81}$$

The following analog of Theorem 5.7 is immediate from the definitions.

THEOREM 8.8. *For* $v_1 \in \mathbf{U}_{m_1} \otimes \mathbf{W}_{n_1}$, $v \in \mathbf{U}_m \otimes \mathbf{W}_n$, $g = (g_1, g_2) \in G_1 \times G_2$, *we have*

$$\hat{g} Y(v_1, \zeta) v = \lambda \ Y(\hat{g} v_1, \zeta) \hat{g} v$$

where

$$\lambda = \lambda(m_1, n_1; m, n; g_1, g_2) = \frac{\epsilon(m, n_1) \mu_1(m_1, m) \mu_2(n_1, n)}{\epsilon(g_1 m, g_2 n_1) \mu_1(g_1 m_1, g_1 m) \mu_2(g_2 n_1, g_2 n)}.$$

COROLLARY 8.9. *Let the notation be as in Theorem 8.8.*
   (a) *If* $m_1 = n_1 = 0$, *then* $\lambda = 1$ *for all* $g_1 \in G_1$, $g_2 \in G_2$, $0 \le m, n \le 3$.
   (b) *If* $g_1 = g_2 = \sigma$ *and* $\mu_1(1, 2) = \mu_1(2, 3) = \mu_1(3, 1)$ *(which implies that* $\mu_1(\sigma i, \sigma j) = \mu_1(i, j)$ *for all* $0 \le i, j \le 3$*), then* $\lambda = 1$ *for all* $0 \le m_1, n_1, m, n \le 3$.

For each k, $1 \le k \le 3$, and each choice of Z_1 and Z_2 the set of operators

$$\{Y_m^1(u), Y_n^2(w) \mid u \in (\mathbf{U}_k)_{1/2}, \ w \in (\mathbf{W}_k)_{1/2}, \ m \in Z_1, \ n \in Z_2\} \tag{8.82}$$

represents on

$$\mathbf{CM}^{(k)}(Z_1, Z_2) = \mathbf{CM}_1^{(k)}(Z_1) \otimes \mathbf{CM}_2^{(k)}(Z_2), \tag{8.83}$$

the generators of a Clifford algebra

$$\mathbf{Cliff}^{(k)}(Z_1, Z_2) = \mathbf{Cliff}_1^{(k)}(Z_1; \mu_1) \otimes \mathbf{Cliff}_2^{(k)}(Z_2; \mu_2). \tag{8.84}$$

These are $\mathbb{Z}_2$-graded tensor products. This means that for $a_i(n) \in \mathbf{C}_i^{(k)}(Z_i)$, $b_i \in \mathbf{Cliff}_i^{(k)}(Z_i; \mu_i)$ and $u_i \in \mathbf{CM}_i^{(k)}(Z_i)$, $i = 1, 2$, we have

$$a_1(n)(b_1 \otimes b_2) = a_1(n) b_1 \otimes b_2, \quad a_2(n)(b_1 \otimes b_2) = \pm b_1 \otimes a_2(n) b_2, \tag{8.85}$$

where the sign is $+$ (or $-$) if b_1 is in the even (or odd) part of $\mathbf{Cliff}_1^{(k)}(Z_1; \mu_1)$, and

$$a_1(n)(u_1 \otimes u_2) = a_1(n) u_1 \otimes u_2, \quad a_2(n)(u_1 \otimes u_2) = \pm u_1 \otimes a_2(n) u_2, \tag{8.86}$$

where the sign is $+$ (or $-$) if u_1 is in the even (or odd) part of $\mathbf{CM}_1^{(k)}(Z_1)$. This is equivalent to the usual construction from the space $\mathbf{A}^{(k)}(Z_1, Z_2) = \mathbf{C}_1^{(k)}(Z_1) \oplus \mathbf{C}_2^{(k)}(Z_2)$ with the symmetric bilinear form on each summand as in (6.106), the summands orthogonal to each other, and the polarization given by the polarizations of each summand. The module (8.83) is an irreducible $\mathbf{Cliff}^{(k)}(Z_1, Z_2)$-module with vacuum vector

$$\mathbf{vac}^{(k)}(Z_1, Z_2) = \mathbf{vac}_1^{(k)}(Z_1) \otimes \mathbf{vac}_2^{(k)}(Z_2). \tag{8.87}$$

We have the decomposition

$$\mathbf{CM}^{(k)}(Z_1, Z_2) = \mathbf{CM}^{(k)}(Z_1, Z_2)^0 \oplus \mathbf{CM}^{(k)}(Z_1, Z_2)^1 \qquad (8.88)$$

into even (^{0}) and odd (^{1}) parity subspaces, where

$$\begin{aligned}
\mathbf{CM}^{(k)}(Z_1, Z_2)^0 = {} & \mathbf{CM}_1^{(k)}(Z_1)^0 \otimes \mathbf{CM}_2^{(k)}(Z_2)^0 \\
& \oplus \mathbf{CM}_1^{(k)}(Z_1)^1 \otimes \mathbf{CM}_2^{(k)}(Z_2)^1
\end{aligned} \qquad (8.89)$$

and

$$\begin{aligned}
\mathbf{CM}^{(k)}(Z_1, Z_2)^1 = {} & \mathbf{CM}_1^{(k)}(Z_1)^0 \otimes \mathbf{CM}_2^{(k)}(Z_2)^1 \\
& \oplus \mathbf{CM}_1^{(k)}(Z_1)^1 \otimes \mathbf{CM}_2^{(k)}(Z_2)^0.
\end{aligned} \qquad (8.90)$$

From (8.19) and (8.20) we see that for $1 \leq k \leq 3$,

$$\begin{aligned}
\mathbf{V}_0 \oplus \mathbf{V}_k &= \mathbf{CM}^{(k)}(\mathbb{Z}+\tfrac{1}{2}, \mathbb{Z}+\tfrac{1}{2}) \oplus \mathbf{CM}^{(k)}(\mathbb{Z}, \mathbb{Z}) \\
&= (\mathbf{U}_0 \oplus \mathbf{U}_k) \otimes (\mathbf{W}_0 \oplus \mathbf{W}_k) \ \oplus \ (\mathbf{U}_{k'} \oplus \mathbf{U}_{k''}) \otimes (\mathbf{W}_{k'} \oplus \mathbf{W}_{k''})
\end{aligned} \qquad (8.91)$$

and

$$\begin{aligned}
\mathbf{V}_{k'} \oplus \mathbf{V}_{k''} &= \mathbf{CM}^{(k)}(\mathbb{Z}+\tfrac{1}{2}, \mathbb{Z}) \oplus \mathbf{CM}^{(k)}(\mathbb{Z}, \mathbb{Z}+\tfrac{1}{2}) \\
&= (\mathbf{U}_0 \oplus \mathbf{U}_k) \otimes (\mathbf{W}_{k'} \oplus \mathbf{W}_{k''}) \ \oplus \ (\mathbf{U}_{k'} \oplus \mathbf{U}_{k''}) \otimes (\mathbf{W}_0 \oplus \mathbf{W}_k).
\end{aligned} \qquad (8.92)$$

From Chapter 2 we see that we have a homogeneous representation of the affine algebra $\hat{\mathfrak{o}}_k(16)$ of type $D_8^{(1)}$ on $\mathbf{CM}^{(k)}(\mathbb{Z}+\tfrac{1}{2}, \mathbb{Z}+\tfrac{1}{2})$ and on $\mathbf{CM}^{(k)}(\mathbb{Z}, \mathbb{Z})$, and the even and odd parity subspaces of each one are irreducible, giving the four level 1 $\hat{\mathfrak{o}}_k(16)$-modules. Similarly, from $\mathbf{CM}^{(k)}(\mathbb{Z}+\tfrac{1}{2}, \mathbb{Z})$ and $\mathbf{CM}^{(k)}(\mathbb{Z}, \mathbb{Z}+\tfrac{1}{2})$ we get four level 1 "**k-p**" graded representations of $\hat{\mathfrak{o}}'_k(16)$. The following theorems summarize what we have accomplished in this chapter.

THEOREM 8.10. *For $1 \leq k \leq 3$, we have*

(a) $\mathbf{CM}^{(k)}(\mathbb{Z}+\tfrac{1}{2}, \mathbb{Z}+\tfrac{1}{2})$ *is a vertex operator superalgebra,*

(b) $\mathbf{CM}^{(k)}(\mathbb{Z}, \mathbb{Z})$ *is a $\mathbb{Z}_2$-twisted $\mathbf{CM}^{(k)}(\mathbb{Z}+\tfrac{1}{2}, \mathbb{Z}+\tfrac{1}{2})$-module, with the twisting given by the order 2 automorphism of $\mathbf{CM}^{(k)}(\mathbb{Z}+\tfrac{1}{2}, \mathbb{Z}+\tfrac{1}{2})$ which is 1 on $(\mathbf{U}_0 \otimes \mathbf{W}_0) \oplus (\mathbf{U}_k \otimes \mathbf{W}_k)$ and -1 on $(\mathbf{U}_0 \otimes \mathbf{W}_k) \oplus (\mathbf{U}_k \otimes \mathbf{W}_0)$,*

(c) $\mathbf{CM}^{(k)}(\mathbb{Z}+\tfrac{1}{2}, \mathbb{Z})$ *is a $\mathbb{Z}_2$-twisted $\mathbf{CM}^{(k)}(\mathbb{Z}+\tfrac{1}{2}, \mathbb{Z}+\tfrac{1}{2})$-module, with the twisting given by the order 2 automorphism of $\mathbf{CM}^{(k)}(\mathbb{Z}+\tfrac{1}{2}, \mathbb{Z}+\tfrac{1}{2})$ which is 1 on $(\mathbf{U}_0 \otimes \mathbf{W}_0) \oplus (\mathbf{U}_k \otimes \mathbf{W}_0)$ and -1 on $(\mathbf{U}_0 \otimes \mathbf{W}_k) \oplus (\mathbf{U}_k \otimes \mathbf{W}_k)$,*

(d) $\mathbf{CM}^{(k)}(\mathbb{Z}, \mathbb{Z}+\tfrac{1}{2})$ *is a $\mathbb{Z}_2$-twisted $\mathbf{CM}^{(k)}(\mathbb{Z}+\tfrac{1}{2}, \mathbb{Z}+\tfrac{1}{2})$-module, with the twisting given by the order 2 automorphism of $\mathbf{CM}^{(k)}(\mathbb{Z}+\tfrac{1}{2}, \mathbb{Z}+\tfrac{1}{2})$ which is 1 on $(\mathbf{U}_0 \otimes \mathbf{W}_0) \oplus (\mathbf{U}_0 \otimes \mathbf{W}_k)$ and -1 on $(\mathbf{U}_k \otimes \mathbf{W}_0) \oplus (\mathbf{U}_k \otimes \mathbf{W}_k)$.*

THEOREM 8.11. *($E_8^{(1)}$ Vertex Operator Para-algebra Theorem) Let $\Gamma = \mathbb{Z}_2^2 \times \mathbb{Z}_2^2$, $\Delta_{(i,j)}$ be the minimal weight of $\mathbf{U}_i \otimes \mathbf{W}_j$, η be given in (8.49), $\mathbf{1} = \mathbf{vac}_1 \otimes \mathbf{vac}_2$ and $\omega = D(-2)\mathbf{1}$. Then, as defined in the Introduction, $(\mathbf{V}_0, Y(\ , z), \mathbf{1}, \omega)$ is a vertex operator algebra, $(\mathbf{V}_k, Y(\ , z))$ is a $\mathbf{V}_0$-module for $1 \leq k \leq 3$, and $(\mathbf{V}, Y(\ , z), \mathbf{1}, \omega, \Gamma, \Delta, \eta)$ is a vertex operator para-algebra.*

References

[B] R.E. Borcherds, *Vertex algebras, Kac-Moody algebras, and the Monster*, Proc. Natl. Acad. Sci. USA **83** (1986), 3068–3071.

[BPZ] A. A. Belavin, A. M. Polyakov, A. B. Zamolodchikov, *Infinite conformal symmetry in two-dimensional quantum field theory*, Nuclear Physics **B241** (1984), 333-380.

[C] C. C. Chevalley, *The Algebraic Theory of Spinors*, Columbia University Press, Morningside Heights, New York, 1954.

[CM] H. S. M. Coxeter, W. O. J. Moser, *Generators and Relations for Discrete Groups*, Springer-Verlag, New York, 1972.

[DL1] C.-Y. Dong, J. Lepowsky, *A Jacobi identity for relative vertex operators and the equivalence of Z-algebras and parafermion algebras*, XVIIth International Colloquium on Group Theoretical Methods in Physics, Ste-Adele, June, 1988, Proceedings (Y. Saint-Aubin, ed.), World Scientific, Singapore, 1989, pp. 235–238.

[DL2] ______, *Generalized vertex algebras and relative twisted vertex operators*, preprint.

[FF] A. J. Feingold, I. B. Frenkel, *Classical affine algebras*, Advances in Mathematics **56** (1985), 117–172.

[FFR] A. J. Feingold, I. B. Frenkel, J. F. X. Ries, *The Exceptional Affine Algebra $E_8^{(1)}$, Triality and Chiral Algebras*, Lie Algebras and Related Topics, Contemporary Mathematics, Vol. 110, Proceedings of a Conference held in Madison, Wisconsin May 22 to June 1, 1988 (Georgia Benkart and J. Marshall Osborn, eds.), American Mathematical Society, Providence, RI, 1989, pp. 67–73.

[FR] A.J. Feingold, J.F.X. Ries, *Vertex Operator Algebras, Triality and $E_8^{(1)}$*, in preparation.

[F1] I.B. Frenkel, *Spinor representations of affine Lie algebras*, Proc. Natl. Acad. Sci. USA **77** (1980), 6303–6306.

[F2] ______, *Two constructions of affine Lie algebra representations and boson-fermion correspondence in quantum field theory*, Journal of Functional Analysis **44** (1981), 259–327.

[FHL] I.B. Frenkel, Y.-Z. Huang, J. Lepowsky, *On axiomatic approaches to vertex operator algebras and modules*, preprint.

[FK] I.B. Frenkel, V.G. Kac, *Basic representations of affine Lie algebras and dual resonance models*, Inventiones Math. **62** (1980), 23–66.

[FLM] I.B. Frenkel, J. Lepowsky, A. Meurman, *Vertex Operator Algebras and the Monster*, Pure and Applied Mathematics, Vol. 134, Academic Press, Boston, 1988.

[FLM2] ———, *Vertex operator calculus*, Mathematical Aspects of String Theory, Proc. 1986 Conference, San Diego (S. -T. Yau, ed.), World Scientific, Singapore, 1987, pp. 150–188.

[GSO] F. Gliozzi, J. Scherk, D. I. Olive, *Supersymmetry, supergravity theories and the dual spinor model*, Nuclear Physics **B122** (1977), 253–290.

[Go] P. Goddard, *Meromorphic conformal field theory*, Infinite Dimensional Lie Algebras and Lie Groups: Proceedings of the CIRM-Luminy Conference, 1988 (Victor G. Kac, ed.), Advanced Series in Mathematical Physics, Vol. 7, World Scientific, Singapore, 1989, pp. 556–587.

[GNORS] P. Goddard, W.Nahm, D.I. Olive, H. Ruegg, A. Schwimmer, *Fermions and octonians*, Commum. Math. Phys. **112** (1987), 385–408.

[GOS] P. Goddard, D.I. Olive, A. Schwimmer, *The heterotic string and a fermionic construction of the E_8 Kac-Moody algebra*, Physics Letters **157B** (1985), 393–399.

[GSW] M. B. Green, J. H. Schwarz, E. Witten, *Superstring Theory*, Volumes 1 and 2, Cambridge University Press, Cambridge, 1987.

[G] R. L. Griess, *The Friendly Giant*, Inventiones Math. **69** (1982), 1–102.

[KP1] V.G. Kac, D.H. Peterson, *Spin and wedge representations of infinite-dimensional Lie algebras and groups*, Proc. Natl. Acad. Sci. USA **78** (1981), 3308–3312.

[KP2] ———, *112 Constructions of the basic representations of the loop group of E_8*, Anomalies, Geometry and Topology, Proc. of the Conf., Argonne 1985, World Scientific, 1985, pp. 276–298.

[L] J. Lepowsky, *Calculus of twisted vertex operators*, Proc. Natl. Acad. Sci. USA **82** (1985), 8295–8299.

[LP] J. Lepowsky, M. Primc, *Standard modules for type one affine Lie algebras*, Lecture Notes in Math., Vol. 1052, Number Theory, New York, 1982, Springer-Verlag, New York, 1984, pp. 194–251.

[LW1] J. Lepowsky, R.L. Wilson, *A new family of algebras underlying the Rogers-Ramanujan identities and generalizations*, Proc. Natl. Acad. Sci. USA **78** (1981), 7254–7258.

[LW2] ———, *The structure of standard modules,I: Universal algebras and the Rogers-Ramanujan identities*, Inventiones Math. **77** (1984), 199–290.

[LW3] ———, *The structure of standard modules,II: The case $A_1^{(1)}$, principal gradation*, Inventiones Math. **79** (1985), 417–442.

[S] R. D. Schafer, *An Introduction to Nonassociative Algebras*, Academic Press, New York, 1966.

[T]   H. Tsukada, *Vertex operator superalgebras*, Communications in Algebra **18** (1990), 2249–2274.

[W]   G. C. Wick, *The evaluation of the collision matrix*, Physical Review **80** (1950), 268–272.

DEPARTMENT OF MATHEMATICAL SCIENCES, THE STATE UNIVERSITY OF NEW YORK, BINGHAMTON, NEW YORK 13902-6000

DEPARTMENT OF MATHEMATICS, YALE UNIVERSITY, NEW HAVEN, CONNECTICUT 06520

DEPARTMENT OF MATHEMATICAL SCIENCES, THE STATE UNIVERSITY OF NEW YORK, BINGHAMTON, NEW YORK 13902-6000